9444

GENERAL BIOLOGY LAB MANUAL

# *Investigations Into Life's Phenomena*

GENERAL BIOLOGY LAB MANUAL

# Investigations Into Life's Phenomena

RUSSELL SKAVARIL

MARY FINNEN

STEVEN LAWTON

*The Ohio State University*

SAUNDERS COLLEGE PUBLISHING

*Harcourt Brace Jovanovich College Publishers*

Fort Worth   Philadelphia   San Diego   New York   Orlando   Austin
San Antonio   Toronto   Montreal   London   Sydney   Tokyo

Text Typeface: Palatino
Compositor: Progressive Typographers, Inc.
Acquisitions Editor: Julie Levin Alexander
Developmental Editor: Lori Weaber
Managing Editor: Carol Field
Project Editor: Nancy Lubars
Copy Editor: Zanae Rodrigo
Manager of Art and Design: Carol Bleistine
Art Director: Doris Bruey
Art Coordinator: Caroline McGowan
Text Designer: Rebecca Lloyd Lemna
Cover Designer: Louis Fuiano
Text Artwork: Rolin Graphics, Inc.
Director of EDP: Tim Frelick
Production Manager: Charlene Squibb

Cover Credit: Front cover: © John Dudak / Phototake NYC
Front cover inset: © 1992 Carolina Biological
Company / Phototake NYC
Back cover: © Russell Skavaril

Printed in the United States of America

GENERAL BIOLOGY LAB MANUAL

ISBN 0-03-032612-5

Library of Congress Catalog Card Number: 92-063111

3456   082   987654321

# Preface

The laboratory portion of a general biology course is vitally important since it is in the laboratory that students directly encounter much of the wonder of biology. Before doing the exercises detailed in this book, the student will be familiar, of course, with the overall aspects of our surrounding biological world, the environment in which we all live, affect, and share. However, many, many wonderful and fascinating details of the biological world, particularly those aspects which cannot be seen with the unaided eye, remain to be understood, appreciated, and explored. It is the adventure of that exploration and acquisition of knowledge from such investigations which are the subject of this book, an exploration of the phenomena of life. It is, in fact, from experiments performed in the general biology laboratory that the student learns first hand about phenomena which, otherwise, exist only as vague concepts.

## TEACHING OPTIONS

This laboratory manual is based upon the collective experience of the authors with, literally, thousands of students in a general biology course at The Ohio State University and a combined professional experience of in excess of a half century of serving such students. From our having worked with those students, we have come to an understanding and appreciation of the students who will use this book. We are indebted to our former students for helping us to learn what would work well with such students in the laboratory, what would be effective, and perhaps most importantly, what would *not* work. We have attempted to present the material of this book in a manner which will be suitable both to the undergraduate non-

science major, as well as to the student who is majoring in one of the science disciplines.

We have intentionally avoided gearing this book to any particular general biology text. Rather, we have attempted to write this book in such a way that it may be used with any of the rich variety of general textbooks which are currently on the market. The manual provides a complete and comprehensive laboratory experience through exercises which we have designed to demonstrate the major concepts and principles in the study of life.

We have attempted to build the laboratory experiences around the use of standard laboratory equipment such as microscopes, glassware, chemicals, easily available biological materials, the spectrophotomoter, etc. A very few of the topics involve the use of a microcomputer, but that utilization has been held to a minimum since we are well aware of the reality that not all schools have general biology laboratories equipped with microcomputers.

## SPECIAL FEATURES

This manual provides complete and comprehensive lab experience through 32 Topics with exercises that have been designed to demonstrate the major concepts and principles in the study of life. To facilitate student learning by requiring and encouraging participation and analyzation of data, each chapter consists of the following features:

OBJECTIVES — Provide clear and concise statements of the learning objectives for each Topic.

INTRODUCTION — Gives a succinct summary of the major details of the subject material of each

Topic and highlights key terms. The introduction also serves as a study guide and review for the student.

PRELAB QUESTIONS — Provide the student with an assessment of his/her readiness and preparation to perform the laboratory work at hand.

DIRECTIONS FOR THE LABORATORY PROCEDURES:

MATERIALS REQUIRED — Lists completely, all materials required for conducting experiments.

PROCEDURES — Supplies clearly written, detailed accounts and step-by-step instructions for each laboratory exercise to be conducted, observations to be made, and data to be gathered. Illustrations enhance the text and space for students to record observations and answers are provided.

POSTLAB QUESTIONS — Give each student a chance to apply information which has been gained from each exercise to its application to concrete concepts and principles. These questions reinforce what the student has learned from the laboratory work.

FOR FURTHER READING — Provides an introduction to the most recent and technically sophisticated reference texts on the subjects of the laboratory exercises for those students who wish to explore the phenomena concerned more comprehensively.

The chapters are independent so that instructors may select those specific exercises which, in the view of the instructor, serve best to accomplish the overall objectives of the laboratory portion of the course.

An Instructor's Manual to the Laboratory Manual containing information for lab instructors as they prepare and set up laboratory exercises is also available. Information on lab equipment and materials, time requirements for each lab exercise and sources for supplies are some of the additional information this manual provides.

## ACKNOWLEDGMENTS

We wish to thank all of those who have contributed in one way or another towards our efforts in producing this work. The following individuals have offered constructive comments on the manuscript at various stages: Alice Jacklet, SUNY at Albany; Carl Hoagstrom, Ohio Northern University; Peter Ducey, SUNY at Cortland; Salvatore Tavormina, Austin Community College; Mark Gromko, Bowling Green State University; Carol Ann Kearns, University of Colorado, Mountain Research Station; Phillip Snider, University of Houston, University Park; C. Michael Bell, Collin County Community College; David Inouye, University of Maryland; Jim Harris, Utah Valley Community College; William Jensen, Ohio State University; S. C. McDonald, Henry Ford Community College; Frank Seabury, The Citadel; T. H. David Ho, Washington University; Elizabeth Waldorf, Mississippi Gulf Coast Community College; Gwen Scottgale, Albertus Magnus College; Ronald Kirk, Indiana University, Purdue University; and, Steve Badger, Southwestern Assemblies of God College.

We also include among those who warrant recognition, our respective spouses, children, families, and friends, fellow workers, editors, and colleagues, and, most particularly, our students who provided us with the inspiration necessary for the production of this work.

**Russell V. Skavaril**
**Mary M. Finnen**
**Steven M. Lawton**
**Columbus, Ohio**
**February, 1993**

# Contents Overview

# Contents

Preface *vii*

GENERAL BIOLOGY LAB MANUAL

# *Investigations Into Life's Phenomena*

# PART I

❑

# The Science of Biology

**TOPIC 1**
Microscopy

**TOPIC 2**
Scientific Method

# Microscopy

1. Identify the components of the compound light microscope and stereomicroscope and state the functions of those parts.

2. Compare and contrast the features of the compound light microscope with those of both the stereomicroscope and the electron microscope.

3. Give the definitions of the following terms and be able to use them correctly in reference to the microscope: resolution (resolving power), magnification, working distance, depth of field, and field of view.

## INTRODUCTION

The microscope is an instrument used to see very small objects (the term "specimen" is commonly used for the term "object") in biology. The Dutch spectacle maker, Zacharias Janssen, is credited with having discovered the principle of the compound microscope in about 1590. By 1674 Anton van Leeuwenhoek had developed microscope lenses that could magnify up to nearly 300 times. The microscope has made it possible to utilize better methods of studying biological material, particularly cells.

There are two general types of microscopes based on the kind of energy source used by the instrument. **Light microscopes** operate by visible light from the sun or an artificial light source. The light passes through glass lenses resulting in a magnified image seen by the eye. **Electron microscopes** do not function by light. The electron microscopes operate by a stream of electrons originating from an electron source within the instrument. Electrons are negatively charged particles that orbit the nucleus of atoms. The electron stream is focused by magnetic lenses resulting in a magnified image appearing on photographic film or on a fluorescent screen, similar to a television picture tube. Other characteristics of the electron microscope will be presented later in Lab 1.F.

You will be using two kinds of light microscopes in this lab exercise. The **compound light microscope** (Fig. 1–1), which we will examine first, is the one you will use most frequently. The compound microscope is probably the image that comes to your mind with the word "microscope." This instrument is called "compound" because it contains two or more lenses that collect and focus light passing through a transparent specimen producing a magnified image. Because light must pass through the specimen for its magnification, the specimen must be extremely thin or transparent for viewing under a compound microscope. In other words, the compound microscope works by **transmitted** light passing through the specimen.

There are times when a compound light microscope is not suitable for magnifying a larger object such as a flower or butterfly. These specimens are best observed by a **stereomicroscope** (Fig. 1–4). This microscope is called "stereo" because it has two eyepieces that give a three-dimensional view of the specimen. The stereomicroscope is useful for examining specimens that are relatively large and opaque and through which light cannot pass. The stereomicroscope works by **reflected** light; that is, the magnified image is formed as the lenses gather and focus light that strikes and bounces off the specimen. For this reason, specimens do not have to be thinly sectioned or transparent in order for the stereomicroscope to magnify a specimen as they do with the compound microscope.

The labs that follow have been designed to introduce you to the use of these microscopes.

## PRELAB QUESTIONS

1. What is the source of energy for an electron microscope?

2. What type of light microscope is used to view large objects such as flowers?

3. What is the difference between transmitted light and reflected light?

4. Why is reflected light used with a stereomicroscope?

## LAB 1.A
## ANATOMY OF THE COMPOUND MICROSCOPE

### MATERIALS REQUIRED
compound light microscope

### PROCEDURE
The microscopes in your lab are not indestructible. You can prevent costly scope damage by following a few simple precautions.

Always carry a microscope with both hands, one supporting the base and the other gripping the arm of the scope. Place the microscope gently on the table.

Store the microscope properly. The shortest objective (Fig. 1–1) should be in place to prevent possible damage to higher power objectives and the electrical cord should be wound neatly around the base. Never leave slides on the stage (Fig. 1–1) during storage. Always place the dust cover (if available) over the microscope when stored.

Use *only* lens paper to clean the objectives and eyepiece lenses (Fig. 1–1). Lens paper should be available within your lab room. Clean the lenses by gently wiping them with a circular motion of the lens paper. If the lens is still soiled, notify your instructor.

In order to use any instrument properly it is necessary to know the various components of its structure. Carefully bring a microscope to your work space if one is not already present. Refer to your own microscope and Figure 1–1 as you follow through this part of the exercise that will acquaint you with microscope anatomy. There are slight variations among different scope models and brands, but they are all similar in basic structure.

1. Position the microscope so that the **ocular lens,** or eyepiece, is pointing toward you. The ocular is the upper lens where you observe the magnified image of the specimen. Your ocular lens may have a movable pointer within it that is useful in locating items within the image. The pointer is moved by rotating the ocular lens. However, do not remove the eyepiece from the body tube.
2. Locate the **arm** of your microscope. The arm is the upper body structure that supports the assembly of lenses and is the proper place to grasp the scope for carrying.
3. The **objectives** are essentially tubes that contain lenses of various magnification power. The magnification power is engraved or painted on each objective. The shortest objective has a magnification power of 4× (pronounced "four times," meaning four times the actual diameter of the specimen). The 4× objective is sometimes called the **scanning lens** because its lower magnification gives an overall view of the specimen. The short 4× objective should always be in place (pointing straight down) when the microscope is stored. This lessens the chance of someone ramming an objective through a microscope slide as they begin focusing. The next longer objective is 10×, also known as **low power.** The third objective after low power may be either 43× or 45×, known as **high power.** Some microscopes may have a fourth objective, which is the longest in length. This is an oil immersion lens that has a magnification of 100×. An oil immersion objective requires the use of immersion oil on the slide, which covers the objective lens when it is in place. The immersion oil increases the resolving power because the oil has a similar index of refraction as glass. This practice is somewhat messy and

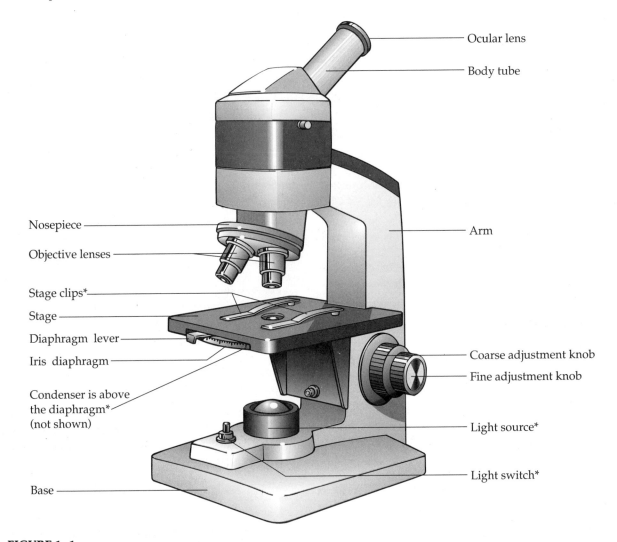

Ocular lens

Body tube

Nosepiece

Objective lenses

Stage clips*

Stage

Diaphragm lever

Iris diaphragm

Condenser is above
the diaphragm*
(not shown)

Base

Arm

Coarse adjustment knob

Fine adjustment knob

Light source*

Light switch*

**FIGURE 1-1**

Your compound microscope may be similar to the one shown in this diagram. Some of the
parts that may not be present on your microscope are marked with an asterisk (*).

beyond the needs of this course. Therefore, you will not be using the oil immersion objective if you have one on your microscope.

The total magnification of a specimen is determined by multiplying the power of the objective by the power of the ocular lens. The ocular lens typically has a magnification power of 10×. With the scanning lens in position, a specimen would be magnified a total of 4 × 10, or 40×.

What is the total magnification of a specimen under low power? ___ High power? ___

The objectives rotate on the **nosepiece** which supports them. Change objectives by turning the nosepiece either to the right or to the left until you feel the next objective click into place and it points straight down.

4. The **stage** is the horizontal surface that supports

the specimen for viewing. The specimen is always placed on a glass microscope slide, which is easy to handle. Notice the hole in the stage. The specimen must always be positioned over this hole for observation. Most scopes are usually equipped with **stage clips** that hold the slide in place by slipping it under the clips.

5. Below the stage is the **condenser,** which does just that — it collects and condenses light coming from below into a focused beam that shines through the hole and specimen on the stage thereby enhancing the resolving power. The light then passes through the objective and other internal lenses forming a magnified image visible through the ocular.

6. The **iris diaphragm** associated with the condenser functions similarly to the iris of your eye — it regulates the amount of light passing through the mi-

croscope lenses, producing various degrees of image brightness. Since many cellular specimens are nearly transparent, there may be a need to reduce the light level to observe structures. Your microscope may have a small lever below the stage that moves side to side that opens and closes the iris diaphragm. The iris diaphragm may also be a rotating wheel with various size holes that allow more or less light to pass through the specimen. Determine which kind of iris diaphragm your microscope has.

7. Focusing knobs are located on both sides of the microscope. The larger knob next to the body of the scope is the **coarse adjustment knob** used for general focus upon the specimen. The smaller knob is the **fine adjustment knob** used for focusing on small details.

8. The **illuminator** is located directly below the condenser. It contains a light bulb, which is the light source of your scope. The on–off switch may be just behind the illuminator, or it may be found on the electrical cord. Some microscope models have a small circular mirror mounted below the condenser instead of an illuminator. The mirror is adjusted to reflect light from a window or another source into the condenser.

9. The **base** is the lower foundation of the microscope that sits on the table.

## POSTLAB QUESTIONS

1. If an ocular lens has a magnifying power of 15× and the objective lens is 10×, what is the total magnification?

2. Identify the part of the microscope with the following functions:
   a. lens that further magnifies the image formed by the objective lens;

   b. concentrates the light before it passes through the specimen;

   c. gives the microscope a firm, steady support;

   d. regulates how much light goes through the specimen;

   e. adjustment knob that gives the microscopist

more precise control over the position of the lens;

f. source of light; and

g. projects a magnified image of the specimen just beneath the ocular.

## LAB 1.B
## USE OF THE MICROSCOPE

### MATERIALS REQUIRED
compound microscope

prepared slide with letter "e"

prepared slide with crossed-colored threads

### PROCEDURE

1. Obtain a slide that has a letter "e" mounted on it as a specimen for focusing practice. Position the microscope so that the ocular is pointing toward you.

2. Plug in the electrical cord and turn on the illuminator. If the microscope was properly stored, the lowest power objective should already be in position. Slowly turn the coarse adjustment knob to move the 4× objective and stage as close as possible to one another while watching the scope from the side. NEVER force an adjustment knob after it stops turning! Depending on the microscope model you have, either the nosepiece or the stage will move when the coarse adjustment knob is turned.

3. Push aside the stage clips. Place the slide on the stage so that the letter is positioned right side up and centered over the hole. Place the stage clips over the ends of the slide to hold it in position.

4. When you look into the ocular lens, try to keep BOTH eyes open. This may seem difficult at first, but you will soon learn to concentrate only on the image in the ocular lens. Keeping both eyes open when working with the microscope reduces eye strain. If you wear glasses, you may find viewing more convenient without them.

5. Begin focusing by looking into the ocular lens and carefully turning the coarse adjustment knob to move the objective and stage away from one an-

other. Continue turning the coarse adjustment until the image is clear. The fine adjustment knob is usually required less at low magnifications such as this.

*Remember:* Always begin focusing with the lowest power objective in place and as close to the stage as possible, then slowly move the objective and stage away from one another. This focusing method is effective and avoids the possibility of jamming an objective into the slide.

Look again at the magnified image of the letter. Is it positioned the same as the letter on the slide?

_____

If not, how is it different? _____

_____

6. While looking into the ocular lens, gently push the letter slide to the left. Which direction does the letter move? _____

Once again look into the ocular lens as you gently pull the slide toward you. Which direction does the letter move? _____

7. This is a good time to observe the effect of the iris diaphragm. Experiment with the iris diaphragm to produce various degrees of brightness, then adjust it to what is comfortable for your eyes.

8. With the letter slide in focus, notice the considerable space between the slide and the objective. This space is called **working distance.** Change objectives to the next higher power and adjust the focus. Now note the working distance between the objective and slide.

Did working distance increase or decrease with the move to a higher power objective? _____

How would working distance be affected if you moved to another objective of higher power?

_____

_____

9. The **field of view** is the area of the slide or specimen that you can see in the ocular lens. Move the scanning objective back into place and center the letter "e" within the field of view.

Can you see the entire letter within the field?

_____

Without moving the slide, change to the next higher power and focus on the letter using the fine adjustment knob. Can you see as much of the letter as you did under the previous objective?

_____

If available, change to the next higher power and again focus on the letter using the fine adjustment knob. Can you see as much of the letter as you did under the previous objective? _____

10. Move the lowest power objective into position and focus on the letter "e." Note the brightness of the field. Now change to the high power objective and note the brightness of the field.

Is the field of view brighter under low or high power? _____

This is an important point to remember when using high power—the field becomes dimmer with each increase in magnification. It is usually good to use more light under high power by adjusting the iris diaphragm. However, too much light with high power can wash out fine details in the image. With a little practice, you will soon be able to determine the proper amount of light to use. Remove the letter "e" slide from your microscope and return it to the proper place.

11. Depth of field is another characteristic of microscopy that changes with magnification. **Depth of field** is the vertical depth of an image that remains in sharp focus under a particular magnification. As an example, a magnified image with great depth of field allows the viewer to observe several layers of cells on a slide, all appearing in clear focus. A magnified image with lesser depth of field forces the viewer to focus up and down through the cells, in order to see each layer in clear focus. Obtain a slide of three colored threads. Using the scanning objective, place the slide on the stage so that the threads intersect at the center of the field of view. Observe the three threads in the area where they intersect.

Are all three threads equally in focus under lowest power? _____

Change to the next higher power and focus on the point of intersection. Are all three threads equally in focus? _____

If available, change to the next higher power and focus on two intersecting threads. Are both threads equally in focus? _____

In order to see the individual threads clearly, is more focusing required under high or low power? _____

## POSTLAB QUESTIONS

1. Using these observations, what is the relationship between working distance and magnification?

2. Using these observations, what is the relationship between field of view and magnification?

3. Using these observations, what is the relationship between depth of field and magnification?

## LAB 1.C
## PREPARATION OF A WET MOUNT

### MATERIALS REQUIRED

slides and coverslips

*Elodea* leaves

dropper bottle of water

pond water or live protozoan cultures

### PROCEDURE

One of the major advantages of the light microscope is that it permits the observation of microscopic *living* things. The microscope enables us to sit and watch a unicellular organism swim about in the same way as we might watch a duck swim on a pond. Living cells or any fresh material is observed under the microscope by making a **wet mount** on a slide.

1. Obtain a clean slide and coverslip.
2. Using forceps, remove a small leaf from the growing tip of the water plant, *Elodea,* and lay it flat on the center of the slide.
3. Add two drops of water to the leaf.
4. Hold the coverslip at an angle by two of its corners and touch the free edge to the slide as in Figure 1 – 2.
5. Slowly pull the coverslip edge across the slide until it meets the water. The water will then spread along the edge of the coverslip.
6. Carefully lower the coverslip onto the slide using your fingers or a needle probe (Figure 1 – 2). If all went well, the water should be spread evenly under the coverslip with no air bubbles. If you have any air bubbles under your coverslip, prepare another wet mount of this same leaf by drying your slide and starting over. With a little practice, you will become efficient at making wet mounts.
7. Place the wet mount on your microscope stage and observe it with the lowest power objective by using the focusing technique described above in Lab 1.B.
8. Observe the leaf under low and high power. Adjust the iris diaphragm to a brightness that reveals the most detail in the leaf. The *Elodea* leaf is only a few cell layers thick, making it ideal for observing plant cells. Note that the leaf resembles a brick wall with each ''brick'' being a cell. The numerous green circles within the cells are **chloroplasts.** They enable the plant to produce its own food by photosynthesis using the energy of the sun. The obvious boundary that outlines each cell is the **cell wall** found in all plant cells, but which is absent in all animal cells.
9. Practice making wet mounts by exploring the pond water or live cultures available on the lab bench. Use a dropper to draw up water and debris from the *bottom* of the container. This ensures that you will capture a good sample of the microorganisms for observation. Place one or two drops of water on a slide and add a coverslip. Ask your instructor for assistance if you have trouble locating specimens.
10. Clean your wet mount slides when finished and

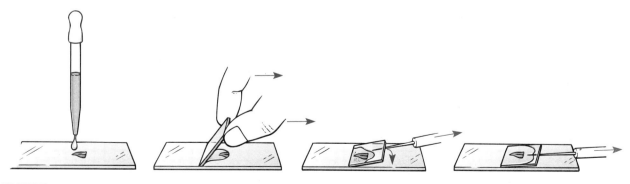

**FIGURE 1 – 2**

Wet mount your living specimen as shown to avoid trapping air bubbles under the coverslip.

dispose of the coverslip according to your instructor's directions.

## POSTLAB QUESTIONS

1. Sketch the *Elodea* leaf observed under the compound microscope. Label the chloroplasts and cell wall.

2. Sketch the various microorganisms observed in the pond water or live cultures provided. If instructors have provided a key, identify these microorganisms.

## LAB 1.D
## THE ADVANTAGE OF STAINING

### MATERIALS REQUIRED

slides and coverslips

forceps

onion pieces

dropper bottle of water

dropper bottle of acetocarmine stain

## PROCEDURE

Cells are crammed with a great variety of specialized structures that have various functions such as the chloroplasts you saw in *Elodea*. These cellular structures are generally referred to as **organelles,** or "little organs," and their functions maintain the cell in a way similar to the organs that keep your body functioning. A great majority of organelles are colorless making them essentially invisible under the microscope. Chloroplasts are an exception because they contain the green pigment chlorophyll which gives plants their green color. Many kinds of special stains have been developed that color specific cell organelles making them visible under the microscope.

1. Obtain a slide, coverslip, and a small piece of onion.
2. Use forceps and carefully remove the thin epidermis, or "skin," from the inner curved side of the small piece of onion layer as shown in Figure 1–3.
3. Place the thin epidermis on the slide avoiding any folds or wrinkles.
4. Make a wet mount of the onion epidermis by adding a drop of water and coverslip to the slide as described in Lab 1.C.
5. Observe the onion epidermis under low and high power. Do you see cell walls in the epidermis? _____ Do you see chloroplasts as in the *Elodea* cells? _____
6. Carefully lift the coverslip from the onion epidermis and add to it a drop of acetocarmine stain. Replace the coverslip and allow the stain to penetrate the onion skin for three minutes. Observe the onion cells again under low and high power. The red acetocarmine stain is absorbed by the cell **nucleus** making it appear as a small red sphere within the cell. The nucleus contains the cell's genetic material that ultimately controls all of the cell functions.
7. Clean your slide and return it to its proper place.

## POSTLAB QUESTIONS

1. If chloroplasts are typical of plant cells, explain why they are not present in onion cells.

2. Your textbook describes many kinds of cell organelles. Why do you think these numerous organelles are not visible in your stained slide of the onion epidermis?

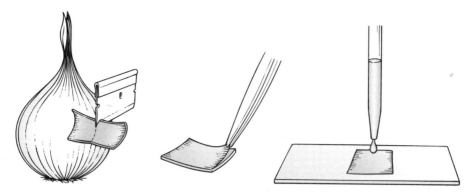

**FIGURE 1–3**

Onion skin epidermis is useful for learning staining technique. The desired epidermis is extremely thin and transparent, and must be removed carefully.

## LAB 1.E
## THE STEREOMICROSCOPE

### MATERIALS REQUIRED

stereomicroscope

large biological specimens such as mold culture, live insects, moss, etc.

### PROCEDURE

As you can see in Figure 1–4, the stereomicroscope has basically the same structure as the compound microscope and should be treated with the same care and precautions. There are some minor differences between the stereo and compound microscopes. The greater working distance between the stage and objective of the stereomicroscope is an advantage because it accommodates larger objects. The stereomicroscope also has a much greater field of view and depth of field than the compound microscope, making it very useful for examining larger specimens. The stereomicroscope is not designed for greater magnifications. Maximum magnification in your stereomicroscope is about 30X. The objective is not moved on the stereomicroscope to change magnification. The zoom knob continually increases or decreases magnification as it is turned. The stereomicroscope has only one focus adjustment knob. The movable ocular lenses can be adjusted to fit your eyes by gently pulling them outward or inward.

Your stereomicroscope may have a built-in light source as in Figure 1–4, or the light source may be a separate, movable lamp. If the light source is separate, the stereomicroscope may have slots in the upper body of the instrument and behind the base for holding the lamp. Turning the illumination control knob on the base of the stereomicroscope creates various background and lighting effects regardless of what kind of light source you have.

Practice using the stereomicroscope by examining containers of specimens such as bread mold, live beetles, and moss. Be sure to use the zoom knob for various powers of magnification. Compare the optical characteristics of the stereomicroscope with the compound scope.

### POSTLAB QUESTIONS

1. As you moved the specimen to the right on the stage, did the image move in the same direction?

2. Was this observation true of the compound microscope?

3. Did the image of the specimen appear in the same position as the specimen on the stage?

4. Was this observation true of the compound microscope?

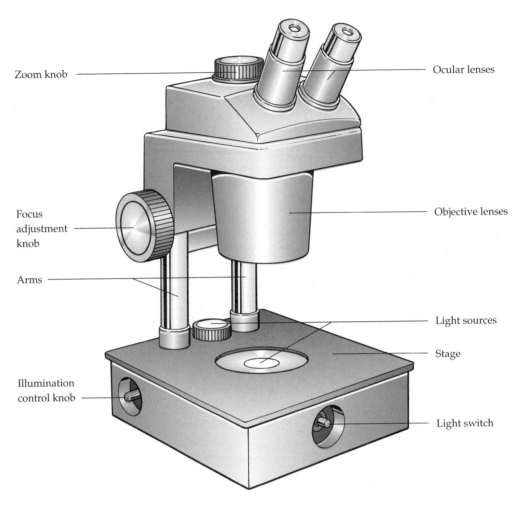

Zoom knob

Ocular lenses

Focus
adjustment
knob

Objective lenses

Arms

Light sources

Stage

Illumination
control knob

Light switch

**FIGURE 1–4**

Your stereomicroscope may be similar to the one shown in this diagram. If not, your instructor
may be able to provide a diagram that exactly represents your microscope.

## LAB 1.F
## THE ELECTRON MICROSCOPE

### MATERIALS REQUIRED

scanning and transmission electron micrographs of
biological specimens

### PROCEDURE

The invention of the electron microscope made a tremendous impact on biology, perhaps as much as the
invention of the original light microscope. Both the
light and electron microscopes have their advantages
and disadvantages. Light microscopes are simple to
use, relatively inexpensive, and portable. Electron microscopes (Fig. 1–5) require a trained technician for
their operation, are costly, and are so large in size that
they usually have a room to themselves. Living organisms can be observed unharmed with a light microscope. An electron microscope requires special preparation of a specimen before it can be observed. The
preparation process is harmful to cells, so living organisms cannot be viewed with the electron microscope.
However, the fantastic magnification of the electron
microscope greatly outweighs any of its disadvantages. It is actually possible to observe *molecules* with
the power of an electron microscope!

In addition to magnification, there is another quality of the electron microscope that makes it ideal for
observing extremely small objects such as cell organelles. The electron microscope has extremely high
**resolution,** the property that determines whether

small objects very close together can be seen separately instead of blurring into one point. Even though the images produced by an electron microscope are in black and white, never color, the exceptional resolution of the electron microscope far exceeds that of even the best light microscope. The resulting images are of crisp detail and amazing clarity.

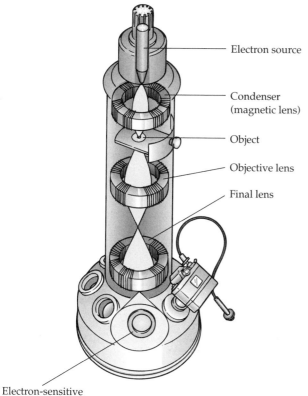

**FIGURE 1–5**

An electron beam is focused by the condenser lens onto the specimen. The objective lens first forms a magnified image of the specimen, which is further magnified by the projector lens onto a fluorescent screen in the electron microscope. The lenses in the electron microscope are actually magnets that bend the beam of electrons.

There are two basic types of electron microscopes. The **scanning electron microscope** scans the specimen with a spray of electrons and forms an image by detecting electrons coming back from the specimen. Images (Fig. 1–6) of the scanning electron microscope have amazing texture and a three-dimensional quality.

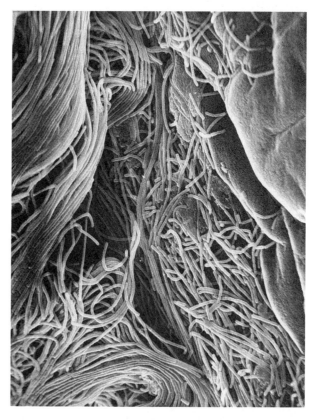

**FIGURE 1–6**

A scanning electron micrograph of a *Bacillus subtilis* colony growing on the surface of a nutrient agar plate, 3000X. *(Scanning electron micrograph by Dr. Robert A. Pfister.)*

1. Observe the scanning electron micrographs (Figures 1–7 to 1–9). (**Micrograph** is the term for a photograph taken by means of a microscope.)

   The **transmission electron microscope** works in a manner similar to the compound light microscope—the specimen must be thinly sectioned, which allows electrons to pass through it. The electrons passing through the specimen are scattered in patterns similar to the specimen structure. The electron patterns then determine the characteristics of the magnified image (Fig. 1–10).

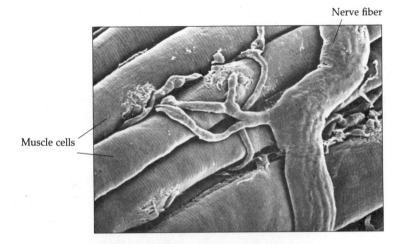

Nerve fiber

Muscle cells

**FIGURE 1-7**

This scanning electron micrograph shows a nerve fiber communicating with several muscle cells (approximately 900X). *(From Desaki, J., Biomedical Research Supplement 139-143, 1981)*

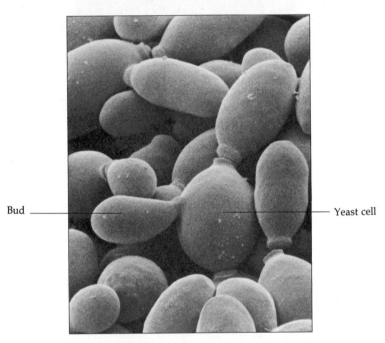

Bud

Yeast cell

**FIGURE 1-8**

A scanning electron micrograph of budding yeast cells (approximately 1800X). *David Scharf)*

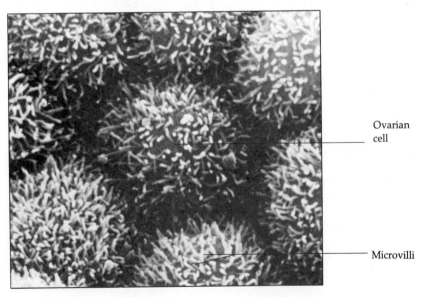

Ovarian cell

Microvilli

**FIGURE 1-9**

A scanning electron micrograph of mouse ovarian epithelial cells showing microvilli (approximately 65,000X). Microvilli occupy the free surfaces of many kinds of cells, greatly increasing the surface area for the absorption of materials. *(Courtesy of Dr. E. Anderson, Harvard Medical School)*

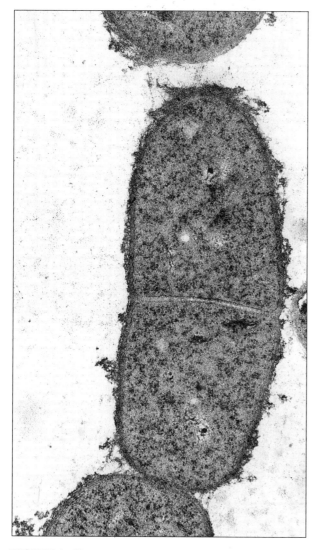

**FIGURE 1–10**

A thin section transmission electron micrograph of *Bacillus cerus*, 80,000X, showing a cross wall and products formed by the cell on the exterior wall. *(Electron micrograph by Dr. Robert A. Pfister.)*

2. Observe the transmission electron micrographs (Figures 1–11 to 1–13).

**POSTLAB QUESTIONS**

1. Based on your observations of the electron micrographs, which type of electron microscope is best for observing internal cell structure?

2. Which electron microscope is best for observing surface features and overall appearance?

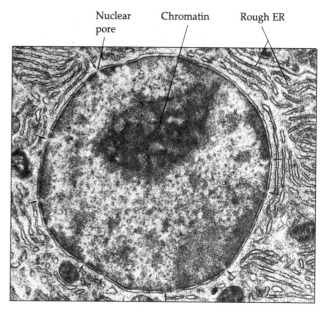

Nuclear pore    Chromatin    Rough ER

**FIGURE 1–11**

Electron micrograph of the nucleus of a pancreatic acinar cell (approximately 40,000X). Note the two membranes that form the nuclear envelope. Arrows indicate the nuclear pores.

3. Identify the appropriate microscope or microscopes:

a. uses light as an energy source,

b. image is neither reversed nor inverted,

c. may view living material,

d. largest depth of field,

e. must use fixed or dead material,

f. uses a beam of electrons as a source of illumination,

g. can observe the internal structure of a nucleus, and

h. highest resolving power.

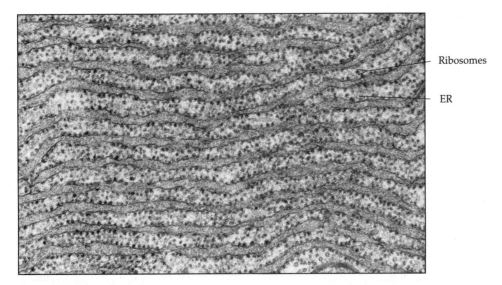

**FIGURE 1-12**

Electron micrograph (magnification approximately 70,000X) of the rough endoplasmic reticulum from a secretory cell of the sea anemone *Metridium.*

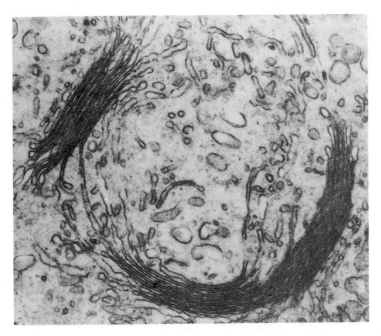

**FIGURE 1-13**

Electron micrograph of a section through the Golgi complex from the sperm cell of a ram. *(Don    Fawcett/Photo Researchers, Inc.)*

## FOR FURTHER READING

Burrells, W. 1977. *Microscope Technique: A Comprehensive Handbook for General and Applied Microscopy.* New York: Wiley.

Freeman, J. A. 1983. *Laboratory Medicine/Urinalysis and Medical Microscopy.* Philadelphia: Lea and Febiger.

Griffin, R. L. 1990. *Using the Transmission Electron Microscope in the Biological Sciences.* New York: Ellis Horwood.

Harrison, F. W. 1991. *Microscopic Anatomy of Invertebrates.* New York: Wiley-Liss.

Heckner, F. 1988. *Practical Microscopic Hematology: A Manual for the Clinical Laboratory and Clinical Practice.* Baltimore: Urban and Schwarzenberg.

James, J. 1991. *Biomedical Light Microscopy.* Boston: Kluwer Academic Publishers.

Locquin, M. 1983. *Handbook of Microscopy.* Boston: Butterworths.

Marmasse, C. 1980. *Microscopes and Their Uses.* New York: Gordon and Breach Science Publishers.

Robards, A. W. 1985. *Botanical Microscopy 1985.* New York: Oxford University Press.

Roos, N. 1990. *Cryopreparation of Thin Biological Specimens for Electron Microscopy: Methods and Applications.* New York: Oxford University Press.

Simpson, D. 1988. *An Introduction to Applications of Light Microscopy in Analysis.* London: Royal Society of Chemistry.

Smith, R. F. 1982. *Microscopy and Photomicrography: A Practical Guide.* New York: Appleton-Century-Crofts.

Tryon, A. F. 1991. *Spores of the Pteridophyta: Surface, Wall Structure, and Diversity Based on Electron Microscope Studies.* New York: Springer-Verlag.

# Scientific Method

## OBJECTIVES

1. Define "scientific method" and list the general sequence whereby scientists build on observations in the formulation of laws.

2. Analyze an experiment and its features in terms of the scientific method.

3. Compare and give examples of the concepts of hypothesis, theory, and principle.

## INTRODUCTION

All fields of science have one unifying principle that is a common tie among these diverse scientific disciplines. That unifying tie is the **scientific method.**

The scientific method is simply an organized, methodical, and structured way of observing and/or investigating a situation in an effort to find information about what is being observed. There are six steps to the scientific method.

The first step is the identification of the situation to be investigated. This is vital because no progress can be made towards understanding the situation unless one knows exactly what is being investigated. Let's consider an example. Suppose that you notice a list of essential nutrients on the label of a box of fertilizer. You wonder how plant growth might be affected if the plant is deprived of just one of those essential nutrients. You decide to investigate the effect of the lack of nitrogen on tomato plants.

The second step of the scientific method is to obtain information about that which is being investigated. One of the biggest advantages in problem solving is knowing the background information about what is being investigated. This is why researchers do searches of the scientific literature when writing a paper or conducting research. Accordingly, you go to the library and read as much as you can about plant nutrition. Let us assume that your literature research was particularly thorough concerning information about nitrogen.

Step three is the formulation of a hypothesis that can be tested to prove or disprove its validity. A **hypothesis** is a possible explanation of the problem or situation based only on what one knows about it so far. Just how good or bad this initial hypothesis really is as a true answer to the problem or as a correct accounting of the situation will be determined by testing, which is a later step. Your first hypothesis might well be "Plants grown in a medium lacking in nitrogen will show definite signs of malnutrition."

The fourth step is to predict the results of a test of the hypothesis. In other words, assuming your hypothesis is correct, you ought to be able to predict the outcome of a situation where your hypothesis was actually applied to the problem. You might now try to imagine how a tomato plant would look when grown in a nitrogen-free medium. Perhaps there would be obvious changes in the leaves and/or the stem height.

We need to define a few terms. An **experiment** is an investigation conducted under very specific conditions wherein all variables are controlled except the one being studied. A **variable** is an event or a condition subject to change. In the nitrogen study, the lack of nitrogen is the variable being investigated.

Step five is to design and conduct an experiment to test the hypothesis. If the hypothesis should be wrong, then it can be modified and further tested. The scientific method commonly results in a long series of repeated testing and hypothesis modification. A hypothesis can never be proven right unequivocally. The result of this is that the hypothesis gradually evolves into becoming more and more valid for the situation or problem. The evolution of a hypothesis is based on conducting experiments, making observations, gathering data, etc., all of which are done to investigate the validity and to challenge the hypothesis under consideration.

You conduct your experiment based on a technique discovered in the literature search. You grow your tomato plants hydroponically (in solutions of plant nutrients with no soil). In your experiment, you have two groups of plants, each group consists of six tomato plants of the same variety and all are the same age, size, and general state of health. In addition, both groups of plants are grown under exactly the same environmental conditions of heat, light, container size, etc. It is important that all of the conditions (except the one being investigated, nitrogen) be exactly the same for both groups. The only difference between the two groups is that one is grown with complete nutrients, the other has complete nutrients except nitrogen.

When your experiment is run, the plants are allowed to grow for two weeks, after which time the plants are compared. In this experiment, the plants growing in the complete nutrient solution serve as the **control group,** which is the group forming the basis for judging any differences that may appear in the **experimental** group, the group grown without nitrogen. A control is essential in any experiment because it reveals any differences in the experimental situation.

Step six is to form a conclusion based on the results of the experiment. The validity of the hypothesis may or may not be determined. Either the results of the experiment support the hypothesis or the results show that the hypothesis needs modification. If you found the control plants to be lush and green with a height increase of three inches since the experiment began and the experimental plants to have no increase in height, to have weak stems, and to have yellowish leaves with brown spots, you would have supported your hypothesis.

In the strictest sense, of course, you cannot say that nitrogen is necessary for plant growth. The experiment does not prove the hypothesis to be correct beyond all shadow of doubt. What the experiment does show, is that under the conditions of the experiment, nitrogen appears necessary, and the hypothesis is supported.

If a hypothesis is continually supported by extensive investigation, it eventually earns the status of a **theory.** A theory is a far more substantial account than a hypothesis because the theory has come to be accepted as a tried-and-true explanation for a situation and has a degree of probability or predictive value. If a theory stands the test of time and becomes widely accepted, it is known as a **scientific principle** or **law.**

There is no question that the scientific method works as a means of research and problem solving. Virtually all of today's technological, medical, and agricultural advances that have improved our lives are a direct outgrowth of discoveries made through the use of the scientific method.

The following laboratory exercises have been devised to provide you with some direct experience in using the scientific method.

## PRELAB QUESTIONS

1. What is a hypothesis?

2. What is an experiment?

3. What is a variable?

4. What is the difference between the control group and the experimental group?

5. Why is a theory more widely accepted than a hypothesis?

## LAB 2.A
## THE SCIENTIFIC METHOD

### MATERIALS REQUIRED
zodiac table

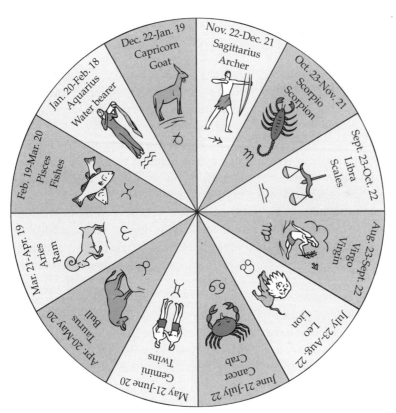

**FIGURE 2-1**

Signs of the Zodiac.

## PROCEDURE

Are our lives influenced and/or affected by the signs of the zodiac? Is what we are to become in life determined by the zodiac sign (Fig 2-1) under which we were born? If that is so, then isn't it reasonable that a group of people, such as the students in your class, all of whom, generally, are pursuing a major in one of the sciences, were more likely born under one or a few of the zodiac signs? Or is it the case that the birth dates of the students in this class are randomly distributed among the signs of the zodiac? Let us apply the scientific method to the problem defined by those questions.

The zodiac consists of 12 divisions or signs. Astrologers—individuals who tell fortunes by studying the stars—believe that a person comes under the influence of the sign under which he or she was born. Our intent here is not to prove or to discredit definitively the contention that the zodiac influences our lives but rather to use the scientific method to investigate the possibility of such contentions. For discussion and demonstration purposes, it might be interesting to have a class discussion concerning these questions before completing the balance of this laboratory exercise in which we will use the scientific method to investigate these questions.

We have identified the situation and are ready to perform the second step of the scientific method, which is the collection of information about what is being investigated. To do this, we need to have every member of the class identify the sign of the zodiac under which that individual was born. The signs of the zodiac and the dates that are generally accepted as defining those signs are as follows:

January 20 through February 18—Aquarius
February 19 through March 20—Pisces
March 21 through April 19—Aries
April 20 through May 20—Taurus
May 21 through June 20—Gemini
June 21 through July 22—Cancer
July 23 through August 22—Leo
August 23 through September 22—Virgo
September 23 through October 22—Libra
October 23 through November 21—Scorpio
November 22 through December 21—Sagittarius
December 22 through January 19—Capricorn

To begin, the instructor should write the names of the signs of the zodiac on the board, and each student should then place a check mark after the sign under which the student was born. This will result in a frequency distribution, which is the observed or raw data. If the class is small, the course instructor may

**Table 2-1** Worksheet

| Zodiac Sign | Observed Number O | Expected Number E | (O − E) | (O − E)²/E |
|---|---|---|---|---|
| Aquarius | | | | |
| Pisces | | | | |
| Aries | | | | |
| Taurus | | | | |
| Gemini | | | | |
| Cancer | | | | |
| Leo | | | | |
| Virgo | | | | |
| Libra | | | | |
| Scorpio | | | | |
| Sagittarius | | | | |
| Capricorn | | | | |
| *Sum* | | | | |

decide to supply the class with a frequency distribution obtained from other sources, for example, from several other classes or several previous courses, so that there will be adequate data with which to work. The raw data should be recorded as the observed numbers in the worksheet provided in Table 2-1.

The hypothesis to be tested is that the observed data are distributed at random among the divisions of the zodiac. There are a number of other ways in which that hypothesis can be stated. For example, the hypothesis can be stated that the observed number of births in the zodiac classes are essentially equal. Can you think of alternative statements of that hypothesis? Our hypothesis is also called the null hypothesis.

With the raw data, we are ready to test the null hypothesis. To do so, we need to use a statistical analysis known as a goodness of fit test that provides, for our purposes, a relatively simple and adequate test. The details of the analysis can best be understood by an example.

When we have data consisting of the numbers of individuals falling within a set of classes (such as our signs of the zodiac), the data and the analysis can be set out in the form of a table such as that found in Table 2-2.

**Table 2-2** Example of Goodness of Fit Test

| Class | Observed Number O | Expected Number E | (O − E) | (O − E)²/E |
|---|---|---|---|---|
| A | 30 | 40 | −10 | 2.5 |
| B | 50 | 40 | 10 | 2.5 |
| C | 20 | 40 | −20 | 10.0 |
| D | 40 | 40 | 0 | 0 |
| E | 60 | 40 | 20 | 10.0 |
| *Sum* | 200 | 200 | 0 | 25.0 |

The first column, the left-most column, lists the five classes (A, B, C, D, and E) in our example. For the class zodiac data, there are 12 classes. The second column gives the raw or observed data and the total number of observations. The expected number, the middle column, is calculated as the total number of observations divided by the number of classes. The expected number calculated in this way is based upon the assumption that the numbers in the classes are equally frequent.

The fourth column from the left is the difference between the observed and the expected. The rightmost column gives the quotient of the square of the observed minus expected divided by the expected number. Finally, the sum of the numbers in the rightmost column is the calculated test statistic. For the analysis we are discussing here, the test statistic is a chi-square.

The chi-square distribution is a family of theoretical curves, each identified by a parameter known as degrees of freedom. In our case, the degrees of freedom are one less than the number of classes in the analysis. These theoretical curves supply the critical values used to interpret the significance of the calculated test statistic, and that interpretation depends upon the chosen level of significance.

The level of significance is formally defined as being the maximum tolerable probability of accepting a false null hypothesis. Commonly, either 5% or 1% is selected as the level of significance; the critical value of chi-square can then be obtained from Table 2–3.

**Table 2–3** Critical Chi-Square Values

| Degrees of Freedom | Level of Significance | |
| --- | --- | --- |
| | 5% | 1% |
| 1 | 3.841 | 6.635 |
| 2 | 5.991 | 9.210 |
| 3 | 7.815 | 11.341 |
| 4 | 9.488 | 13.277 |
| 5 | 11.070 | 15.086 |
| 6 | 12.592 | 16.812 |
| 7 | 14.067 | 18.475 |
| 8 | 15.507 | 20.090 |
| 9 | 16.919 | 21.666 |
| 10 | 18.307 | 23.209 |
| 11 | 19.675 | 24.725 |
| 12 | 21.026 | 26.217 |
| 13 | 22.362 | 27.688 |
| 14 | 23.685 | 29.141 |
| 15 | 24.996 | 30.578 |
| 16 | 26.296 | 32.000 |
| 17 | 27.587 | 33.409 |
| 18 | 28.869 | 34.805 |
| 19 | 30.144 | 36.191 |
| 20 | 31.410 | 37.566 |

If we select 5% as our level of significance, then the critical value of chi-square with four degrees of free-

dom, from Table 2–3, is 9.488, which we then compare with our calculated test statistic. Our calculated test statistic was 25.0, which exceeds our critical value. Therefore, we reject the hypothesis that the five classes are equally frequent, which is another way of saying that the observations are not randomly distributed among the five classes.

Had the calculated test statistic been less than or equal to the critical value, we would have been unable to reject the null hypothesis and would have concluded that the observations were randomly distributed among the classes.

Perform the chi-square goodness of fit test for the zodiac data of your class (or the data supplied by your instructor) by completing Table 2–1. Determine whether or not there is evidence in support of, or against, the hypothesis that the observed data are randomly distributed among the classes defined by the signs of the zodiac.

If time allows, apply the scientific method and the chi-square goodness of fit test to determine whether the digits 0 through 9 occur randomly in the first digit of the Social Security Numbers of the students in your class. Repeat the investigation with respect to the last digit of the students' Social Security Numbers. If there is a lottery in your area, apply the scientific method to determine whether or not the digits of the winning lottery numbers for, say, the past six weeks (data to be supplied by the instructor) are random. Can you think of other instances where one might investigate whether or not the digits 0 through 9 randomly occur?

## POSTLAB QUESTIONS

Answer the following questions for the zodiac data analysis and for each assigned analysis in this laboratory exercise.

1. Write the specific steps of the scientific method that were used in the exercise.

2. What hypothesis was tested?

3. What observations were made?

4. Did the observations support the hypothesis?

5. What variable was studied?

6. What conclusion was made from the analysis?

The following questions are not associated with any one specific data analysis but are general in nature.

7. Identify, apart from the situations mentioned in this exercise, three other possible investigations wherein the chi-square goodness of fit test might be used.

8. In a chi-square goodness of fit analysis, there were 200 observations among four classes (A, B, C, and D) and the expected numbers in the classes were in a $9:3:3:1$ ratio for the classes A, B, C, and D, respectively. What would be the expected numbers?

9. Under what circumstances might it be more desirable or appropriate to use a 1% level of significance rather than a 5% level of significance?

10. Does the chi-square goodness of fit test provide an absolute test of the null hypothesis involved?

11. How would increasing sample size (total number of observations) affect the confidence one has in the conclusion from an analysis?

## LAB 2.B
## SNOW ON CHOLERA

In the 1800s, Dr. John Snow applied the scientific method in his classical study of cholera. Cholera is a bacterial infection of the small intestine that causes severe diarrhea and vomiting. Untreated, more than half of its victims will die due to dehydration.

In 1855, Snow recorded several cases of cholera. The following quote depicts a typical case of cholera. "A man came from Hull (where cholera was prevailing), by trade a painter; his name and age are unknown. He lodged at the house of Samuel Wride, at Pocklington; was attacked on his arrival on the 8th of September, and died on the 9th. Samuel Wride himself was attacked on the 11th of September and died shortly afterwards. . . ." (Snow, 1936)

Snow also noted,

*Nothing has been found to favour the extension of cholera more than want of personal cleanliness, whether arising from habit or scarcity of water, although the circumstance till lately remained unexplained. The bed linen nearly always becomes wetted by the cholera evacuations, and as these are devoid of the usual colour and odour, the hands of persons waiting on the patient become soiled without their knowing it; and unless these persons are scrupulously cleanly in their habits, and wash their hands before taking food, they must accidentally swallow some of the excretion, and leave some on the food they handle or prepare, which has to be eaten by the rest of the family, who, amongst the working classes, often have to take their meals in the sick room: hence the thousands of instances in which, amongst this class of the population, a case of cholera in one member of the family is followed by other cases; whilst medical men and others, who merely visit the patients, generally escape. (Snow, 1936)* ❏

In analyzing the cholera epidemic of the 1850s, Dr. Snow noted that the people of London received their water supply from two companies: the Lambeth Company and the Southwark and Vauxhall Company. In 1852, the Lambeth Company had changed their source of water to a new water supply, which was free from the sewage of London. The Southwark and Vauxhall Company received water that was contaminated by the sewage of London. Some districts of London were solely supplied by the Lambeth com-

Deaths from Cholera from August 1853 through January 1854 by Source of Water*

| Sub-Districts | Population in 1851 | Deaths from Cholera in 1853 | Deaths by Cholera in each 100,000 Living | Water Supply |
|---|---|---|---|---|
| First 12 sub-districts | 167,654 | 192 | 114 | Southwk. & Vaux. |
| Next 16 sub-districts | 301,149 | 182 | 60 | Both Companies |
| Last 3 sub-districts | 14,632 | — | — | Lambeth Comp. |

\* Only partially reproduced.

pany, other districts were solely supplied by the Southwark and Vauxhall Company, and in a third group of districts, the residents had a choice of water companies. The above data compiled by Snow clearly indicated the effect of water companies on the transmission of cholera.

Regarding the table data, Snow stated the following.

*Although the facts shown in the above table afford very strong evidence of the powerful influence which the drinking of water containing the sewage of a town exerts over the spread of cholera, when that disease is present, yet the question does not end here; for the intermixing of the water supply of the Southwark and Vauxhall Company with that of the Lambeth Company, over an extensive part of London, admitted of the subject being sifted in such a way as to yield the most incontrovertible proof on one side or the other. In the sub-districts enumerated in the above table as being supplied by both Companies, the mixing of the supply is of the most intimate kind. The pipes of each Company go down all the streets, and into nearly all the courts and alleys. A few houses are supplied by one Company and a few by the other according to the decision of the owner or occupier at that time when the Water Companies were in active competition. In many cases a single house has a supply different from that on either side. Each company supplies both rich and poor, both large houses and small; there is no difference either in the condition or occupation of the persons receiving the water of the different Companies. Now it must be evident that, if the diminution of cholera, in the districts partly supplied with the improved water, depended on this supply, the houses receiving it would be the houses enjoying the whole benefit of the diminution of the malady, whilst the houses supplied with the water from Battersea Fields [Southwark and Vauxhall Company's source of water] would suffer the same mortality as they would if the improved supply did not exist at all. As there is no difference whatever, either in the houses or the people receiving the supply of the two Water Companies, or in any of the physical conditions with which they are surrounded, it is obvious that no experiment could have been devised which would more thoroughly test the effect of water supply on the progress of cholera than this, which circumstances placed ready made before the observer.*

*The experiment, too, was on the grandest scale. No fewer than three hundred thousand people of both sexes, of every age and occupation, and of every rank and station, from gentlefolks down to the very poor, were divided into two groups without their choice, and in most cases, without their knowledge; one group being supplied with water containing the sewage of London, and amongst it, whatever might have come from the cholera patients, the other group having water quite free from such impurity.*

*To turn this grand experiment to account, all that you required was to learn the supply of water to each individual house where a fatal attack of cholera might occur (Snow, 1936).* ❏

In July of 1854, another outbreak of cholera in London afforded Dr. Snow the chance to test his hypothesis.

*The following is the proportion of deaths to 10,000 houses, during the first 7 weeks of the epidemic, in the population supplied by the Southwark and Vauxhall Company, in that it supplied by the Lambeth Company, and in the rest of London.*

| | Number of Houses | Deaths from Cholera | Deaths in each 10,000 houses |
|---|---|---|---|
| Southwark and Vauxhall Company | 40,046 | 1,263 | 315 |
| Lambeth Company | 26,107 | 98 | 37 |
| Rest of London | 256,423 | 1,422 | 59 |

*The mortality in the houses supplied by the Southwark and Vauxhall Company was therefore between eight and nine times as great as in the houses supplied by the Lambeth Company. . . . (Snow, 1936)* ❏

## POSTLAB QUESTIONS

1. State the problem confronting Dr. Snow with regard to cholera.

2. Identify one of Snow's early observations in the transmission of cholera.

3. State Snow's hypothesis on the transmission of cholera.

4. What did Snow predict would be the outcome of his experiment?

5. What was Snow's experiment and what did he observe?

6. What was the variable in Snow's experiment?

7. What did Snow conclude from his experiment?

8. Snow's experiment was a natural experiment. What component of an experiment was it lacking?

9. Did Snow's observations support his hypothesis?

10. Was Snow's conclusion bias due to sampling error? Why or why not?

11. Was Snow's hypothesis proven to be true? Why or why not?

12. Why, later in the 1854 outbreak, did the difference between the number of cholera cases associated with each water supply decrease?

13. What were the ethical implications of this natural experiment?

14. Design an experiment to test the effect of adding fluoride to the drinking water to the amount of dental caries (cavities) in the people who drink fluoridated water.

## FOR FURTHER READING

Ausubel, Frederick M., ed. 1990. *Current Protocols in Molecular Biology.* New York: Green Pub. Associates and Wiley-Interscience.

Baker, J., and G. Allen. 1968. *Hypothesis, Prediction and Implication in Biology.* Reading, Mass.: Addison-Wesley.

Berra, Tim M. 1990. *Evolution and the Myth of Creationism: A Basic Guide to the Facts in the Evolution Debate.* Stanford, Calif.: Stanford University Press.

Gardner, E. 1972. *History of Biology.* Minneapolis: Burgess.

Harre, R. 1983. *Great Scientific Experiments.* Oxford: Oxford University Press.

Hillis, David M., and Craig Moritz, eds. 1990. *Molecular Systematics.* Sunderland, Mass.: Sinauer Associates.

Linskens, H. F., and J. F. Jackson, eds. 1990. *Physical Methods in Plant Sciences.* New York: Springer-Verlag.

Lloyd, G. E. R. 1991. *Methods and Problems in Greek Science.* New York: Cambridge University Press.

Manly, Bryan F. 1991. *Randomization and Monte Carlo Methods in Biology.* New York: Chapman and Hall.

Pittendrigh, C. S. 1965. In *Life: An Introduction to Biology,* edited by G. G. Simpson et al. New York: Harcourt Brace Jovanovich.

Rickwood, D., and B. D. Hines, eds. 1990. *The Practical Approach Series Cumulative Methods Index.* New York: Oxford University Press.

Salmon, W. C. 1990. *Four decades of Scientific Explanation.* Minneapolis: University of Minnesota Press.

Schreck, Carl B., and Peter B. Moyle, eds. 1990. *Methods for Fish Biology.* Bethesda, Md.: American Fisheries Society.

Smith, J. M. 1986. *The Problems of Biology.* London: Oxford University Press.

Snow, J. 1936. "On the mode of communication of cholera." In *Snow on Cholera.* New York: The Commonwealth Fund, pp 1–175.

Stedman, Thomas L. 1990. *Medical Dictionary.* Baltimore: Williams and Wilkins.

Uebel, T. E. 1991. *Rediscovering the Forgotten Vienna Circle: Austrian Studies on Otto Neurath and the Vienna Circle.* Boston: Kluwer Academic Publishers.

Zalucki, Myron P., ed. 1991. *Heliothis, Research Methods and Prospects.* New York: Springer-Verlag.

# PART II

□

# The Organization of Life

**TOPIC 3**
Organic Compounds

**TOPIC 4**
Cell Structure and Function

**TOPIC 5**
Diffusion, Osmosis, and Biological Membranes

# TOPIC 3

❏

# Organic Compounds

## OBJECTIVES

1. Compare the major groups of biologically important organic compounds—carbohydrates, lipids, proteins, and nucleic acids—with respect to their chemical composition and function.

2. Distinguish among monosaccharides, disaccharides, and polysaccharides, giving examples of each.

3. Describe neutral fats and give the biological function of this type of molecule.

4. Describe the functions and chemical structure of proteins.

5. Describe the chemical structure of nucleotides and nucleic acids and explain the importance of these compounds in living organisms.

6. Name the specific laboratory test used to identify (a) reducing sugar, (b) starch, (c) neutral fat, (d) protein, and (e) nucleic acid.

---

## INTRODUCTION

The various molecules found in all life forms are called **organic compounds** because they are produced by organisms. Organic molecules are carbon based and range from small molecules to ones that are enormous in size. These large molecules are not as complex as they first seem because they are composed of simpler, smaller molecules or **monomers** linked together into long chains or **polymers.**

There are four major classes of biologically important organic compounds. Within the living cell, each of these four major classes has specific important functions.

1. **Carbohydrates** are energy sources and provide structural support as in the cell wall of plants. Carbon, hydrogen, and oxygen are the elements found in carbohydrates. Carbohydrates may be classified as **monosaccharides, disaccharides,** or **polysaccharides** (Fig 3–1). Monosaccharides are simple sugars such as fructose and glucose. Glucose, for example, is a readily usable energy source. Two

monosaccharides bonded together form a disaccharide. A common disaccharide is sucrose (table sugar), which consists of a glucose and fructose molecule. Three or more bonded monosaccharides form a polysaccharide. Starch, cellulose, and glycogen are polysaccharides. Starch, produced by plants, and glycogen, produced by animals, are storage forms of energy. Cellulose is the structural component of plant cell walls.

2. **Lipids** are a varied group of molecules most of which are insoluble in water. Like carbohydrates, carbon, hydrogen, and oxygen are the principal elements of lipids although the oxygen content is much reduced. Lipids are essential components of membranes, a good means of storing energy and some are essential hormones. Neutral fats, phospholipids, steroids, carotenoids, and waxes are lipids. Fatty acids are the simplest lipids. Neutral fats are the most abundant group of lipids in the biological world. Neutral fats are composed of three fatty acid molecules bonded to a molecule of the alcohol **glycerol.** Neutral fats have at least

twice the energy-storing capacity per unit weight as carbohydrates do. Cholesterol is an important steroid that is a part of some hormones. Steroids differ from most other lipids by virtue of the structure of steroids consisting of carbon rings instead of chains.

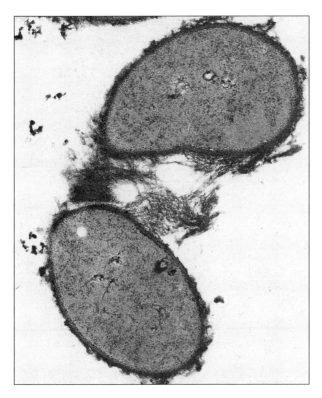

**FIGURE 3-1**

Thin section electron micrograph of *Bacillus cereus* showing polysaccharide produced by the cells, 50,000X. *(Electron micrograph by Dr. Robert A. Pfister.)*

3. **Proteins** have numerous important roles in the living organism. Proteins form many structural features such as hair, hooves, and tendons. The elements in proteins are carbon, hydrogen, oxygen, nitrogen, and, in some proteins, sulfur. All proteins are polymers of **amino acids** covalently bonded in long chains that subsequently coil and fold into complex shapes that determine the function of the resulting protein molecule. The chemical bond that holds amino acids together is called a **peptide bond.** The bond forms between the functional groups of two amino acids. A functional group is the portion of a molecule that takes part in a chemical reaction. Specifically, a peptide bond occurs between the **carboxyl group** ($-COOH$) of one amino acid and the **amino group** ($-NH_2$) of its neighboring amino acid. Many amino acids linked together form a **polypeptide.** Proteins are polypeptides. A particularly important class of proteins or polypeptides in biology are the enzymes that control the many chemical reactions that keep the cell alive.

4. **Nucleic acids** are extremely large and complex molecules that have a variety of important biological roles. The elements of nucleic acids are carbon, hydrogen, oxygen, nitrogen, and phosphorus. One general type of nucleic acid (DNA, deoxyribonucleic acid) is the genetic information (Fig. 3–2), the

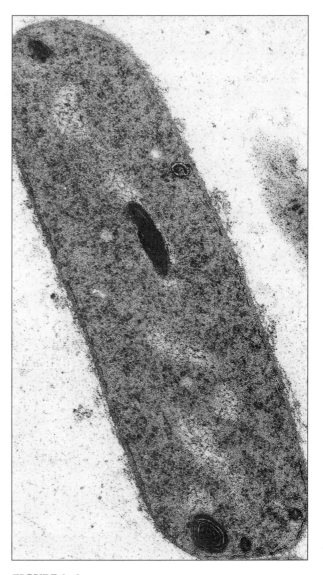

**FIGURE 3-2**

Thin section electron micrograph of *Escherichia coli*, 50,000X, showing condensed DNA. *(Electron micrograph by Dr. Robert A. Pfister.)*

genetic code of cells. A second type (RNA, ribonucleic acid) is involved in the expression of the genetic code when amino acids are assembled into proteins. **Nucleotides** are the monomers in the nucleic acid polymers. **ATP** (adenosine triphosphate) is a nucleotide existing in nonpolymer form. ATP supplies energy for the many chemical reactions of the cell.

In the laboratory exercises of this chapter, you will gain some useful knowledge about the four major biologically important molecules — carbohydrates, lipids, proteins, and nucleic acids — and experience in performing laboratory tests to identify those types of molecules.

## PRELAB QUESTIONS

1. What is (are) the function(s) of carbohydrates?

2. What classification of carbohydrates is cellulose?

3. What are the major components of a neutral fat?

4. What smaller molecules or monomers bond together to form a protein?

5. What type of organic compound is DNA?

## LAB 3.A
## TESTING FOR CARBOHYDRATES: REDUCING SUGARS

### MATERIALS REQUIRED

| | |
|---|---|
| 4 test tubes | Benedict's reagent |
| test tube rack | distilled water |
| test tube holder | glucose solution |
| boiling water bath | sucrose solution |
| wax pencil | starch solution |
| small ruler | |

### PROCEDURE

Some carbohydrates called reducing sugars have an aldehyde functional group as part of their molecular structure (Fig. 3–3) which makes them react with Benedict's reagent when heated. The reaction produces a color change in the Benedict's reagent from blue to green or orange-red, depending on how much reducing sugar is present. Therefore, this color change is a positive test for reducing sugar. Reducing sugars are so named because they accept an oxygen atom from the Benedict's reagent causing the reagent to become reduced. You will test three kinds of carbohydrates with Benedict's reagent to determine which of them are reducing sugars (See Fig. 3–4).

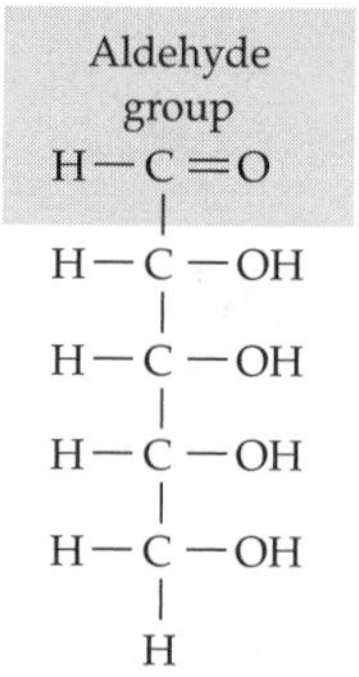

**FIGURE 3–3**

Structural formula of the reducing sugar ribose showing the aldehyde functional group.

*Note:* In all the chemical tests of this exercise, you will always test a substance, such as distilled water, that gives a negative result. This is known as a control and it serves as an unchanging standard for judging positive reactions.

1. Use the ruler and wax pencil to mark 1 cm and 3 cm from the bottom of 4 clean test tubes. Label the top end of the test tubes as 1, 2, 3, and 4.
2. Fill each test tube up to the 1 cm mark with one of the following: distilled water (tube 1), glucose solution (tube 2), sucrose solution (tube 3), and starch solution (tube 4).
3. Add Benedict's reagent up to the 3 cm mark of all tubes.
4. Place all four test tubes in a boiling water bath and heat for 3 minutes.
5. Remove the test tubes from the hot water bath with the test tube holder and place them in the test tube rack. Record the color of each tube in Table 3–1.

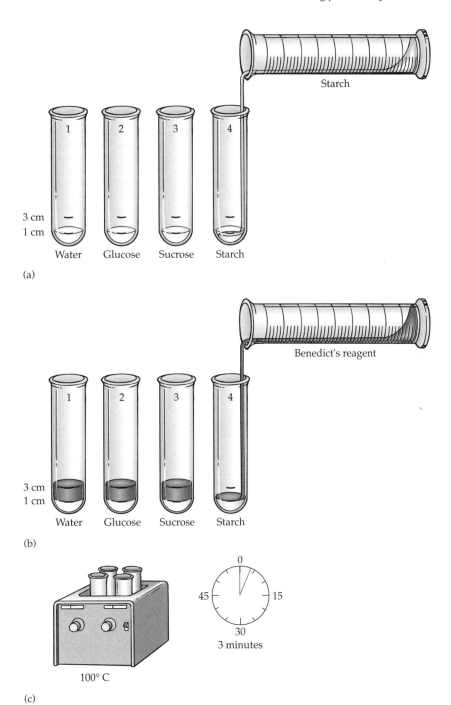

**FIGURE 3–4**

Procedure for determining the presence of reducing sugars.

**Table 3-1**   Results of Benedict's Test for Reducing Sugars

| Tube Number | Tube Contents | Color of Tube After Boiling |
|---|---|---|
| 1 | water + Benedict's reagent | |
| 2 | glucose + Benedict's reagent | |
| 3 | sucrose + Benedict's reagent | |
| 4 | starch + Benedict's reagent | |

## POSTLAB QUESTIONS

1. Which carbohydrate solution contains reducing sugar?

2. What functional group present in this carbohydrate molecule produces a positive reaction with Benedict's reagent?

3. If onion juice mixed with Benedict's reagent gives a mustard yellow color, in what form is sugar stored in onions?

Discard the contents of the test tubes and rinse them for use in the next exercise. Shake out excess water from the tubes.

## LAB 3.B
## TESTING FOR CARBOHYDRATES: STARCH

### MATERIALS REQUIRED

| | |
|---|---|
| 4 test tubes | sucrose solution |
| test tube rack | starch solution |
| wax pencil | fresh potato |
| small ruler | razor blade |
| iodine solution | clean slide |
| distilled water | coverslip |
| glucose solution | |

## PROCEDURE

Iodine reacts with starch resulting in a dark, blue-black color that is a positive test for starch. The reaction occurs because starch is a coiled polysaccharide made of repeated glucose molecules (Fig. 3–5). The coiled shape of the starch molecule binds with the iodine molecules and produces the color change (See Fig. 3–6).

1. Use the ruler and wax pencil to mark 1 cm from the bottom of 4 clean test tubes. Label the top end of the test tubes as 1, 2, 3, and 4.
2. Fill each test tube up to the 1 cm mark with one of the following: distilled water (tube 1), glucose solution (tube 2), sucrose solution (tube 3), and starch solution (tube 4).
3. Add 3 drops of iodine solution to each test tube and swirl to mix. Record the color of each tube in Table 3–2.

**Table 3-2**   Results of Iodine Test for Starch

| Tube Number | Tube Contents | Color of Tube |
|---|---|---|
| 1 | water + iodine | |
| 2 | glucose + iodine | |
| 3 | sucrose + iodine | |
| 4 | starch + iodine | |

Discard the contents of the test tubes and rinse them for use in the following exercises. Shake out excess water from the tubes.

Starch is stored in the seeds, roots, and tubers (fleshy underground stems) of plants as an energy reserve. The starch is converted to simple sugars and is used by the plant during times of growth or when low levels of photosynthesis may not meet its energy requirements. The potato tuber is an example of starch storage in plants.

1. Use a razor blade to cut a small, extremely thin slice of fresh potato. The potato slice should be thin enough to be nearly transparent.

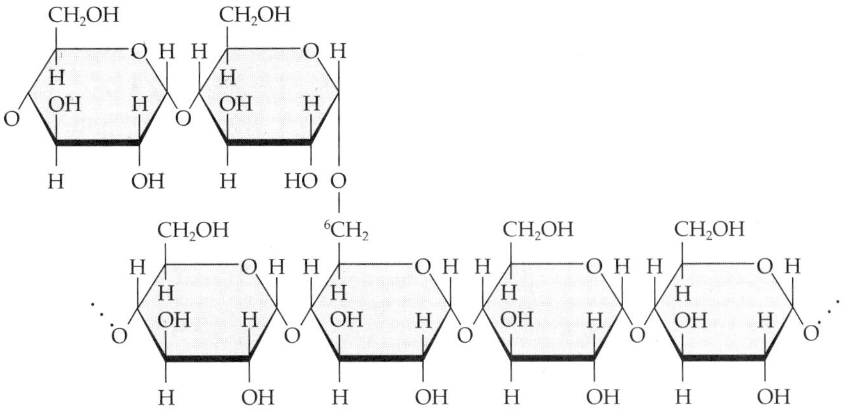

**FIGURE 3-5**

Molecular structure of starch consisting of repeated glucose molecules bonded together into long chains. This figure represents only a tiny segment of a large starch molecule.

2. Place the potato slice on a clean microscope slide (see Lab 1.C for wet mount preparation). Add a drop of water to the potato and a coverslip. Observe the potato slice under low and high power of your microscope.

3. Remove the coverslip and add a few drops of iodine solution. Replace the coverslip, wait a few moments for the iodine to penetrate, and again observe the potato under low and high power. Draw the cells and any details you see in the box provided.

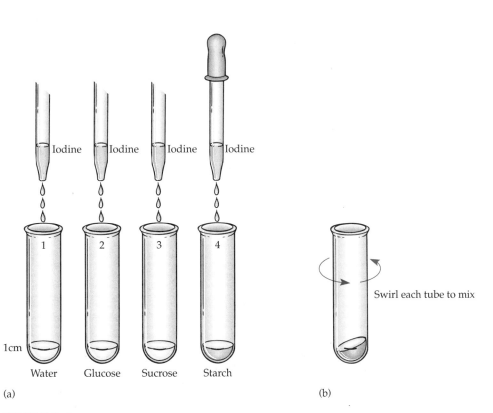

(a)                                      (b)

**FIGURE 3-6**

Procedure for determining the presence of starch.

## POSTLAB QUESTIONS

1. Do you see a blue-black color in any of the carbo-hydrate solutions other than the starch?

2. Does the iodine test help you distinguish starch from sucrose? Why or why not?

3. Starch and cellulose are both polymers of glucose molecules, but starch gives a positive iodine test and cellulose does not. What is a plausible expla-nation for this difference?

4. Did you see small dark bodies within the potato cells? If you did not, review your slide and, if need be, contact your instructor for assistance.

5. What molecule is obviously contained in these dark bodies?

6. If you mix potato juice with the Benedict's reagent in the previous experiment, would a color change occur? Why or why not?

Discard the potato slice. Clean your microscope slide and return it and the scope to their proper place.

## LAB 3.C
## TESTING FOR PROTEINS: BIURET TEST

### MATERIALS REQUIRED

| | |
|---|---|
| 4 test tubes | honey solution |
| test tube rack | corn oil |
| small ruler | egg white solution |
| wax pencil | 10% NaOH solution |
| distilled water | 1% $CuSO_4$ solution |

## PROCEDURE

Protein molecules consist of multiple amino acid mol-ecules linked together in long chains (Fig. 3–7). The amino acids are linked by peptide bonds that react with copper sulfate that is used in the biuret test pro-ducing a violet color. Therefore, the violet color indi-cates a positive biuret test for protein (see Fig. 3–8).

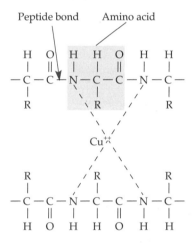

### FIGURE 3–7

In the Biuret test, a complex is formed between the copper atom and four nitrogen atoms.

1. Mark 4 test tubes at 2 cm and 4 cm measured from the bottom of the test tube. Label the top end of the test tubes as 1, 2, 3, and 4.

2. Fill each test tube up to the 2 cm mark with one of the following: distilled water (tube 1), honey solu-tion (tube 2), corn oil (tube 3), and egg white solu-tion (tube 4).

3. Add 10% NaOH (sodium hydroxide) solution to each test tube up to the 4 cm mark.

**C A U T I O N**  Sodium hydroxide can cause skin burns. Thoroughly wash with soap and water if contact occurs with skin or clothing.

4. Add 5 drops of 1% $CuSO_4$ (copper sulfate) solution to each test tube and swirl to mix. Examine the tubes for the appearance of any violet color and record your results in Table 3–3.

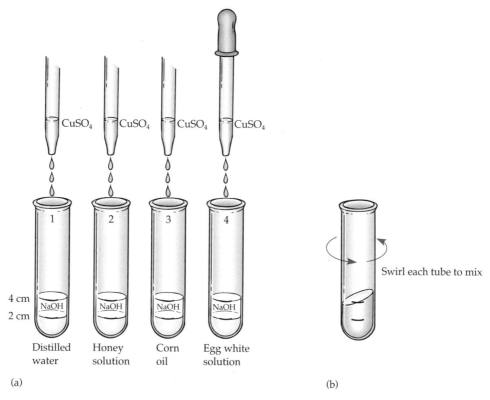

**FIGURE 3-8**

Procedure for determining the presence of proteins.

**Table 3-3**   Results of Biuret Test for Protein

| Tube Number | Tube Contents | Color of Tube |
|---|---|---|
| 1 | water + NaOH + $CuSO_4$ | |
| 2 | honey + NaOH + $CuSO_4$ | |
| 3 | corn oil + NaOH + $CuSO_4$ | |
| 4 | egg white + NaOH + $CuSO_4$ | |

3. Would you expect a solution containing an enzyme to yield a positive biuret test? Why or why not?

4. What color change would occur if iodine was added to the egg white?

Discard the contents of the tubes, wash thoroughly with detergent, and rinse them well for use in the next exercises. Shake out excess water from the tubes.

**POSTLAB QUESTIONS**

1. Which solution(s) contain(s) protein?

2. What type of chemical bond causes the biuret reagent to react and to give the characteristic violet color?

# LAB 3.D

## TESTING FOR NUCLEIC ACIDS: DISCHE DIPHENYLAMINE TEST

### MATERIALS REQUIRED

3 test tubes          test tube holder

test tube rack        distilled water

boiling water bath

wax pencil

Dische diphenylamine reagent

small ruler

DNA solution

RNA solution

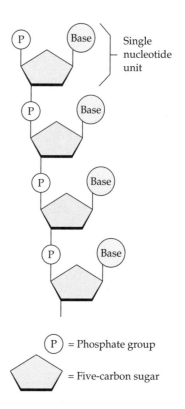

## PROCEDURE

Nucleic acids, RNA and DNA, are large molecules composed of many nucleotides bonded together in long strands (Fig. 3–9). Each nucleotide contains a five-carbon sugar as part of its structure. Specifically, DNA contains the sugar deoxyribose that reacts with the diphenylamine reagent when heated forming a blue color—a positive test for DNA (Fig. 3–10).

1.  Mark 3 test tubes at 2 cm and 4 cm measured from the bottom of the test tube. Label the top end of the test tubes as 1, 2, and 3.

2.  Fill each test tube up to the 2 cm mark with one of the following: distilled water (tube 1), DNA solution (tube 2), and RNA solution (tube 3).

3.  Add Dische diphenylamine reagent to each test tube up to the 4 cm mark.

(P) = Phosphate group

= Five-carbon sugar

**FIGURE 3–9**

Generalized diagram of a nucleic acid molecule composed of multiple nucleotides. Each nucleotide consists of three parts: a five-carbon sugar, a phosphate group, and a base.

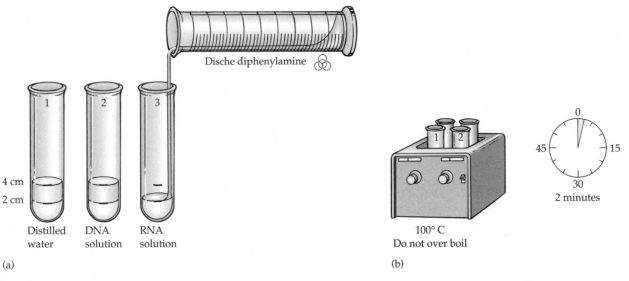

(a)

(b)

**FIGURE 3–10**

Procedure for determining the presence of DNA.

**C A U T I O N** Dische diphenylamine reagent contains strong acids. Use great care when handling the reagent. If the reagent contacts skin or clothing, flush the area under running water for several minutes.

4. Boil the test tubes in the hot water bath for 2 minutes or until one of the tubes turns blue. Do not over boil the test tubes — this causes the contents to boil out of the tubes.

5. After boiling, use the test tube holder to place the tubes into the test tube rack. Record the color of each tube in Table 3 – 4.

**Table 3-4** Results of Dische Diphenylamine Test

| Tube Number | Tube Contents | Color of Tube |
|---|---|---|
| 1 | water + diphenylamine reagent | |
| 2 | DNA + diphenylamine reagent | |
| 3 | RNA + diphenylamine reagent | |

**POSTLAB QUESTIONS**

1. Did the test tube containing DNA yield a positive test for this nucleic acid?

2. Based on the chemical composition of DNA and RNA nucleotides, explain why the nucleic acid RNA did not yield a positive diphenylamine test.

Discard the contents of the tubes in the container designated for diphenylamine reagent disposal — DO NOT pour liquids containing diphenylamine down the sink drain! Wash the test tubes well for use in the next exercise. Shake out excess water from the tubes.

## LAB 3.E
## TESTING FOR LIPIDS: SUDAN III TEST

### MATERIALS REQUIRED

| | |
|---|---|
| 4 test tubes | egg white solution |
| test tube rack | honey solution |
| Sudan III dye | corn oil |
| wax pencil | distilled water |
| small ruler | |

### PROCEDURE

Lipids are nonpolar compounds composed of a glycerol molecule bonded to three fatty acid molecules (Fig. 3 – 11). Because lipids are nonpolar, they do not dissolve in water or other polar liquids. Lipids are soluble in nonpolar solvents such as ether. Liquid Sudan III dye is made with ether as its solvent. Therefore, solubility in Sudan III dye can be used as a posi-

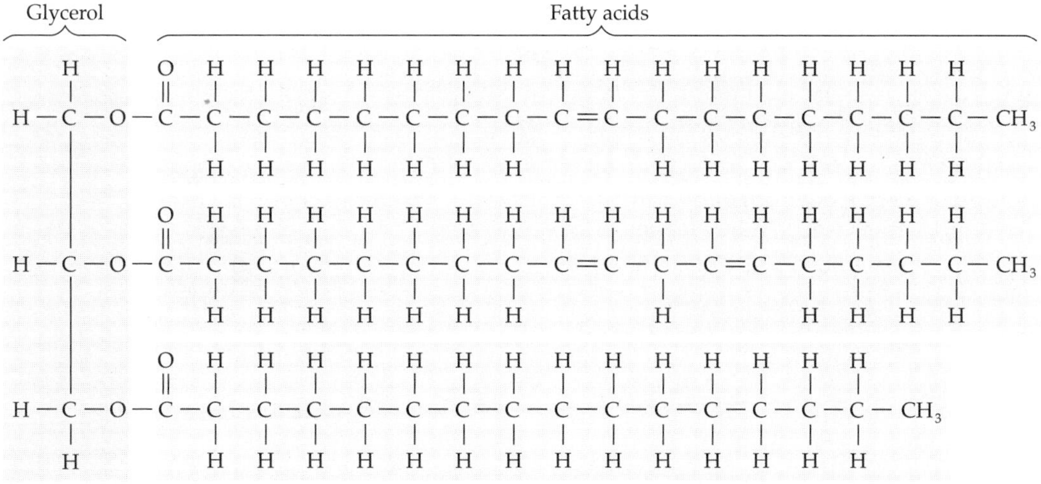

**FIGURE 3-11**

A lipid (fat) molecule consisting of three fatty acids bonded to one glycerol molecule.

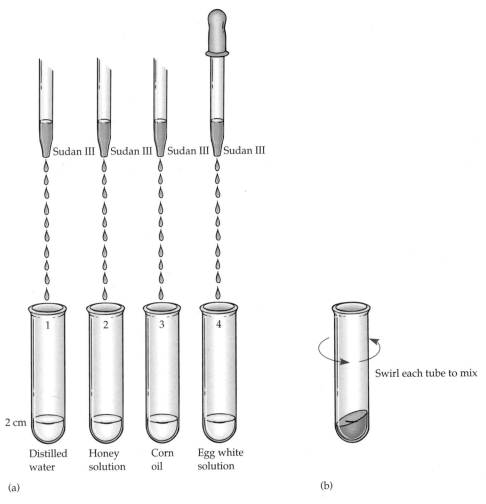

**FIGURE 3–12**

Procedure for determining the presence of lipids.

tive test for lipids. Nonlipid substances will not dissolve in Sudan III dye and form two distinct layers in a test tube (Fig. 3–12).

1. Mark 4 test tubes at 2 cm measured from the bottom of the tube. Label the top end of the test tubes as 1, 2, 3, and 4.
2. Fill each test tube up to the 2 cm mark with one of the following: distilled water (tube 1), honey solution (tube 2), corn oil (tube 3), and egg white solution (tube 4).
3. Add ten drops of Sudan III dye to each test tube.

---

**C A U T I O N** Sudan III is a powerful stain. Avoid contact with skin and clothing.

---

Swirl the tubes to mix. Examine the tubes for results. Record your results in Table 3–5 as ''lipids present'' (even mixing of liquid and Sudan III dye) or ''lipids absent'' (layers of liquid and Sudan III dye form).

**Table 3–5**  Results of Sudan III Test for Lipids

| Tube Number | Tube Contents | Lipids Present/Absent |
|---|---|---|
| 1 | water + Sudan III dye | |
| 2 | honey + Sudan III dye | |
| 3 | corn oil + Sudan III dye | |
| 4 | egg white + Sudan III dye | |

## POSTLAB QUESTIONS

1. Which compound(s) contain(s) lipids?

2. Which tube served as the control?

3. Which compound contains more calories, honey or corn oil? Why?

Discard the contents of all tubes in the container designated for Sudan III disposal — DO NOT pour Sudan III down the sink drain! Thoroughly brush your test tubes with detergent and water, particularly the tubes containing corn oil. Shake out any excess water remaining in the tubes. This will ensure good test results for those students in the next lab period.

## FOR FURTHER READING

Alberts, B., et al. 1989. *Molecular Biology of the Cell.* New York: Garland Publishing.

Atkins, P. W. 1987. *Molecules.* New York: Scientific American Library.

Bettelheim, F.A., and J. March. 1990. *Introduction to General, Organic and Biochemistry.* Philadelphia: Saunders College Publishing.

Bloomfield, M. 1980. *Chemistry and the Living Organism.* New York: John Wiley and Sons.

Coleman, David C., and Brian Fry, eds. 1991. *Carbon Isotope Techniques.* San Diego: Academic Press.

Creighton, T. E. 1990. *Protein Function: A Practical Approach.* New York: IRL Press.

Denyer, S. P., and W. B. Hugo, eds. 1991. *Mechanisms of Action of Chemical Biocides: Their Study and Exploitation.* Boston: Blackwell Scientific Publications.

Glick, David M. 1990. *Glossary of Biochemistry and Molecular Biology.* New York: Raven Press.

Hames, B. D., and D. Rickwood. 1990. *Gel Electrophoresis of Proteins: A Practical Approach.* New York: IRL Press at Oxford University Press.

Harris, E. L. V., and S. Angal, eds. 1990. *Protein Purification Applications: A Practical Approach.* New York: IRL Press.

Lieth, H., and B. Market, eds. 1990. *Element Concentration Cadasters in Ecosystems.* New York: VCH Publishers.

Moody, Peter C. E. 1990. *Protein Engineering.* New York: IRL Press at Oxford University Press.

Moore, Geoffrey R. 1990. *Cytochrome c: Evolutionary, Structural, and Physiochemical Aspects.* New York: Springer-Verlag.

Morimoto, Richard I., et al., eds. 1990. *Stress Proteins in Biology and Medicine.* Cold Spring Harbor, N.Y.: Cold Spring Harbor Laboratory.

Reddy, C. Channa, et al., eds. 1990. *Biological Oxidation Systems.* San Diego: Academic Press.

Revzin, Arnold, ed. 1990. *The Biology of Nonspecific DNA-Protein Interactions.* Boca Raton, Fl.: CRC Press.

Siddle, K. and T. E. Creighton, eds. 1990. *Peptide Hormone Action: A Practical Approach.* New York: IRL Press at Oxford University Press.

Sorensen, Elsa M. B. 1991. *Metal Poisoning in Fish.* Boca Raton, Fl.: CRC Press.

Stein, Stanley. 1990. *The Fundamentals of Protein Biotechnology.* New York: M. Dekker.

Stein, Wilfred D. 1990. *Channels, Carriers, and Pumps: An Introduction to Membrane Transport.* San Diego: Academic Press.

Stryer, L. *Biochemistry.* 1988. New York: Freeman.

❑

# Cell Structure and Function

O B J E C T I V E S

1. State and describe the significance of the cell theory.

2. Describe the general characteristics of cells, including their size, range, and shape.

3. Contrast prokaryotic and eukaryotic cells. Draw and label a diagram of a prokaryotic cell and a eukaryotic cell.

4. Contrast plant and animal cells. Sketch a diagram of a plant cell and an animal cell.

5. Describe and list the functions of the principal cell organelles and structures: nucleus, chloroplast, nucleolus, vacuole, mitochondrion, smooth endoplasmic reticulum, rough endoplasmic reticulum, lysosome, cell wall, centriole, and ribosome.

## INTRODUCTION

The **cell theory** states that all organisms consist of one or more cells, that cells are the smallest and the most fundamental living entity, and that a cell comes only from division of a previously existing cell. This theory also implies that cells are the biological structures within which the biological processes of metabolism occur and that cells contain hereditary material (nucleic acids).

The cells that make up unicellular and multicellular organisms are highly varied. Cells come in a wide range of sizes — from the smallest, which is approximately 10 micrometers in diameter (2500 of these cells placed end to end would measure approximately 1 inch in length) to the largest cells, which are bird eggs. Different types of cells vary considerably in shape, from the round shape of normal red blood cells of humans to the long thread-like shape of human nerve cells. Nevertheless, all cells are of two basic types: prokaryotic and eukaryotic. **Prokaryotic cells** (eubacteria and archaebacteria) are simpler in structure than eukaryotic cells and are usually surrounded by a thick cell wall that provides protection and shape. The major differences are the lack of a membrane-bound nucleus and organelles in the prokaryotes. DNA is present as a single circular molecule folded into a **nuclear area.** The plasma membrane is the only membrane found in most prokaryotes. Many bacteria are capable of movement by rotating long, thread-like **flagella** (Fig. 4 – 1) that extend from the cell wall.

**Eukaryotic cells** (protists, fungi, plants, and animals) have membranes that surround an organized nucleus and various organelles with distinct functions. In general, cells of eukaryotic organisms are larger and more complex than prokaryotes.

DNA is contained within multiple linear **chromosomes** that carry genetic information and are contained within the **nucleus,** (Fig. 4 – 2), which segregates the chromosomes from the cytoplasm. The nucleus also holds the **nucleolus,** that produces granule-like **ribosomes** which are involved in protein synthesis. Prokaryotes also have ribosomes, but of smaller size. The nucleus is surrounded by a porous double membrane called the **nuclear envelope,** which controls the passage of materials into and out of the nucleus.

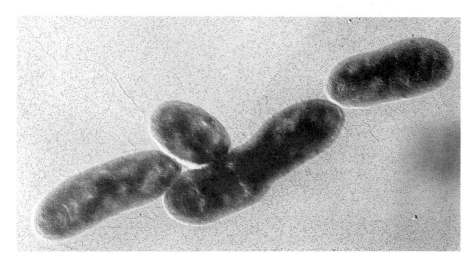

**FIGURE 4-1**

A shadowed electron micrograph preparation of *Pseudomonas sp.* showing flagella, 22,000X. *(Electron micrograph photo by Dr. Robert A. Pfister.)*

The outer nuclear membrane is continuous with the **endoplasmic reticulum** (ER), a system of flattened membranes throughout the cytoplasm. There are two types of ER. The first type, known as rough

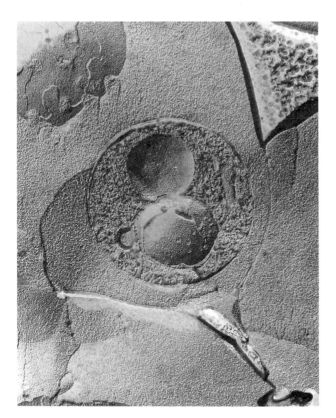

**FIGURE 4-2**

Freeze etching of *Plasmodium gallinaceum*, a blood parasite of the chicken, inside a chicken red blood cell. The electron micrograph, taken at a magnification of nearly 27,000X, shows, at upper left, the nucleus of the red blood cell and, at center, the nucleus of the parasite. Both nuclei show nuclear pores. *(Electron micrograph by Drs. Thomas Seed, Robert A. Pfister, and Julius P. Kreier.)*

endoplasmic reticulum (RER), is associated with ribosomes and is involved with protein synthesis. The second type, known as smooth endoplasmic reticulum (SER), is not associated with ribosomes and is involved in lipid metabolism. The RER stores and segregates proteins prior to secretion from the cell. Before leaving the cell, proteins are frequently carried by vesicles from the RER to a **Golgi complex,** another membranous structure. Within the Golgi complex, proteins are modified and packaged prior to secretion or transportation to other organelles, such as **lysosomes.** Lysosomes are membrane-bound sacs containing enzymes that digest food molecules and worn-out cell components. The enzymes travel by vesicles from the Golgi complex and become deposited in the lysosome membrane.

**Mitochondria** are bound by an inner and outer membrane and function in eukaryotic cellular respiration wherein adenosine triphosphate (ATP) is produced. ATP leaves the mitochondria and provides energy for many cytoplasmic chemical reactions. Mitochondria have their own DNA, RNA, and ribosomes, which are involved in the production of new mitochondria.

Those organelles described thus far (chromosome, nucleus, nucleolus, ribosome, nuclear envelope, endoplasmic reticulum, Golgi complex, lysosome, and mitochondrion) are found in essentially all plant and animal cells. Some organelles are characteristic of plants only. **Plastids** are similar to mitochondria in having a double membrane and their own nucleic acids and ribosomes, which are involved in self-reproduction. Plastids occur only in photosynthetic eukaryotes. Some plastids contain pigments that produce bright colors in plant parts and other plastids store food. **Chloroplasts** are membrane-bound organelles that contain green chlorophyll. Chlorophyll is a special pigment that gives plants the capability of photosynthesis, a process that produces food (sugars) for the plant and, consequently, for any animal that eats the plant.

The **cell wall** is another organelle absent in animal cells. It provides external support and protection for the plant cell. The wall, composed of cellulose and other fibers, is porous, allowing an exchange of materials between the plant cell and its environment. Plant cells also typically have a large **vacuole,** which functions in the storage of fluids and many dissolved substances. The vacuole is surrounded by a membrane that keeps stored materials separate from the cytoplasm, thereby increasing available surface area. Fluid within the vacuole exerts pressure against the cell wall, which helps support the bodies of softer plants.

Eukaryotic cells have a **cytoskeleton** that establishes cell shape and is involved in cell movement. The cytoskeleton is a network of protein filaments attached to the plasma membrane and various organelles. The thickest filaments of the cytoskeleton are the **microtubules**. Microtubules compose the **centrioles** that function in cell division. Centrioles are absent in higher plants. Short **cilia** and the longer flagella also consist of microtubules. They both function in cell movement or sweep liquid past a cell surface in immobile cells. Cilia and flagella do not occur in higher plants.

In the laboratory work of this chapter, we shall examine some of these interesting and important organelles.

## PRELAB QUESTIONS

1. What are the three basic components of the cell theory?

2. Compare the size, shape, and function of cells from one organism to another.

3. Define the major differences between eukaryotic and prokaryotic cells.

4. What organelle functions in eukaryotic cellular respiration wherein ATP is produced?

5. What is the difference between rough and smooth endoplasmic reticulum?

# LAB 4.A
## CELLULAR DIVERSITY

### MATERIALS REQUIRED

motor neurons, prepared slide
live *Elodea* plants
*Spirostomum* culture
forceps
dropper
slides and coverslips
methyl cellulose solution in dropper bottle

### PROCEDURE

In this exercise, you will examine the tremendous variation in the shape, size, and function of cells from different organisms.

**Motor neurons—animal cells shaped for specialized function.**

1. Obtain a prepared slide of motor neurons. Place the slide on your microscope and examine it under both low and high power. Motor neurons are nerve cells that carry nerve impulses to muscles or glands resulting in some kind of action. The numerous cells on your slide probably came from the spinal cord of a mammal. The motor neurons are the larger, spider-like cells with a prominent **nucleus** and appendages that branch from the cell. Sketch a motor neuron in the space provided below.

*Elodea*—a representative of plant cells.

2. Obtain a blank slide and forceps. Use the forceps to remove a small leaf from the growing tip of an *Elodea* strand (see Topic 1). Place the leaf on the slide, then add a drop of water and coverslip. Place

the slide on your microscope and observe under both low and high power. One of the most striking features you'll notice in these plant cells are the small green bodies called **chloroplasts.**

Notice that the cells are rectangular in shape and resemble bricks in a wall. The well-defined boundary you see around each plant cell is the **cell wall,** a rigid box-like structure that completely encloses the cell.

Using high power, carefully focus on one *Elodea* cell. Try not to focus on any cells in lower layers. Notice the distribution of chloroplasts within the cell.

Are the chloroplasts located mostly along the cell wall or are they evenly distributed throughout the cell? _____

_____

Any unequal distribution of chloroplasts you may see is due to the presence of a **vacuole** within the plant cell. Vacuoles frequently are large and occupy much of the cell volume, crowding the other contents of the cell out towards the cell wall. Sketch a few *Elodea* cells in the space provided below.

*Spirostomum* — **the cell as an organism.**

3. *Spirostomum* belongs to a group of unicellular organisms known as protozoa. *Spirostomum* is a relatively large protozoan. You may even see *Spirostomum* with the unaided eye. It may look like lint moving through the culture water. *Spirostomum* moves by means of cilia, which are short hair-like strands that cover the surface of the cell. The cilia beat rapidly in a regulated rhythm, thereby propelling the cell through the water.

Obtain a blank slide and use a dropper to draw a few drops of water from the bottom of a *Spirosto-*

*mum* culture. It is important to take a sample from the bottom because it increases your chances of capturing organisms since they tend to settle. Place a drop of the culture on your slide and add a drop of methyl cellulose solution. This is a harmless solution that thickens the water and slows the rapid swimming of small organisms for easier viewing under a microscope. Add a coverslip and place the slide on your microscope. Look for *Spirostomum* using the scanning objective. After locating an individual, observe under both low and high power.

Carefully focus on the protozoan and look for the whirling movement of the cilia along the outer edge of the cell. Cilia and internal cellular details are best observed by adjusting the iris diaphragm of your microscope to lower the light intensity. Bright light can wash out small details under high power. If you cannot see cilia or other details of *Spirostomum,* ask your instructor for help.

Closely observe the interior of the protozoan. Does the cytoplasm appear to be uniform throughout the cell or is there some evidence that different parts of the cell may be specialized? _____

_____

Although *Spirostomum* is a single cell, it has various organelles that function in water balance, food capture, and digestion similar to the function of organ systems in multicellular animals. Sketch *Spirostomum* in the space below.

## POSTLAB QUESTIONS

1. What do you think is the function of the appendages extending from the nerve cells?

2. What three organelles (or structures) did you view in the *Elodea* leaf cells, but not in the mammalian nerve cells?

3. Does *Spirostomum* have a cell wall? How did you know?

4. How do nerve cells, *Elodea* leaf cells, and *Spirostomum* compare in size?

5. Even before examining eukaryotic versus prokaryotic cell structure, would you predict that the mammalian nerve cells, *Elodea* leaf cells, and the protozoan *Spirostomum* are eukaryotic or prokaryotic cells? Why?

## LAB 4.B
## A LOOK AT PROKARYOTIC CELLS

### MATERIALS REQUIRED

| | |
|---|---|
| plain yogurt | dropper bottle of water |
| flat toothpicks | *Anabaena* culture |
| slides and coverslips | |

### PROCEDURE

Prokaryotic cells, such as bacteria and cyanobacteria, may be simple in structure, but many species are capable of biochemical activities which make them very important to both the ecosystem and to humans. An important ecological biochemical activity is the decomposition of dead plant and animal bodies by bacteria, which release trapped nutrients back into the ecosystem where they are again used by other organisms. An important biochemical activity of humans is the use of certain bacteria in the processing of cheese and yogurt.

Cyanobacteria are a group of prokaryotes capable of making their own food by photosynthesis. Being prokaryotes, the photosynthetic pigment is not contained within chloroplasts, but is embedded in extensions of the cell membrane.

1. Use the flat end of a toothpick to place a tiny dab of yogurt onto a clean slide. Add a drop of water and mix thoroughly with the yogurt. Place a coverslip over the mixture and place the slide on your microscope. Examine under both low and high power. Looking carefully among the small bits of yogurt, you should see numerous bacteria that resemble tiny, thin filaments. These bacteria are *Lactobacillus.* They convert milk into yogurt. Note the size of these bacteria.

2. Use a dropper to draw up a bit of *Anabaena* culture, a representative of the cyanobacteria. Make a wet mount on a slide. Place the slide on your microscope and observe under both low and high power. As you can see, the individual cells of *Anabaena* grow in a chain resembling a string of beads.

What color are the *Anabaena* cells? _____

### POSTLAB QUESTIONS

1. Which cells are the smallest-bacteria, *Elodea* or *Spirostomum?*

2. Based on size, which of these cells is the simplest in structure?

3. Why is the plasma membrane of the various cell types you have observed not visible with your microscope?

4. Were nuclei visible in the bacterium *Lactobacillus* and the cyanobacterium *Anabaena?*

5. Knowing that *Anabaena* is photosynthetic, do you think that the color you observed was due to the presence of chloroplasts? Explain your answer.

## LAB 4.C
## A LOOK AT EUKARYOTIC CELLS

### MATERIALS REQUIRED

onion pieces

forceps

slides and coverslips

dropper bottle of water

acetocarmine stain

human cheek epithelium, prepared slide

## PROCEDURE

By far, most organisms are eukaryotes, which possess a membrane-bound nucleus and other membrane-bound organelles within the cytoplasm of their cells.

1. Obtain a small piece of onion layer and a clean slide. Carefully remove the delicate, transparent epidermis or "skin" from the inner curved side of the onion layer (see Fig. 1–3). Avoiding any wrinkles, place the onion epidermis on the slide. Add a drop of water and a coverslip. Place the slide on your microscope and observe under both low and high power.

    Do you see any structures within the onion epidermis cells? _____

    Carefully lift the coverslip from the slide and apply two drops of acetocarmine stain, which is specific for nucleic acids. Replace the coverslip and wait a moment for the stain to penetrate the epidermis. Then examine the stained wet mount. Do you see any more details in the cells after staining?

    _____

    If so, try to identify the structure(s) visible in the cells after staining. _____

    Using the experience with cells that you have gained so far, place a check by the structures you see present in the stained onion epidermis.

    ____ chloroplasts ____ nucleus

    ____ cell wall ____ plasma membrane

    ____ appendages

2. Obtain a prepared slide of human cheek epithelium. Observe the cells under high and low power. You can see that the cells on the slide have been stained for better visibility. These thin, flat epithelial cells form a protective lining of the mouth and other areas of the digestive tract. Observe several individual cells, noting as many details as you can.

    Using your experience with cells gained thus far, place a check by the structures that you see present in the stained human epithelial cells.

    ____ chloroplasts ____ nucleus

    ____ cell wall ____ plasma membrane

    ____ appendages

## POSTLAB QUESTIONS

1. In the first exercise, you examined a wet mount of *Elodea* leaf cells. In this exercise, you examined a stained slide of onion epidermis. Since onion cells and *Elodea* leaf cells are both plant cells, why was the nucleus visible only in the onion cells?

2. Do you agree or disagree with the statement: All plant cells contain chloroplasts. Why or why not?

3. What organelles (or structures) do onion epidermal cells possess that are NOT found in human cheek epithelial cells?

4. What organelle (or structure) did you see in the onion epidermal cell that was also visible in the human cheek epithelial cell?

## LAB 4.D
## IDENTIFICATION OF UNKNOWN CELLS

### MATERIALS REQUIRED

3 numbered microscopes on display with slides of animal, plant, and prokaryotic cells

### PROCEDURE

The various cells that you observed in the previous labs have given you enough experience to view unknown cells and to then classify them as prokaryotic, eukaryotic plant, or eukaryotic animal cells. Carefully look at the three unknown slides on display and try to identify them as representatives of one of the three general categories. The following clues should help you make a decision.

Are chloroplasts, or photosynthetic pigments, present or absent? (If observing a prepared slide, do not mistake green staining for photosynthetic pigment.)

Is a cell wall present or absent?

Is the cell relatively large or small?

Does the cell interior appear modified and complex

(organelles present) or does it seem to be unmodified and simple?

Does the cell have an organized nucleus?

Record your decisions below by checking the appropriate answers.

Slide #1 ___ prokaryotic ___ eukaryotic

___ plant ___ animal

Slide #2 ___ prokaryotic ___ eukaryotic

___ plant ___ animal

Slide #3 ___ prokaryotic ___ eukaryotic

___ plant ___ animal

## POSTLAB QUESTIONS

1. Based on their appearance, which cells appear more primitive, eukaryotic or prokaryotic? How do you think this relates to the theory of evolution?

2. How do you think the presence of a cell wall affects an organism?

## LAB 4.E
### EXPLORATION OF CELL DIVERSITY

### MATERIALS REQUIRED

pond water or mixed protozoa culture

slides and coverslips

dropper

dropper bottle of methyl cellulose solution

### PROCEDURE

Pond water or a mixed culture of microscopic organisms provides an excellent hunting ground for exploring a variety of cell types, shapes, and structures.

1. Use a dropper to draw up debris from the bottom of a container of pond water or other culture for microscopic observation. If green alga is present, be sure to obtain some in your sample. Place a few drops of the sample on a slide, add a coverslip, and observe under low power.
2. Scan the slide for movement or anything that catches your eye. If the organisms are moving too

fast for good observation, carefully lift the coverslip and add a drop of methyl cellulose. Observe any object of interest under high power if it improves visibility. In the space provided below, draw at least four different specimens you find on the slide.

## POSTLAB QUESTIONS

1. Do most of the specimens on your slide appear to be prokaryotic or eukaryotic? What criteria do you use to make this distinction?

2. Did you find mostly unicellular or multicellular organisms? Is it easy to distinguish between them?

3. Are any of the specimens you drew photosynthetic? How can you tell?

## LAB 4.F
### CELLS AND ORGANELLE FUNCTION

### PROCEDURE

Many of the organelles within eukaryotic cells are too small to be easily seen with your microscope even though they are abundant in most of the cells you saw today. Use the table below and the electron micrographs in Figures 4–3 to 4–11 to familiarize yourself with some of these major cell organelles. With the help of your text, complete Table 4–1 on page 50 by supplying the location (nucleus vs. cytoplasm), function, and origin (plant, animal, both) of the listed organelles.

Cell wall

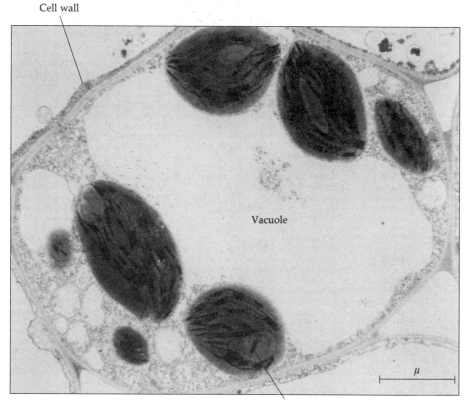

Vacuole

Chlorophasts

**FIGURE 4-3**

Electron micrograph of a grass cell showing the vacuole, cell wall, and chloroplasts typical of plant cells. *(Biophoto Associates/Science Source/Photo Researchers)*

Centrioles

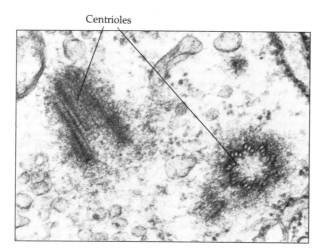

**FIGURE 4-4**

Electron micrograph of two centrioles (approximately 80,000X), one sectioned lengthwise (left), the other in cross section (right). *(B.F. King, School of Medicine, Univ. of CA-Davis/BPS)*

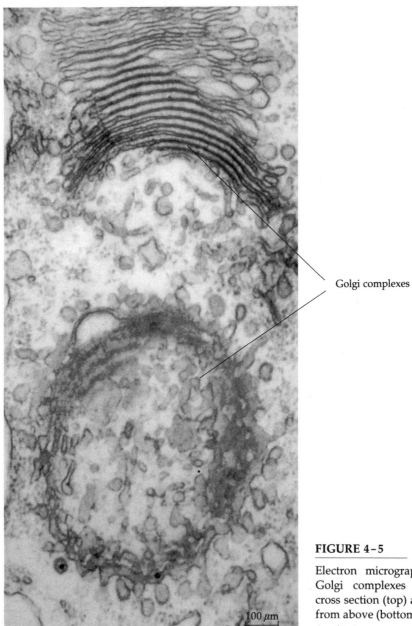

Golgi complexes

100 μm

**FIGURE 4–5**

Electron micrograph of two Golgi complexes shown in cross section (top) and viewed from above (bottom).   *(Biophoto Associates)*

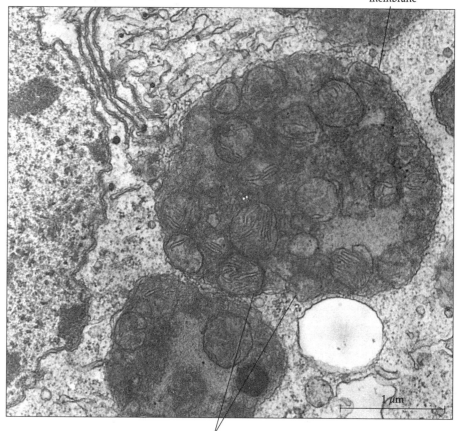

Lysosomal
membrane

Mitochondria
being digested

**FIGURE 4-6**

Electron micrograph of a lysosome digesting a starving cell's own mitochondria as a food source. *(Biophoto Associates)*

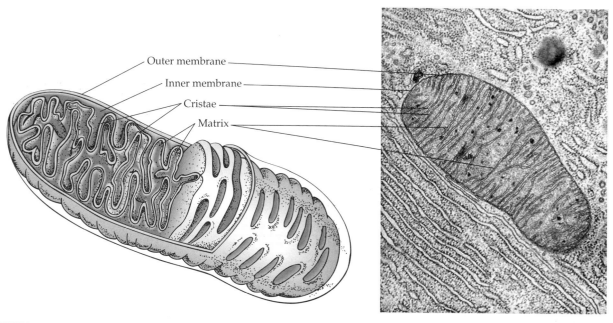

Outer membrane
Inner membrane
Cristae
Matrix

**FIGURE 4-7**

Drawing and electron micrograph of a mitochondrion (approximately 80,000X). *(Keith Porter/ Photo Researchers, Inc.)*

Nucleus

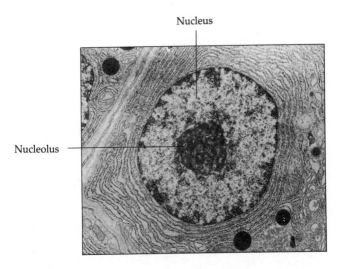

Nucleolus

**FIGURE 4-8**

Electron micrograph of a nucleus (approximately 11,000X) and a darker nucleolus within its interior.    *(Dr. Susumu Ito, Harvard Medical School)*

Ribosomes

Endoplasmic reticulum

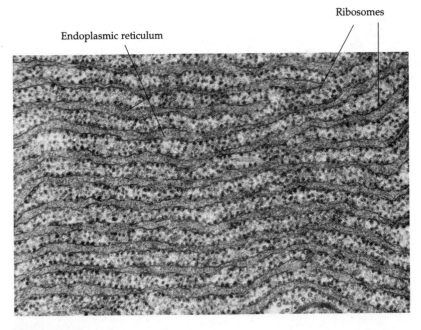

**FIGURE 4-9**

Electron micrograph of rough endoplasmic reticulum (approximately 70,000X). Note the many ribosomes associated with the endoplasmic reticulum membranes that give the "rough" appearance. Smooth endoplasmic reticulum appears similar but lacks the attached ribosomes.    *(Courtesy of Dr. E. Anderson, Harvard Medical School)*

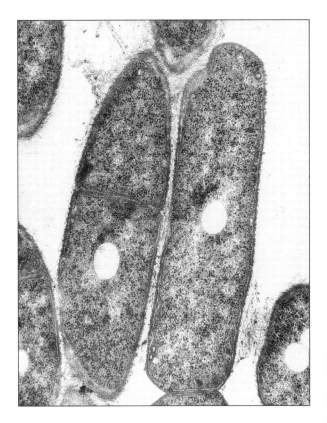

**FIGURE 4–10**

Thin section transmission electron micrograph of *Bacillus subtilis,* 30,000X, showing cell wall, cytoplasmic membrane, ribosomes, and polybetahydroxy butyric (PHB) acid granules. PHB is a cellular storage product and might possibly be important in the potential development of nonpetroleumbased plastic. *(Electron micrograph by Dr. Robert A. Pfister.)*

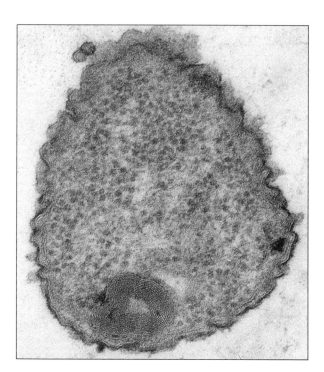

**FIGURE 4–11**

Thin section electron micrograph of *Pseudomonas sp.,* 60,000X, showing mesosome. The mesosome is an infolding of the cytoplasmic membrane. The exact function of the mesosome is unknown. *(Electron micrograph by Dr. Robert A. Pfister.)*

**POSTLAB QUESTIONS**

1. What is similar about the functions of RER, SER, and the Golgi complex? How might they interact?

2. Which organelles are exclusive to plants? To animals? How does this affect the functions of these organisms?

**Table 4-1**  Cell Organelles

| Organelle | Location | Function | Origin |
|---|---|---|---|
| Cell wall | | | |
| Centriole | | | |
| Chloroplast | | | |
| Golgi complex | | | |
| Lysosome | | | |
| Mitochondrion | | | |
| Nucleolus | | | |
| Nucleus | | | |
| Ribosome | | | |
| Rough endoplasmic reticulum | | | |
| Smooth endoplasmic reticulum | | | |
| Vacuole | | | |

## FOR FURTHER READING

Adams, R. L. P. 1990. *Cell Culture for Biochemists.* New York: Elsevier.

Bard, Jonathan B. L. 1990. *Morphogenesis: The Cellular and Molecular Processes of Developmental Anatomy.* New York: Cambridge University Press.

Becker, W. 1986. *The World of the Cell.* Menlo Park, Calif.: Benjamin Cummings.

Bellvue, Anthony R., and Henry J. Vogel, eds. 1991. *Molecular Mechanisms in Cellular Growth and Differentiation.* San Diego: Academic Press.

Bloom, W., and D. Fawcett. 1986. *A Guided Tour of the Living Cell.* New York: Freeman.

Crane, Frederick L., et al., eds. 1991. *Oxidoreduction at the Plasma Membrane: Relation to Growth and Transport.* Boca Raton, Fla.: CRC Press.

Darnell, J., H. Lodish, and D. Baltimore. 1986. *Molecular Cell Biology.* New York: Scientific American.

DeDuve, Christian. 1991. *Blueprint for a Cell: The Nature and Origin of Life.* Burlington, N. C.: N. Patterson.

Doyle, R. J., and Mel Rosenberg, eds. 1990. *Microbial Cell Surface Hydrophobicity.* Washington, D.C.: American Society for Microbiology.

Hill, Walter E., et al., eds. 1990. *The Ribosome: Structure, Function, and Evolution.* Washington, D.C.: American Society of Microbiology.

Holtzman, E., and A. B. Novikoff. 1984. *Cells and Organelles.* Philadelphia: Saunders College Publishing.

Malacinski, George M., ed. 1990. *Cytoplasmic Organizational Systems.* New York: McGraw-Hill.

Muramatsu, Takashi. 1990. *Cell Surface and Differentiation.* New York: Chapman and Hall.

Prescott, D. M. 1988. *Cells.* Boston: Jones and Bartlett.

Schlesinger, M. J., et al., eds. 1990. *Stress Proteins: Induction and Function.* New York: Springer-Verlag.

Thorpe, N. 1984. *Cell Biology.* New York: John Wiley and Sons.

Wolfe, S. 1985. *Cell Ultrastructure.* Belmont, Calif.: Wadsworth.

❏

# Diffusion, Osmosis, and Biological Membranes

## O B J E C T I V E S

1. Discuss the plasma membrane of the cell, describing the various functions it performs.

2. Describe the currently accepted model for the structure of the plasma membrane.

3. Contrast the physical (for example, diffusion and osmosis) with the physiological processes (for example, facilitated diffusion and active transport) by which materials are transported across cell membranes.

4. Solve simple problems involving osmosis. Predict whether cells will swell or shrink under various osmotic conditions.

---

## INTRODUCTION

Surrounding the contents of a cell is the **plasma membrane,** a physical barrier that separates the cell interior from the environment and that regulates substances passing into or out of the cell. Lipids (mostly phospholipids) and proteins are the main components of the plasma membrane and of membranes surrounding cell organelles. The bilayer arrangement of the lipid molecules results from an interaction between the lipids and the aqueous environment of the cell. The polar hydrophilic portion of the phospholipid molecules faces the outer and inner cell surfaces due to their attraction for water molecules. The long, nonpolar hydrophobic portion of the phospholipids turns away from water and aligns between the hydrophilic ends. Protein molecules of various types are distributed throughout the lipid bilayer. The nature and function of the plasma membrane are determined by both the proteins and lipids present in the membrane.

Normal cellular function is characterized by many substances moving into and out of the cell. Simple diffusion is one means of exchange between a cell and its environment. **Diffusion** is the process where molecules of two or more substances move about and become evenly mixed or dispersed. When the molecules remain uniformly distributed an **equilibrium** exists. As the molecules move away from their area of highest density, there is a gradual decrease in molecule concentration as the distance increases from the point of highest molecule concentration. This is called a **concentration gradient.** Molecules diffuse *down* a concentration gradient toward the area of least concentration. Simple diffusion is the mechanism by which small, uncharged molecules enter or leave a cell.

Not all molecules will diffuse through the plasma membrane. The membrane is **selectively permeable,** which means that some substances pass through more readily than others. In fact, the membrane may be **impermeable** to some molecules, which means that it completely prevents the passage of those molecules into or out of the cell.

Proper water balance is crucial for a cell. Water passes easily through the plasma membrane. This diffusion of water through a selectively permeable membrane is called **osmosis.** Like all other substances, water diffuses down its concentration gradient toward an area where there is less of it. The **osmotic potential** of a solution is its tendency to gain water by osmosis when separated from pure water by a selectively permeable membrane. The osmotic potential is determined by the number of dissolved particles (solutes) in the solution. The more particles there are in a solution, the greater the solution's tendency to gain water by osmosis.

If a cell is placed in a solution where the solute concentration is equal to the cell's solute concentration (equal osmotic potential), there is no net movement of water into or out of the cell. Such a solution is termed **isotonic.** A **hypertonic** solution contains a higher concentration of solutes than is found within the cell. If a cell is placed in a hypertonic environment, water moves out of the cell. A **hypotonic** solution contains a lower concentration of solutes than the cell interior. If a cell is placed in a hypotonic solution, water moves into the cell. These effects can best be remembered if one recalls that, in such systems, water always moves toward the area of lower concentration.

Liquids are not the only substances that diffuse through a membrane. **Dialysis** is the diffusion of a dissolved solute through a selectively permeable membrane. Consider two salt solutions of different concentrations separated by a membrane that is permeable to water and salt molecules. Salt molecules will diffuse by dialysis from the solution of greatest concentration into the solution of least concentration until the concentration of salt is equal on both sides of the membrane. Of course, diffusion of water (osmosis) across the membrane will also occur until equilibrium is reached.

In addition to diffusion, there are other processes and mechanisms by which materials pass through the plasma membrane. **Endocytosis** is the ingestion of large molecules or particles into a cell that are too large to pass directly through the plasma membrane. **Phagocytosis** is a specific type of endocytosis wherein the cell engulfs an object by surrounding it with arm-like extensions of the plasma membrane. The plasma membrane fuses around the object forming a vacuole that is pulled into the cell's cytoplasm. Enzymes within the vacuole then digest the material.

**Membrane carrier proteins** transport lipid insoluble materials such as small ions, glucose, and amino acids by three different mechanisms: facilitated diffusion, protein channels, and active transport. **Facilitated diffusion** occurs when a carrier protein in the membrane combines with a particular substance and moves it across the membrane down its concentration gradient to an area of least concentration. Some membrane carrier proteins form **protein channels** through the membrane. These channels allow the passage of molecules that are not lipid soluble. The transfer of materials with the utilization of energy is called **active transport,** which can move substances against or with a concentration gradient. An example is the **sodium-potassium pump,** which is a set of membrane proteins that use energy from ATP to transfer sodium ($Na+$) out of the cell and potassium ($K+$) into the cell. Electrical energy can also be used in active transport. The plasma membrane has an electrical charge, or **membrane potential,** because the fluid outside a cell contains a higher concentration of positively charged ions than the cell interior. The membrane potential is produced by the sodium-potassium pump, which continually maintains the proper balance of sodium and potassium ions on both sides of the plasma membrane. The electrical energy of the membrane potential powers many important cellular activities. The pump also enables the cell to control its water content indirectly as water follows the higher ion concentration.

The following laboratory exercises will demonstrate some of these principles.

## PRELAB QUESTIONS

1. What are the principal components of plasma membranes and membranes surrounding cell organelles?

2. What process involves the movement of molecules down a concentration gradient toward the area of least concentration?

3. What process is the diffusion of water through a selectively permeable membrane?

4. If a cell is placed in a solution where the solute concentration is greater than the solute concentration within the cell, what is the direction of the net movement of water?

5. What process involves the transfer of materials across a membrane with the use of energy from ATP?

## LAB 5.A
## MOLECULAR MOTION—THE MECHANISM OF DIFFUSION

### MATERIALS REQUIRED

slides and coverslips

toothpicks

container of carmine dye powder

### PROCEDURE

The molecules of all matter, particularly gases and liquids, are in constant, vibrating motion. As a result of this motion, molecules are continually colliding with one another, much like balls scattering over a pool table. Using a microscope, you can observe these molecular collisions that produce the effects of diffusion and osmosis.

1. Obtain a slide and place two drops of water on it. Set the slide down on the table.
2. Stick a toothpick vertically down into the container of carmine dye powder to a depth of about one half inch. Hold the toothpick vertically over the water drop on your slide with the point just above the water. Gently tap the toothpick to shake a light sprinkling of red dye powder onto the water drop.
3. Add a coverslip over the drop of water. Place the slide on your microscope and observe under low and high power. Carefully examine an area where the powder granules are evenly dispersed throughout the field of view. Note that the vibrating motion of the red granules is caused by vibrating water molecules colliding with each other and the powder granules.

### POSTLAB QUESTIONS

1. When will an equilibrium be reached? Do the powder granules stop moving at equilibrium?

2. Predict how a change in medium (that is, water to gelatin) would affect the rate of diffusion?

## LAB 5.B
## THE EFFECT OF TEMPERATURE ON DIFFUSION

### MATERIALS REQUIRED (per pair of students)

| | |
|---|---|
| 2 test tubes | forceps |
| 10ml graduated cylinder | potassium |
| test tube rack | permanganate |

### PROCEDURE

Diffusion is the movement of a substance from a region of higher concentration to a region of lower concentration. It is facilitated by the molecular motion you observed in Lab 5.A. Several factors affect diffusion such as temperature and the molecular weight of the diffusing molecules.

1. Working with a partner, obtain two test tubes of equal size. Using a graduated cylinder, pour 10 ml of cold tap water into one of the tubes. Pour 10 ml of hot tap water into the other tube.
2. Using forceps, add *one* crystal of potassium permanganate into each test tube and set them in a test tube rack. Without disturbing the tubes in any way, observe the diffusion of the purple potassium permanganate in both test tubes for three minutes.

### POSTLAB QUESTIONS

1. In which test tube is diffusion the greatest after three minutes?

2. Knowing that diffusion is facilitated by molecular collisions, how does the temperature of the tube with greatest diffusion affect the molecular motion of the water and potassium permanganate?

3. With time, would the tube with least diffusion eventually reach the same state as the tube with greatest diffusion?

## LAB 5.C
## THE EFFECT OF MOLECULAR WEIGHT ON DIFFUSION

### MATERIALS REQUIRED
glass tube supported by clamps on ringstands

rubber stoppers

dropper bottle of HCl

dropper bottle of $NH_4OH$

meter stick

### PROCEDURE

The instructor has set up a demonstration that shows the diffusion of two compounds, HCl (hydrochloric acid) and $NH_4OH$ (ammonium hydroxide). Observe the set up from time to time during the lab period for results. A few drops of each compound were placed in opposite ends of the glass tube, which was then plugged with stoppers (Fig. 5–1). The liquid compounds quickly evaporate, and the gases diffuse throughout the tube. The gases react with one another when they meet, forming a new compound, $NH_4Cl$ (ammonium chloride), which appears as a white band within the tube.

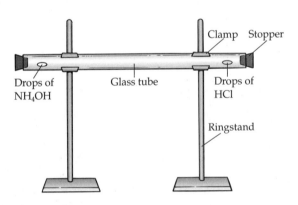

Drops of $NH_4OH$    Glass tube    Drops of HCl

Clamp  Stopper

Ringstand

**FIGURE 5–1**

Glass tube demonstration set up to show diffusion rates of two compounds of different molecular weights, HCL (hydrochloric acid) and $NH_4OH$ (ammonium hydroxide).

When the white band forms in the tube, record the following measurements in centimeters, using a meter stick:

1. Distance from HCl end to white band _____.

2. Distance from $NH_4OH$ end to white band _____.

### POSTLAB QUESTIONS

1. HCl has a molecular weight of 36, and $NH_4OH$ has a molecular weight of 35. Based on your measurements, how does molecular weight affect the rate of diffusion?

2. In summary, what factors effect the rate of diffusion and in what way do they effect the diffusion rate?

## LAB 5.D
## MEMBRANE PERMEABILITY

### MATERIALS REQUIRED
dialysis tubing in water

large test tube

2 regular test tubes

large beaker

test tube rack

test tube holders

boiling water bath

rubber bands

starch solution in squeeze bottle

glucose solution in squeeze bottle

Benedict's reagent in dropper bottle

iodine in dropper bottle

2 droppers

### PROCEDURE

This exercise simulates a selectively permeable plasma membrane by using dialysis tubing, an artificial membrane with microscopic pores throughout its surface. Molecules smaller than the diameter of the pores will diffuse through the membrane, whereas molecules larger than the pores will not. This selectivity regulates the diffusion of substances into and out of the cell based on molecular size, although other factors are involved.

1. Obtain a large test tube and fill it half full with tap water. Set the tube in a large beaker so that it is supported.

2. Pick up a strip of dialysis tubing that has been soaking in water. Carefully tie a knot at one end of the dialysis tubing. Make sure the knot is secure enough to be water tight.

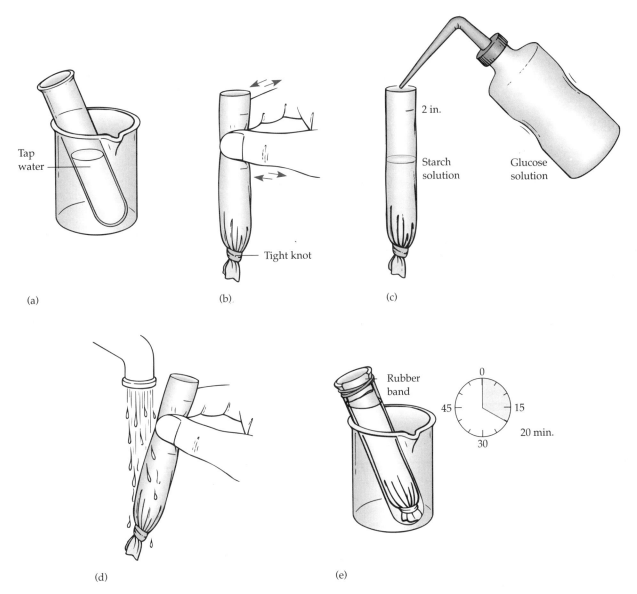

(a)    (b)    (c)

(d)    (e)

**FIGURE 5-2**

(a–e) Preparation of dialysis bag containing starch and glucose solutions. (Figure continued on page 56)

3. Open the free end of the dialysis tubing by slipping it back and forth between your thumb and forefinger. Continue separating the walls of the dialysis tubing along its entire length in this same manner. Always keep the dialysis tubing wet during the process and handle it carefully.

4. Using the squeeze bottles, fill half of the dialysis tubing with starch solution and the remaining half with glucose solution to within about two inches of the top of the dialysis tubing.

5. Hold the filled dialysis tubing closed with your fingers and rinse it in running water to wash off any solutions on the outside of the tubing.

6. Place the filled dialysis tubing into the large test tube with water. Fold the top of the dialysis tubing over the lip of the test tube and keep it in place with a rubber band around the mouth of the tube (Fig. 5-2). Let the dialysis tubing sit undisturbed for 20 minutes. While you are waiting, proceed with the next exercise.

After 20 minutes have passed, your next step is to determine if any of the starch and glucose diffused out of the dialysis tubing. The water in the test tube can be tested for the presence of starch and glucose with two simple chemical tests: Benedict's test and the starch/iodine test. Benedict's test

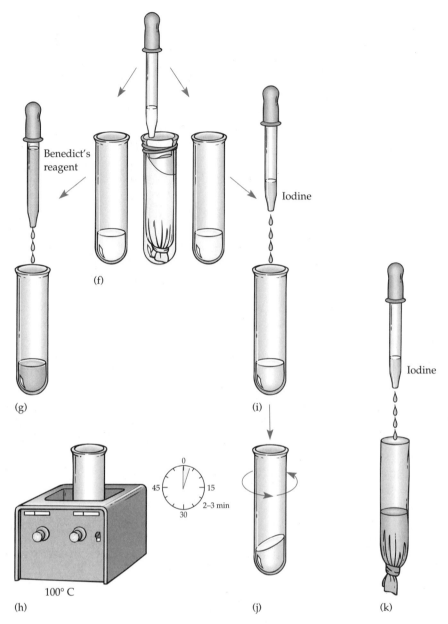

**FIGURE 5-2**

(*continued*) (f-k) Procedure to detect the diffusion of starch and/or glucose.

uses Benedict's reagent. When heated with glucose a reaction occurs that changes the blue color of the reagent to green, yellow, orange, or dark red, depending on how much glucose is present. Any color resulting in green to dark red indicates a positive test for glucose. The starch/iodine test uses iodine. In the presence of starch, iodine turns a dark, blue-black color. This is a positive test for starch.

7. Place two full droppers of the water surrounding the dialysis tubing into a clean test tube. *Note:* Use *only* a clean dropper. Rinse well with water if nec-

essary. Add one dropper of Benedict's reagent and boil the tube in a hot water bath for 2-3 minutes. Did you get a positive test for glucose? _____

8. Again, place two full droppers of the water surrounding the dialysis tubing into a clean test tube. Add four drops of iodine into the tube and swirl to mix. Did you get a positive test for starch? _____

9. Carefully open the top of the dialysis tubing and add four drops of iodine into the starch/glucose mixture. Did you get a positive test for starch? ____

**POSTLAB QUESTIONS**

1. Did glucose diffuse out of the dialysis tubing? Substantiate your answer with evidence.

2. Did starch diffuse out of the dialysis tubing? Support your answer with evidence.

3. What can you say about the relative size of glucose molecules as compared to those of starch molecules?

## LAB 5.E
## OSMOSIS

### MATERIALS REQUIRED

potato cylinders in containers of:

   20% salt (NaCl) solution

   .9% salt (NaCl) solution

   distilled water

forceps

small rulers

paper towels

### PROCEDURE

Osmosis is a particular kind of diffusion that involves only water diffusing through a selectively permeable membrane, such as the plasma membrane. Because water is a major component of living things, osmosis is an important process that determines the water content of cells.

   In this exercise, the direction of water movement in osmosis is determined by the solute (dissolved substance) concentration of a salt solution as compared to the solute concentration within the potato cells. The potato cylinders you see in the salt solutions and in the distilled water were all originally 7 cm in length. Over time, water osmotically moved into or out of the potato tissue according to the concentration gradient of water. Remember that the solute concentration of the solution determines the concentration gradient. In other words, more salt in a solution means less water. The osmotic gain or loss of water causes a change in size of the potato cylinders. You will record these changes by taking measurements.

1. Using forceps, carefully remove a potato cylinder from the 20% salt solution container and place it on a paper towel.

2. Measure the length of the cylinder in centimeters as accurately as you can with a small ruler. Record the measurements in Table 5–1. Return the potato cylinder to its original container.

**Table 5–1**  Demonstration of Osmosis

| Conc. of Solution | Length (cm) of Cylinder |
| --- | --- |
| 20% salt solution | |
| .9% salt solution | |
| distilled water | |

3. Measure one potato cylinder from the other two containers. Also record the measurements in Table 5–1. *Note:* Make sure you replace the potato cylinders in the correct container after you measure them.

### POSTLAB QUESTIONS

1. Which potato cylinders increased in length? Decreased in length? How do you explain these changes in length?

2. Which solution was hypertonic to the solute concentration within the potato cells?

3. In which solution would there be no net movement of water into or out of the potato cells?

4. Which solution contained a lower concentration of solutes than the interior of the potato cells?

5. A solution of 0.9% NaCl is isotonic to red blood cells. What would happen if you placed the red blood cells in a solution of 10% NaCl? In distilled water?

6. Imagine that three dialysis tubes filled with varying concentrations of sucrose are placed in a beaker containing a 25% sucrose solution. Tube A contains a 1% sucrose solution. Tube B contains a 25% sucrose solution and Tube C contains a 50% sucrose solution. Will water move into or out of Tube A? Tube B? Tube C? Which solution is hypertonic to the solution in the beaker? Hypotonic? Isotonic?

## LAB 5.F
## PLASMOLYSIS IN *ELODEA*

### MATERIALS REQUIRED

slides and coverslips

live *Elodea* plants

forceps

20% salt (NaCl) solution in dropper bottles

### PROCEDURE

Plant cells are completely surrounded by a rigid **cell wall.** The plasma membrane lies just inside the cell wall much like a box lined with thin plastic wrap. Water and all other materials that enter and leave the cell pass through the plasma membrane and cell wall.

1. Prepare a wet mount of an *Elodea* leaf taken from the growing tip of the plant. Place the slide on your microscope and observe under both low and high power. Note the appearance of these normal *Elodea* cells. Draw a few of the cells in the space below.

2. Lift the coverslip and place a few drops of 20% salt solution on the *Elodea* leaf. Replace the coverslip and again examine the leaf cells under low and high power. Note the appearance of these cells after treatment with salt solution. Draw a few of these cells in the space below and compare their appearance with normal *Elodea* cells.

The difference in appearance between normal and salt-treated *Elodea* cells is due to water loss. The water loss reduces the volume of the cell contents and the plasma membrane pulls away from the rigid cell wall as the cell shrinks. This process is known as *plasmolysis.*

### POSTLAB QUESTIONS

1. By what process does water leave the salt-treated *Elodea* cells?

2. The 20% salt solution is hypertonic/hypotonic/isotonic to the solute concentration of *Elodea* cells.

3. How would you reverse the process of plasmolysis in *Elodea* cells?

4. Compare the location of chloroplasts in normal versus salt-treated *Elodea* cells. Why the difference in locations?

### FOR FURTHER READING

Alberts, B., et al. 1989. *Molecular Biology of the Cell.* New York: Garland.

Boutilier, R. G., ed. 1990. *Vertebrate Gas Exchange: From Environment to Cell.* New York: Springer-Verlag.

Crane, Frederick L., et al., eds. 1991. *Oxidoreduction at the Plasma Membrane: Relation to Growth and Transport.* Boca Raton, Fla.: CRC Press.

Glaser, R., and D. Gingell, eds. 1990. *Biophysics of the Cell Surface.* New York: Springer-Verlag.

Goodfellow, Julia M. 1990. *Molecular Dynamics: Applications in Molecular Biology.* Boca Raton, Fla.: CRC Press.

Karp, G. 1984. *Cell Biology.* New York: McGraw-Hill.

Kuhn, P. J., ed. 1990. *Biochemistry of Cell Walls and Membranes in Fungi.* New York: Springer-Verlag.

Larson, C., and I. M. Miller, eds. 1990. *The Plant Plasma Membrane: Structure, Function, and Molecular Biology.* New York: Springer-Verlag.

Mato, Jose M. 1990. *Phospholipid Metabolism in Cellular Signaling.* Boca Raton, Fla.: CRC Press.

Michel, Hartmut, ed. 1991. *Crystallization of Membrane Proteins.* Boca Raton, Fla.: CRC Press.

Prescott, D. M. 1988. *Cells.* Boston: James and Bartlett.

O'Day, Danton H., ed. 1990. *Calcium as an Intracellular Messenger in Eucaryotic Microbes.* Washington, D.C.: American Society for Microbiology.

Turner, A. J., ed. 1990. *Molecular and Cell Biology of Membrane Proteins: Glycolipid Anchors of Cell Surface Proteins.* New York: Ellis Horwood.

Vasilescu, D., ed. 1990. *Water and Ions in Biomolecular Systems: Proceedings of the 5th UNESCO International Conference.* Boston: Birkh auser Verlag.

Vigo-Pelfrey, Carmen, ed. 1990. *Membrane Lipid Oxidation.* Boca Raton, Fla.: CRC Press.

Wotton, Roger S., ed. 1990. *The Biology of Particles in Aquatic Systems.* Boca Raton, Fla.: CRC Press.

# PART III

## Life and the Flow of Energy

**TOPIC 6**
Enzymes

**TOPIC 7**
Cellular Respiration and Anaerobic Pathway

**TOPIC 8**
Photosynthesis

# TOPIC 6

❑

# Enzymes

## OBJECTIVES

1. Describe the chemical nature and structural aspects of enzymes.

2. Summarize the role of enzymes as chemical regulators and to describe the biological function of the active sites.

3. Identify the effects of pH and temperature on enzyme activity.

4. Describe the action and effect of an inhibitor on enzyme activity.

## INTRODUCTION

**Proteins** have important roles in living organisms. They are amino acid polymers covalently bonded in long chains that are subsequently coiled and folded into complex shapes. Proteins form structural features such as hair, hooves, and tendons.

Other proteins, known as **enzymes,** affect the many biochemical reactions that keep cells alive. Enzymes act as **catalysts** in biochemical reactions. This means that they speed chemical reactions that would, in the absence of the enzymes, proceed at very slow rates or perhaps not even at all. Enzymes accomplish this effect by lowering the **energy of activation** of a biochemical reaction, which is the energy required to initiate the reaction. Specific **substrates,** or **reactants,** combine with enzymes at their **active site(s),** which induces biochemical reactions that result in products used by the cell (Fig. 6–1).

The activity of a specific enzyme is affected by many factors such as pH, temperature, **allosteric interactions,** and **inhibitors** (substances that prevent normal action of the enzyme).

A temperature and/or pH higher than what are normal for an enzyme will **denature** (alter the molecular shape of) the protein, thereby disabling the enzyme's normal activity. A pH lower than normal will

have a similar effect. Enzymes are not denatured by lower temperatures, but the kinetic energy is lowered to a point where the rate of biochemical reactions is greatly reduced.

Enzyme activity can be decreased by inhibitors that bind to the enzyme's active site. This prevents substrate molecules from binding and no chemical reaction occurs (Fig. 6–2a). Enzyme activity can also be decreased by inhibitors that bind to a site other than the active site of the enzyme. This renders the enzyme inoperative by altering its shape (Fig. 6–2b).

Some enzymes, termed **allosteric enzymes,** have more than one binding site (the portion of the enzyme molecule that normally comes in contact with the substrate molecule). Such enzymes frequently exhibit a dramatic change in activity when one site is occupied by a substrate or, in some cases, by a molecule similar to the substrate.

All of an organism's biochemical reactions are collectively referred to as its **metabolism.** The reactions involved are produced by a series of enzyme-mediated activities called **metabolic pathways.** The phenomenon of **negative feedback** occurs when the concentration of an enzymatically produced product decreases the activity of the enzyme(s) involved. When the product becomes used up, enzyme activity

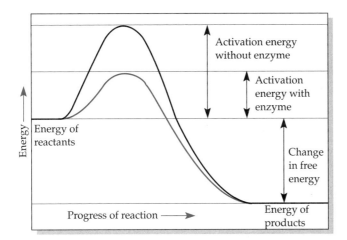

**FIGURE 6-1**

An enzyme speeds up a chemical reaction by lowering its activation energy. A catalyzed reaction proceeds more quickly than an uncatalyzed reaction because the barrier of activation is lowered.

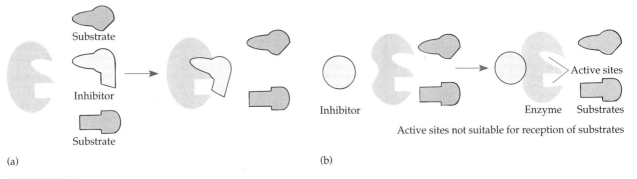

(a)                                                            (b)

**FIGURE 6-2**

(a) The inhibitor competes with the normal substrate for the active site of the enzyme. (b) The inhibitor binds with the enzyme at a site other than the active site, altering the shape of the enzyme and thereby inactivating it.

resumes once again. Negative feedback is one of a variety of methods of metabolic control in an organism.

We have stated that enzymes are proteins; however, it has recently been discovered that certain types of RNA (a nucleic acid) can act as an enzyme, which shows that our knowledge of enzymatic biochemistry is still growing.

The following laboratory exercises demonstrate some of the fascinating properties of enzymes.

## PRELAB QUESTIONS

1. What are the monomers that compose an enzyme?

2. What is the effect of enzymes on chemical reactions?

3. What factors affect enzyme activity?

4. What effect does extreme pH have on normal enzyme activity?

## LAB 6.A
## ENZYME ACTIVITY: THE EFFECT OF TEMPERATURE

### MATERIALS REQUIRED

potato extract

catechol

5 test tubes

test tube rack

test tube holder

boiling water bath, 100°C

warm water bath, 40°C

ice bath, 0°C

wax pencil

small ruler

## PROCEDURE

The enzyme you will be using, catechol oxidase, is found in plant tissues. As its name implies, catechol oxidase oxidizes (removes hydrogen from) the substrate catechol, also found in plants, and converts it to benzoquinone, which is brown in color. This reaction explains the darkening of bruised fruit and vegetables. The injured cells release catechol and catechol oxidase, which do not contact one another in unbroken cells. The benzoquinone product, which is toxic to bacteria, helps prevent decay in damaged plant tissue. You will use potato extract as a source of catechol oxidase and a solution of catechol to investigate the various factors that affect enzyme activity (see Fig. 6–3).

1. Use the ruler and wax pencil to accurately mark both 1 cm and 2 cm from the *bottom* of 5 clean test tubes. Number the top end of the test tubes 1, 2, 3, 4, and 5.
2. Fill tube 1 to the 1 cm mark with distilled water. This tube remains in the test tube rack and will later serve as a control.
3. Fill tubes 2 through 5 up to the 1 cm mark with potato extract. Place tube 2 in the hot water bath, tube 3 in the warm water bath, tube 4 in the ice bath, and tube 5 in the test tube rack (which remains at room temperature). Allow all test tubes to stand for 5 minutes at their various temperatures.
4. After the tubes reach their temperatures, add catechol to the 2 cm mark in all 5 test tubes.

---

**CAUTION** Catechol is toxic! Use caution when filling test tubes. Wash thoroughly with soap and water if catechol contacts your skin.

---

5. Watch for any color change in the 5 test tubes during the next 20 minutes. Draw a table to record color changes at 5-minute intervals for 20 minutes.

Use the following plus and minus color intensity scale presented below for recording color intensity. Mix the solutions at 5-minute intervals by gently tapping the bottom of the tubes. Be sure to use a test tube holder to handle tube 2 in the hot water bath.

**Color Intensity Scale**

---

| | |
|---|---|
| − | = no color change |
| + | = very slight color change |
| ++ | = light color change |
| +++ | = definite color change |
| ++++ | = dark color |

---

**CAUTION** After recording results, *do not* empty test tubes into the drain. Dispose of catechol in the container designated for catechol disposal.

---

## POSTLAB QUESTIONS

1. Identify the substrate, the enzyme, and the product involved in this experiment.

2. Was the control lacking the substrate or the enzyme?

3. In Step 3, why is it necessary to let the test tubes stand for 5 minutes in their various temperatures?

4. Based on your results, what was the optimum temperature for catechol oxidase activity?

5. What effect did a hot water bath have on enzyme activity? Why?

6. Using your results, argue for or against the following statement: Enzymes function equally and efficiently at all temperatures.

7. Why does lemon juice applied to slices of apples or bananas prevent darkening of the fruit?

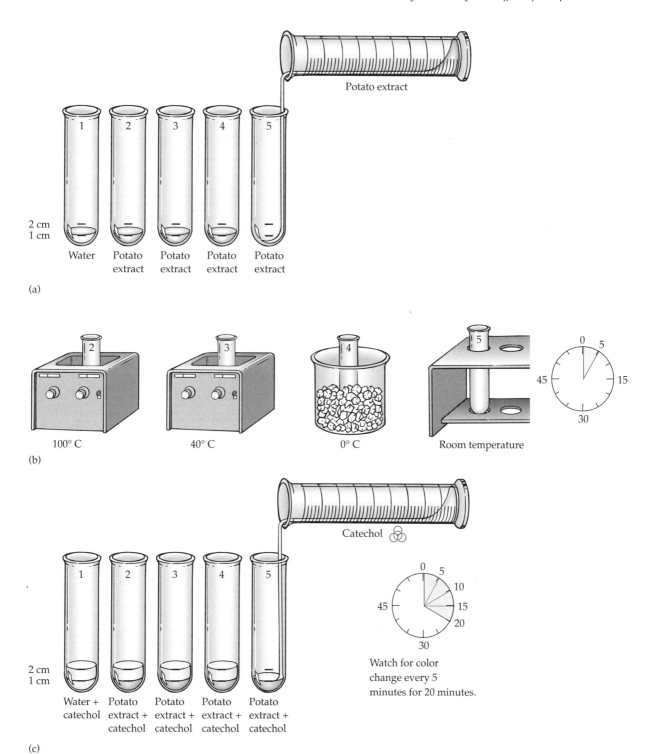

**FIGURE 6-3**

Procedure for determining the effect of temperature on enzyme activity.

## LAB 6.B
## ENZYME ACTIVITY: THE EFFECT OF AN INHIBITOR

### MATERIALS REQUIRED

3 test tubes

test tube rack

wax pencil

small ruler

potato extract

sodium phosphate solution

PTU solution

distilled water in dropper bottles

catechol

### PROCEDURE

Enzyme inhibitors interfere with the function of enzymes by binding with some critical portion of the enzyme molecule. Your objective is to determine which of two solutions, sodium phosphate ($NaPO_4$) or phenylthiourea (PTU), is an inhibitor of the enzyme catechol oxidase (see Fig. 6–4).

1. Use the ruler and wax pencil to accurately mark both 1 cm and 2 cm from the *bottom* of 3 clean test tubes. Number the top end of the test tubes 1, 2, and 3.
2. Fill all tubes to the 1 cm mark with potato extract.
3. Add five drops of distilled water to tube 1, five drops of sodium phosphate solution to tube 2, and five drops of PTU solution to tube 3.

---

C A U T I O N  PTU is toxic! Flush thoroughly with running water if the solution contacts skin or clothing.

---

4. Fill all tubes to the 2 cm mark with catechol and mix well.
5. Watch for any color changes in the 3 test tubes during the next 20 minutes. Mix the tube contents at 5-minute intervals while timing. Record color changes in a table at 5-minute intervals using the plus and minus scale in the previous lab.

---

C A U T I O N  After recording results, *do not* empty PTU solution into the sink drain. Dispose of PTU in the container designated for catechol disposal.

---

### POSTLAB QUESTIONS

1. Identify the substrate, the enzyme, the inhibitor, and the product involved in this experiment.

2. What is the purpose of tube 1?

3. What was the effect of an inhibitor on enzyme activity?

4. The enzyme catechol oxidase has a copper ion that forms part of its active site. Knowing this, what do you suggest as the mechanism of action for the inhibitor to affect enzyme activity?

## LAB 6.C
## ENZYME ACTIVITY: THE EFFECT OF pH

### MATERIALS REQUIRED

4 test tubes

test tube rack

wax pencil

small ruler

buffer solutions of pH 10, 7, 6, 3

potato extract

catechol

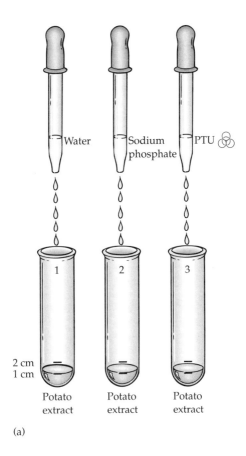

(a)

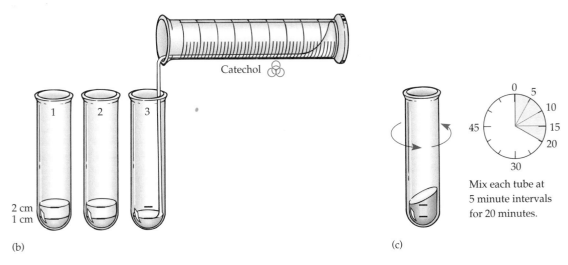

(b)

(c)

Mix each tube at
5 minute intervals
for 20 minutes.

**FIGURE 6–4**

Procedure for determining which solution is an inhibitor of the enzyme catechol oxidase.

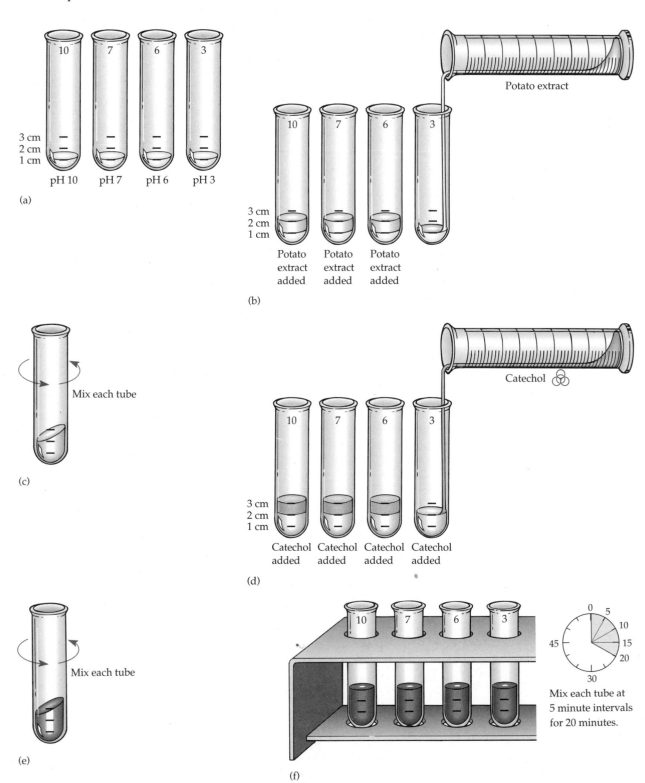

**FIGURE 6-5**

Procedure for determining the effect of pH on enzyme activity.

## PROCEDURE

The acidity, or alkalinity of a substance can be expressed as a numerical value, which is called pH. The pH scale ranges from 0 to 14. A pH value of 7 is neutral—values less than 7 are more acidic, whereas values greater than 7 are more alkaline (see Fig. 6–5).

1. Use the ruler and wax pencil to mark 1 cm, 2 cm, and 3 cm from the *bottom* of 4 clean test tubes. Mark the top end of the test tubes 10, 7, 6, and 3.
2. Fill each test tube up to the 1 cm mark with the appropriate buffer solution of the same pH value as that written on the top of the tube.
3. Fill each test tube up to the 2 cm mark with potato extract. Mix the contents well.
4. Finally, fill each test tube up to the 3 cm mark with catechol. Mix well.
5. Watch for color changes in the four test tubes during the next 20 minutes. Record color changes in a table at 5-minute intervals. Use the plus and minus scale presented in Lab 6.A. Mix the solutions at 5-minute intervals while timing.

## POSTLAB QUESTIONS

1. Based on your results, what was the optimum pH for catechol oxidase activity?

2. What effect does a high or low pH have on enzyme activity? Why?

3. Based on your observations, why do cytoplasm and many body fluids have a pH measure near 7?

## FOR FURTHER READING

Alberts, B., et al. 1983. *Molecular Biology of the Cell.* New York: Garland Publishing.

Atkins, P. W. 1989. *Molecules.* New York: Scientific American Library.

Bettelheim, F. A., and J. March. 1991. *Introduction to General, Organic and Biochemistry.* 3rd ed. Philadelphia: Saunders College Publishing.

Chaplin, M. F. 1990. *Enzyme Technology.* New York: Cambridge University Press.

Creighton, T. E., ed. 1990. *Protein Structure: A Practical Approach.* New York: IRL Press.

Dordick, Jonathan S., ed. 1991. *Biocatalysts for Industry.* New York: Plenum Press.

Fersht, A. 1985. *Enzyme Structure and Function.* San Francisco: W. H. Freeman.

Harris, Harry. 1976. *Handbook of Enzyme Electrophoresis in Human Genetics.* New York: American Elsevier Publishing Company.

Hayashi, Katsuya. 1986. *Dynamic Analysis of Enzyme Systems: An Introduction.* New York: Springer-Verlag.

Jornvall, H., et al. 1991. *Methods in Protein Analysis.* Boston: Birkhauser Verlag.

Lehninger, A. 1982. *Principles of Biochemistry.* New York: Worth.

Lojda, Zdenek. 1979. *Enzyme Histochemistry: A Laboratory Manual.* New York: Springer-Verlag.

Prescott, D. M. 1988. *Cells.* Boston: Jones and Bartlett.

Stauffer, Clyde. 1989. *Enzyme Assays for Food Scientists.* New York: Van Nostrand Reinhold.

Suelter, Clarence H. 1985. *A Practical Guide to Enzymology.* New York: Wiley.

Ternynck, Therese. 1990. *Immunoenzymatic Techniques.* New York: Elsevier.

Wiseman, Alan, ed. 1991. *Genetically Engineered Proteins and Enzymes from Yeasts: Production Control.* New York: Ellis Horwood.

❏

# Cellular Respiration and Anaerobic Pathways

1. Write an overall summary for cellular respiration.

2. Give a brief overview of the four phases of cellular respiration, and indicate where the reactions of each phase take place in the cell.

3. Write a general equation illustrating hydrogen and electron transfer from a substrate to a hydrogen (or electron) acceptor such as $NAD^+$.

4. Describe fermentation in terms of a final hydrogen acceptor and end products.

5. Describe the relationship between overall metabolic rate and the amount of carbon dioxide produced.

## INTRODUCTION

Virtually all of the activities of an organism are accomplished by energy extracted from high energy-containing food molecules. Most organisms do this by a series of complex biochemical reactions known, overall, as **cellular respiration.** The process consists of the stepwise oxidation of high energy-containing food molecules to the low energy-containing molecules carbon dioxide and water. This sequence utilizes oxygen as the final hydrogen (or electron) acceptor in a series of oxidation-reduction reactions by means of which food molecules, most commonly the six-carbon sugar glucose, are completely degraded. Most of the extracted energy is in the form of hydrogen atoms or their electrons which, as glucose is broken down, are removed and passed through the electron transport system where adenosine triphosphate (ATP) is synthesized. The energy contained within ATP molecules is subsequently utilized when cellular work is performed.

Hydrogen atoms are transported throughout the reactions of respiration by **coenzymes.** These are enzymes that require an additional organic molecule in their structure, usually a vitamin. Specifically, the respiratory coenzymes are $NAD^+$ (nicotinamide adenine dinucleotide) and **FAD** (flavin adenine dinucleotide).

The complex reactions of respiration can be considered in four major steps (see Fig. 7–1):

1. **Glycolysis,** which occurs in the cytoplasm, breaks down glucose into two 3-carbon **pyruvate** molecules. A small amount of ATP is also produced and $NAD^+$ is reduced to NADH.

2. Pyruvate enters the matrix of the mitochondrion and loses one carbon atom as carbon dioxide, leaving a two-carbon acetyl group. The acetyl group attaches to coenzyme A forming **acetyl CoA.** More NADH is also produced.

3. Acetyl CoA enters the **citric acid cycle** where the remaining carbon atoms leave as carbon dioxide.

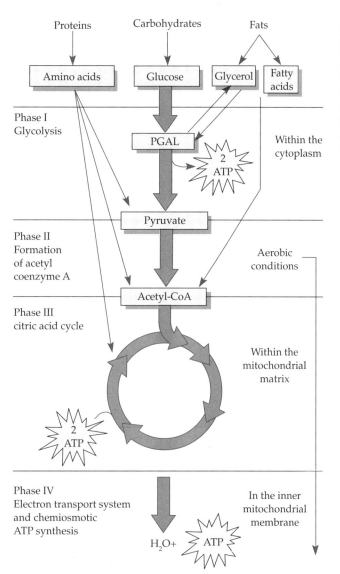

Proteins  Carbohydrates  Fats

Amino acids | Glucose | Glycerol | Fatty acids

Phase I
Glycolysis

PGAL

Within the
cytoplasm

2 ATP

Pyruvate

Phase II
Formation
of acetyl
coenzyme A

Aerobic
conditions

Acetyl-CoA

Phase III
citric acid cycle

Within the
mitochondrial
matrix

2 ATP

Phase IV
Electron transport system
and chemiosmotic
ATP synthesis

In the inner
mitochondrial
membrane

$H_2O+$  ATP

**FIGURE 7–1**

Four main phases in cellular respiration are (I) glycolysis, (II) formation of acetyl coenzyme A, (III) the citric acid cycle, and (IV) the electron transport system with its associated chemiosmotic ATP synthesis.

This cycle occurs in the matrix of the mitochondrion. More hydrogen atoms are removed from the molecule forming additional NADH and reducing FAD to $FADH_2$. A small amount of ATP is also produced.

4. NADH and $FADH_2$ carry hydrogen to the **electron transport system,** a series of carrier proteins in the inner mitochondrial membrane. Here, the hydrogen atoms are separated into hydrogen ions ($H^+$) and electrons. The hydrogen ions are pushed to one side of the membrane forming a hydrogen ion gradient. Electrons arriving at the end of the transport system are rejoined with hydrogen ions and picked up by oxygen atoms. This forms water. **Chemiosmotic ATP synthesis** occurs when hydrogen ions flow back to the other side of the membrane through protein channels. ATP synthe-

tase enzymes associated with the channels use the energy of the moving hydrogen ions to attach inorganic phosphate to adenosine diphosphate (ADP). This produces ATP. Although ATP production occurs in several steps of respiration, most of the cell's ATP is produced at the electron transport system.

As shown above, most of the energy from food molecules is obtained by passing electrons from hydrogen atoms along the electron transport system. The more hydrogen there is in a food molecule, the more energy in that substance. Fats contain an abundance of hydrogen atoms in their molecular structure. Therefore, fat is an efficient medium of energy storage. Food is converted to fat, which accumulates on the body when more food is consumed than utilized. When food is scarce, stored fat and body proteins are metabolized as energy sources. These compounds enter at various steps along the cellular respiration pathway (see Fig. 7–1).

Cellular respiration cannot proceed without oxygen. Some organisms are capable of **fermentation**— the incomplete breakdown of food molecules using organic molecules instead of oxygen as electron acceptors. Fermentation yields only 2 ATPs for each glucose molecule fermented as compared to 38 ATPs for each glucose molecule oxidized by cellular respiration. Despite the difference, fermentation allows cells to live in conditions without oxygen. Fermentation begins by glycolysis, which produces pyruvate. The resulting NADH transfers its hydrogen to pyruvate (or its derivative) as an electron acceptor. NAD+ is freed and can then carry more hydrogen in another glycolysis reaction.

There are two types of fermentation: alcoholic and lactate. **Alcoholic fermentation** occurs when yeast cells break down pyruvate into acetaldehyde and carbon dioxide. The acetaldehyde accepts hydrogen from NADH and forms ethyl alcohol. **Lactate fermentation** occurs in muscle cells. Pyruvate from glycolysis accepts hydrogen from NADH and is converted into lactate. This allows muscles to make a small amount of additional ATP during times of heavy exercise when adequate amounts of oxygen cannot be obtained from the blood supply. After prolonged muscular activity, the accumulated lactate is oxidized back to pyruvate or glucose. The additional oxygen utilized in the oxidation of lactate is called an **oxygen debt.**

The following laboratory exercises have been selected and designed to illustrate the concepts and principles mentioned in this brief discussion of cellular respiration and anaerobic pathways.

## PRELAB QUESTIONS

1. What phase of cellular respiration breaks down glucose into two pyruvate molecules?

2. Where in the cell do the citric acid cycle reactions take place?

3. Identify the two molecules that transport hydrogen to the electron transport system.

4. What is the final hydrogen acceptor in the electron transport system?

5. What is (are) the end product(s) of alcoholic fermentation?

# LAB 7.A
## FERMENTATION AND CARBON DIOXIDE PRODUCTION IN YEAST

### MATERIALS REQUIRED (per group of students)

| | |
|---|---|
| 4 fermentation tubes | yeast suspension |
| wax pencil | starch solution |
| small metric ruler | sucrose solution |
| small beaker | glucose solution |
| 2 small graduated cylinders | distilled water |

### PROCEDURE

Yeast, a unicellular fungus, is extremely important to the baking and brewery industries because of its fermentation byproducts—carbon dioxide, which causes bread dough to rise, and ethyl alcohol, which is found in alcoholic beverages. Unlike many organisms, yeast is capable of metabolizing by both fermentation and aerobic respiration processes, depending on the presence or absence of oxygen in the yeast's environment at the time. In the following exercise, you will observe the degrees to which yeast can ferment various carbohydrates by measuring the amount of carbon dioxide gas produced.

1. Obtain 4 fermentation tubes. Use a wax pencil to label them 1, 2, 3, and 4.

2. Since the volume of various fermentation tubes may vary, you must first determine the volume of your tubes before proceeding further. Do this by filling one of your fermentation tubes with water. The straight portion of the tube should be completely filled with liquid so that there is no air space, as shown in Figure 7–2. Fill the straight portion of the tube by holding your thumb over the tube opening and tipping it backwards. After properly filling the fermentation tube with water, pour the water from the tube into a graduated cylinder and note the volume of liquid.

3. Using a graduated cylinder, add yeast suspension equal to half the tube volume to all fermentation tubes. Be sure to keep the amount of yeast equal in all tubes.

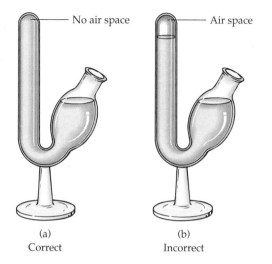

No air space — (a) Correct

Air space — (b) Incorrect

**FIGURE 7-2**

Fermentation tubes should be filled completely with liquid, as shown in the correct illustration.

4. Fill the fermentation tubes with the various solutions as listed below in an amount equal to the yeast suspension.

   Tube 1 — distilled water

   Tube 2 — starch solution

   Tube 3 — sucrose solution

   Tube 4 — glucose solution

   Rinse the graduated cylinders after measuring *each* solution. Be sure to keep the initial amount of solution equal in all tubes. Tip the fermentation tubes as above when you fill them so that no air space remains in the upper end of the tube.

5. Set the tubes aside for one hour as the yeast ferments the carbohydrate solutions and continue with the other exercises. Your results will not be harmed if the tubes stand for more than an hour.

**Table 7-1**   Measurement of Gas Production by Fermentation

| Tube Number | Carbohydrate Solution | Millimeters (mm) of Gas Produced |
|---|---|---|
| 1 | Distilled water | |
| 2 | Starch solution | |
| 3 | Sucrose solution | |
| 4 | Glucose solution | |

After timing, measure in millimeters the amount of gas that has collected in the end of the fermentation tube and record your results in Table 7-1.

**POSTLAB QUESTIONS**

1. What gas is produced by fermentation in yeast?

2. Is fermentation in yeast aerobic or anaerobic respiration?

3. What advantage does fermentation offer yeast?

4. Did the tubes show varying amounts of gas production? Why?

5. What tube showed the greatest rate of gas production? Why?

# LAB 7.B

# DIFFERENCES IN CO₂ PRODUCTION BASED ON METABOLIC RATE

**MATERIALS REQUIRED (per group of students)**

| | |
|---|---|
| 3 small beakers | plastic wrap |
| wax pencil | perforated plastic spoon |
| 2 graduated cylinders | |
| aquarium water | phenolphthalein solution |
| live crayfish | |
| live goldfish | white paper |
| dip net | sodium hydroxide (NaOH) solution |

**PROCEDURE**

Metabolism is a general term for all biochemical processes occurring in an organism that keep it alive. The process of aerobic respiration is certainly a big part of metabolism. Aerobic organisms consume oxygen and produce carbon dioxide at various rates based on several factors such as body temperature, activity level, and complexity of the animal. In other words, their

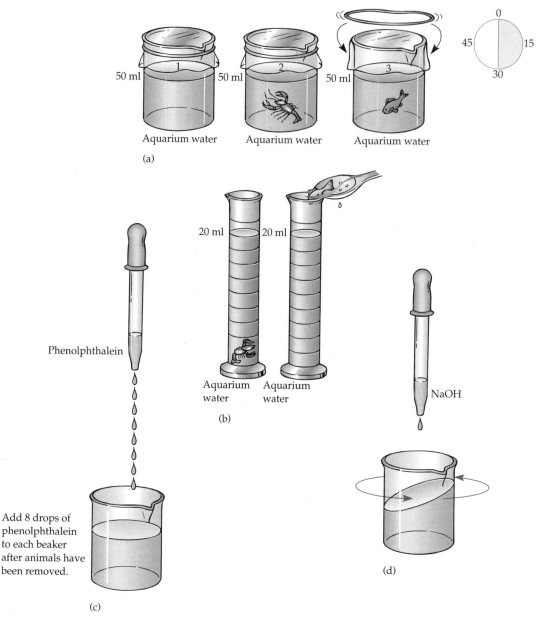

**FIGURE 7-3**

Procedure for comparing metabolic rates of a goldfish with a crayfish.

metabolic rates differ for these reasons as well. In the following exercise, you will compare the metabolic rate of a goldfish, a lower vertebrate, and a crayfish, which is a common freshwater invertebrate related to the lobster.

The metabolic rates of these animals will be measured as the amount of carbon dioxide produced within a specific time. Carbon dioxide reacts with water by forming carbonic acid. By neutralizing the carbonic acid with a determined amount of sodium hydroxide, you can calculate the amount of carbon

dioxide produced by the goldfish and crayfish (see Fig. 7-3).

1. Obtain three 100 ml beakers and label them 1, 2, and 3 with a wax pencil.
2. Using a graduated cylinder, add 50 ml fresh aquarium water to each beaker. Do not use tap water!
3. Beaker 1 serves as a control; add nothing to the water in this beaker. Using a dip net, place a small

crayfish from the holding tank into beaker 2. Place a small goldfish into beaker 3. *Please handle the animals carefully and respectfully!*

4. Tightly seal the top of each beaker with plastic wrap and allow the beakers to stand undisturbed for 30 minutes.

5. After timing, remove the plastic wrap from all beakers. Have ready two 25 ml graduated cylinders filled as accurately as possible with 20 ml aquarium water. You will use the graduated cylinders to determine the body volumes of the goldfish and crayfish, which are important for later calculations.

6. Using a perforated plastic spoon, carefully transfer the crayfish from its beaker into one of the graduated cylinders, removing as little water from the beaker as possible. Carefully transfer the goldfish to the other graduated cylinder in the same manner.

7. Observe the new level of water in each of the graduated cylinders now holding the animals. The difference between the new and previous water levels in each cylinder is the body volume of the animals in milliliters. Record the volumes of the animals in Table 7–2 and return them to their holding tank.

8. Add 8 drops of phenolphthalein solution to each beaker. Phenolphthalein is colorless in acidic conditions and becomes pink in alkaline conditions.

9. Hold beaker 1 over a sheet of paper or some other white background that will help you detect a color change. Add NaOH (0.0025M) drop by drop to beaker 1, swirling it gently to mix after each drop. Count the number of drops required to turn the water in the beaker a slight pink color that lingers as you continue to swirl the beaker. Record in Table 7–2 the number of sodium hydroxide drops necessary for the color change. Repeat the process for beakers 2 and 3.

10. Since 20 drops of NaOH is equal to one milliliter, divide the number of sodium hydroxide drops added to each beaker by 20. Record the amount of sodium hydroxide added in milliliters in Table 7–2.

11. Using the data you have collected, determine the amount of carbon dioxide produced by each animal as $\mu M/ml/hr$ by completing the following calculations.

a. ml NaOH (experimental) − ml NaOH (control) = ml NaOH (adjusted)

b. $\dfrac{\text{ml NaOH (adjusted)} \times 2.5\mu M \text{ NaOH/ml}}{\text{Animal volume (ml)} \times \text{time (hr)}}$

*Note:* Check your fermentation tubes from Lab 7.A for results if sufficient time has passed.

**Table 7–2** Differences in CO₂ Production Based on Metabolic Rate

| Beaker Number | Volume of animal (ml) | NaOH | | CO₂ produced by animal (μM/ml/hr) |
|---|---|---|---|---|
| | | Drops | ml | |
| 1 (control) | | | | |
| 2 (crayfish) | | | | |
| 3 (goldfish) | | | | |

## POSTLAB QUESTIONS

1. Why were the beakers sealed with plastic wrap?

2. Why was phenolphthalein added to each beaker?

3. What did the number of sodium hydroxide drops indicate about carbon dioxide production?

4. Which organism had the greater metabolic rate?

5. Why was aquarium water rather than tap water used in this experiment?

6. Explain the possible results of this experiment if you used a very large crayfish and a small goldfish.

## LAB 7.C
## USE OF A DYE AS AN INDICATOR OF ELECTRON TRANSPORT IN YEAST

### MATERIALS REQUIRED (per group of students)

| | |
|---|---|
| 6 clean test tubes | yeast suspension |
| test tube rack | iodoacetamide solution |
| wax pencil | citrate solution |
| small metric ruler | formaldehyde solution |
| methylene blue dye | citrate + iodoacetamide solution |
| distilled water | |
| rubber stoppers | magnesium chloride solution |

### PROCEDURE

Methylene blue is a dark blue dye commonly used to stain specimens for observation under a microscope. Methylene blue can also be used as an indicator of cellular respiration. In the absence of oxygen, methylene blue will accept electrons, and the hydrogen ions that follow them, from the electron transport system within cells. The dye becomes reduced as it gains hydrogen and changes from blue to colorless. If oxygen is introduced to the dye, methylene blue changes from colorless back to the blue oxidized condition.

In the following exercise, you will watch for electron transport using methylene blue as the final electron acceptor in the cellular respiration of yeast cells. You will observe the effect of various compounds on cellular respiration and the enzymes involved. You also will observe the effect of adding citrate to a tube of yeast cells, which bypasses glycolysis and directly enters the Krebs cycle (see Fig. 7-4).

1. Use a ruler and a wax pencil to mark 1 cm and 2 cm from the *bottom* of 6 clean test tubes. Number the tubes 1, 2, 3, 4, 5, and 6.
2. Add yeast suspension to all test tubes up to the 1 cm mark. Swirl the container to mix the yeast.
3. Fill the tubes to the 2 cm mark with the various solutions as listed below.

---

**C A U T I O N**  Iodoacetamide and formaldehyde are toxic! Handle them carefully. Wash thoroughly with soap and water if these solutions contact your skin or clothing.

---

Tube 1 — magnesium chloride

Tube 2 — iodoacetamide

Tube 3 — citrate

Tube 4 — citrate + iodoacetamide

Tube 5 — formaldehyde

Tube 6 — distilled water

4. Add *one* drop of methylene blue solution to each test tube. Make sure the drops are of equal size in all tubes by holding the dropper straight up and down when you deliver a drop of dye.
5. Mix the tubes by shaking them from side to side and return them to the test tube rack. Observe the tubes for the next 10–15 minutes *without* shaking them. When observing the test tubes for color changes, disregard the thin blue layer that may remain at the surface of the tube contents. Record your results in Table 7-3 using the color scale below.

Color Intensity Scale

---

+ + = original blue color, no change

+   = lighter than original blue, some change

−   = blue color mostly gone, definite change

**Table 7-3**  Methylene Blue Indicator Results of Electron Transport

| Tube Number | Contents | Color Scale |
|---|---|---|
| 1 | Magnesium chloride | |
| 2 | Iodoacetamide | |
| 3 | Citrate | |
| 4 | Citrate + iodoacetamide | |
| 5 | Formaldehyde | |
| 6 | Distilled water | |

### POSTLAB QUESTIONS

1. In this exercise, methylene blue substitutes for what molecule as the final electron acceptor in the electron transport system of yeast cells?

2. What does a decrease in the blue color of methylene blue indicate about the activity of the electron transport system?

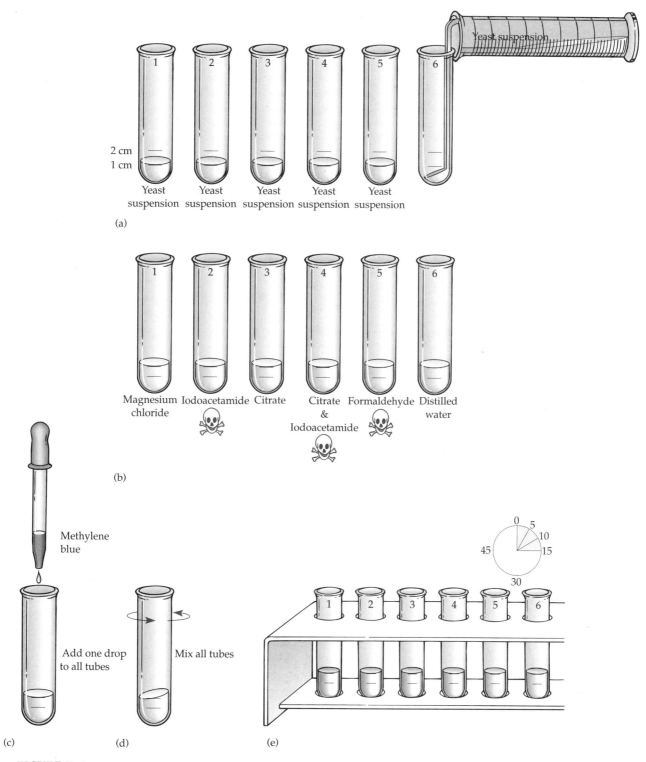

**FIGURE 7-4**

Procedure for using methylene blue as the final electron acceptor in the cellular respiration of yeast cells.

3. Identify the tubes that retained their original blue color. Which compounds deactivated the enzymes of cellular respiration?

4. Why does the reduction of methylene blue occur in tube 4, but not in tube 2?

5. Why do you think formaldehyde was used in the past as a preservative for keeping animal specimens?

## FOR FURTHER READING

Alberts, B., et al. 1989. *Molecular Biology of the Cell.* New York: Garland.

Becker, W. 1986. *The World of the Cell.* Menlo Park, Calif.: Benjamin Cummings.

Bridger, William A. 1983. *Cell ATP.* New York: Wiley.

Brock, T., et al. 1988. *Biology of Microorganisms.* Englewood Cliffs, N.J.: Prentice-Hall.

Darnell, J., et al. 1986. *Molecular Cell Biology.* New York: Scientific American.

Douce, R., and D. A. Day, eds. 1985. *Higher Plant Cell Respiration.* New York: Springer-Verlag.

Dubyak, George R., and Jeffrey S. Fedan, eds. 1990. *Biological Actions of Extracellular ATP.* New York: New York Academy of Sciences.

Hendler, Sheldon S. 1989. *The Oxygen Breakthrough: 30 Days to an Illness-Free Life.* New York: Morrow.

Holtzman, E., and A. B. Novikoff. 1984. *Cells and Organelles.* Philadelphia: Saunders College Publishing.

Lehniger, A. 1982. *Principles of Biochemistry.* New York: Worth.

Luckner, Martin. 1990. *Secondary Metabolism in Microorganisms, Plants, and Animals.* New York: Springer-Verlag.

Malik, Marek. 1990. *Biological Electron Transport Processes: Their Mathematical Modelling and Computer Simulation.* New York: Ellis Horwood.

McNeil, B., and L. M. Harvey, eds. 1990. *Fermentation: A Practical Approach.* New York: IRL Press.

Stryer, L. 1988. *Biochemistry.* San Francisco: W. H. Freeman.

Zehnder, Alexander J. B., ed. 1988. *Biology of Anaerobic Microorganisms.* New York: Wiley.

# TOPIC 8

❏

# Photosynthesis

## OBJECTIVES

1. Write an overall summary of photosynthesis, using only the terms carbon dioxide, chlorophyll, light, oxygen, sugar, and water.

2. Summarize the events of the light-dependent and the light-independent reactions of photosynthesis.

3. Draw and explain the significance of the absorption spectrum of chlorophyll a relative to energy utilization in photosynthesis.

4. Describe the procedure of paper chromatography as it applies to the separation of photosynthetic pigments.

## INTRODUCTION

Photosynthesis is the process whereby plants and most algae capture the sun's energy and store it in the form of chemical bonds in carbohydrate molecules. This energy is made available to all other organisms when they consume algae, plants, or the animals that eat algae or plants. Photosynthesis produces energy-rich carbohydrates from carbon dioxide and water using light energy that has been trapped by chlorophyll. Oxygen is released as a by-product and is used by most organisms in respiration. Overall, the basic summary equation of photosynthesis is

Carbon dioxide + water $\xrightarrow[\text{chlorophyll}]{\text{sunlight}}$

carbohydrate + oxygen.

Photosynthesis uses visible light, which comprises a small portion of the electromagnetic spectrum (Fig. 8–1). Visible light carries enough energy to cause chemical reactions without destroying biological molecules, a reaction that occurs with other spectrum wavelengths. Only light that is absorbed by the plant can be used for photosynthesis. The so-called "white light" visible to our eyes is actually composed of many different wavelengths that we see individually as different colors only when white light is separated into its spectrum by a prism. Only the violet, blue, and red wavelengths are absorbed, or captured, by chlorophyll as an energy source that drives photosynthesis, whereas the green wavelength is mainly reflected. The various wavelengths of light that are characteristically absorbed by a particular substance are known as its **absorption spectrum.**

Photosynthesis occurs within the **chloroplast,** which houses a system of membranes called **thylakoids.** Thylakoids are arranged in stacks **(grana)** and surrounded by **stroma,** a protein-rich solution that fills the chloroplast. The enzymes necessary for **carbon fixation** (carbohydrate production) are located in the stroma. This location makes the enzymes readily available to react with the other materials involved in photosynthesis. Molecules of the green photosynthetic pigment **chlorophyll** are embedded in the thylakoid membranes. The large surface area of the membranes intercepts light which then reacts with the chlorophyll molecules. Chlorophyll a is called the main photosynthetic pigment.

The processes of photosynthesis take place primarily in two stages—during **light-dependent reactions** and **light-independent reactions.** Both stages occur

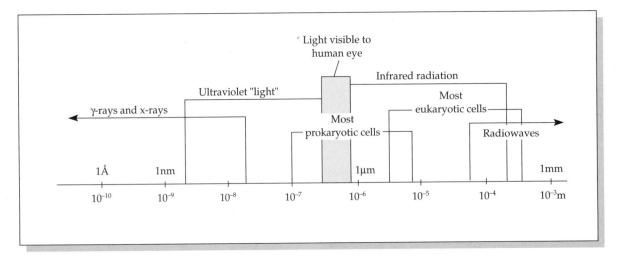

**FIGURE 8-1**

The electromagnetic spectrum. Note how small the proportion of visible light is to the entire spectrum.

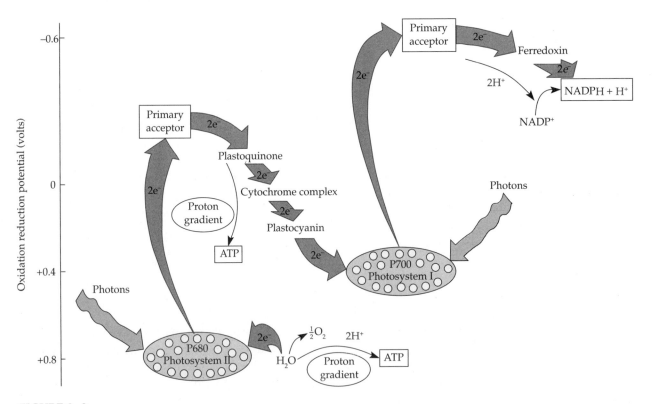

**FIGURE 8-2**

Light energy is used in the formation of ATP and NADPH + H⁺ during the light-dependent reactions of photosynthesis. These reactions, which occur in the thylakoid membranes, also result in the photolysis of water and the subsequent release of oxygen.

within the chloroplast. The light-dependent reactions (Fig. 8–2) result in the production of ATP (adenosine triphosphate) and NADPH + H$^+$ (reduced nicotinamide adenine dinucleotide phosphate) that provide energy and hydrogen, respectively, necessary for carbohydrate production. Molecules of chlorophyll a and b absorb light energy and donate electrons to the electron transport system. The electrons lost from chlorophyll are replaced by electrons donated from water molecules that are subsequently split in the thylakoid interior. This produces the oxygen that is released during photosynthesis.

The **electron transport system** is also located in the thylakoid membrane near the chlorophyll molecules where it can readily accept electrons from them. This position in the thylakoid membrane also enables the electron transport system to produce a hydrogen ion (H$^+$) reservoir within the thylakoid interior, a reservoir that is involved in ATP production. ATP synthetase enzymes are associated with channels in the membrane that allow hydrogen ions to flow back across the membrane. The enzymes use the energy of the moving hydrogen ions to join inorganic phosphate with ADP (adenosine diphosphate), which forms ATP. Another segment of the electron transport system carries electrons from chlorophyll molecules, and reduces NADP to NADPH + H$^+$. With the formation of ATP and NADPH + H$^+$, the fixation of carbon dioxide into a carbohydrate molecule occurs during the light-independent reactions.

The process requires energy from ATP, hydrogen from NADPH + H$^+$, carbon dioxide, and **ribulose bisphosphate** (RuBP), a five-carbon sugar already present in the stroma. Enzymes attach carbon dioxide to RuBP, which forms two 3-carbon molecules of phosphoglycerate (PGA). Hydrogen from NADPH + H$^+$ reduces PGA, which converts it to two 3-carbon molecules of phosphoglyceraldehyde (PGAL). At this point, two 3-carbon molecules of PGAL may be joined, which forms a 6-carbon glucose molecule; or, PGAL may be converted to a number of other biological compounds not directly involved in carbon fixation. Other PGAL molecules are converted into more RuBP, which in turn become involved in yet further carbon fixations. The spent ADP, inorganic phosphate, and NADP$^+$ are recycled into the light-dependent reactions which lead to more ATP and NADPH + H$^+$ production and a continuation of carbon fixation and carbohydrate production.

In the final analysis (Fig. 8–3), photosynthesis may be viewed as a process of energy transfers. Light en-

ergy striking the thylakoid membrane starts a flow of electrons (electrical energy) from water through chlorophyll molecules and the electron transport system. As a result of the electron transport system, energy from the flow of electrons is held as chemical energy in ATP and NADPH + H$^+$. Finally, the chemical energy of ATP and NADPH + H$^+$ is kept in a more permanent form as a carbohydrate molecule that can be stored.

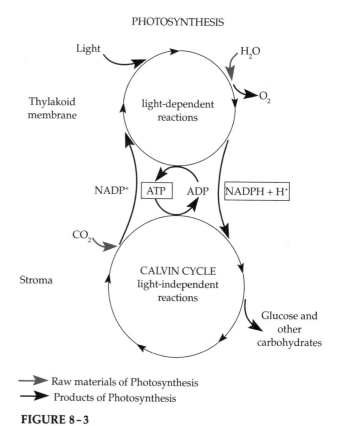

PHOTOSYNTHESIS

——▶ Raw materials of Photosynthesis
▬▬▶ Products of Photosynthesis

**FIGURE 8–3**

A summary of photosynthesis.

The following laboratory exercises are investigations of some of the concepts involved in photosynthesis.

## PRELAB QUESTIONS

1. What are the principal raw materials for photosynthesis and what are their functions in the photosynthetic process?

2. Identify the major pigment involved in photosynthesis.

3. What wavelengths of light play a role in photosynthesis?

4. Identify the high energy compounds produced in the light-dependent reactions of photosynthesis.

5. During what stage of photosynthesis does carbon fixation occur?

## LAB 8.A
## STARCH PRODUCTION IN PHOTOSYNTHESIS

### MATERIALS REQUIRED (per group of students)

| | |
|---|---|
| *Coleus* leaf | forceps |
| hot plate | petri plate |
| beaker | iodine solution |
| ethyl alcohol | |

### PROCEDURE

The abundant amount of simple sugars produced during photosynthesis is often converted into starch. Starch is stored in various places throughout the plant body, such as in the roots, the stems, and even the leaves. The following exercise is a simple illustration of how plentiful starch can be in a single leaf (see Fig. 8–4).

1. Remove a leaf from a healthy, variegated *Coleus* plant that has been growing in bright light for several days.

2. Using forceps, place the leaf in boiling ethyl alcohol. Boil the leaf for several minutes until the leaf is very pale.

---

**C A U T I O N**   Alcohol is very flammable! Make sure there is no open flame in the room during the entire laboratory period.

---

3. Carefully remove the leaf with forceps from the boiling alcohol and briefly rinse it with tap water. Place the leaf on a petri plate.

4. Cover the leaf with iodine solution and wait a few moments. What areas on the leaf changed color?

### POSTLAB QUESTIONS

1. What color change indicates a positive iodine test for starch? Why?

2. Is starch production occurring in the whole leaf or in specific areas of the leaf? Why?

3. Suggest two reasons why boiling the leaf in alcohol is an essential step in this exercise.

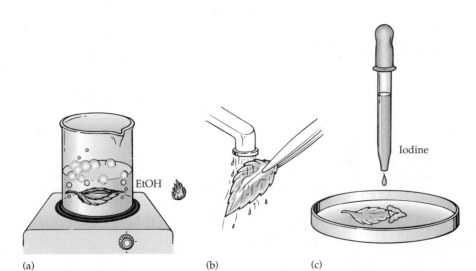

(a)            (b)            (c)

Iodine

**FIGURE 8–4**

Procedure to illustrate starch production in a *Coleus* leaf.

## LAB 8.B
## LIGHT ABSORPTION BY CHLOROPHYLL

### MATERIALS REQUIRED (per group of students)

spectrophotometer          chlorophyll extract

2 cuvettes                 lintless tissue paper

acetone

### PROCEDURE

A spectrophotometer is a machine that can be used to determine the absorption spectrum of a chlorophyll extract. The spectrophotometer can be adjusted to screen out all light wavelengths but the desired one. These wavelengths are then directed through a tube containing the sample that is being tested for light absorbance. The amount of light absorbed by the sample is then displayed on the instrument's meter. By adjusting the spectrophotometer for various wavelengths, the absorption spectrum of a substance is determined by noting light absorbance at each wavelength.

Refer to Figure 8–5 and review all of the following steps for operating the spectrophotometer. These steps will prepare you for the experiment in which you will determine the absorption spectrum of chlorophyll.

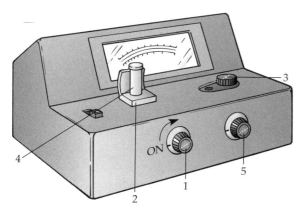

### FIGURE 8–5

*The spectrophotometer,* with numbers referring to parts of the instrument described in LAB 8.B.

1. Turn knob (1) to the right until it clicks on. The instrument must warm for 5 to 10 minutes before proceeding further.
2. Make sure the sample holder (2) is covered by its lid, which should always be closed. Turn knob (3) to the left or right to a particular wavelength.

3. Turn knob (1) to adjust the meter to infinity absorbance, the last line at the lower left of the meter scale.
4. Open the sample holder lid (2) and insert a spectrophotometer cuvette (4) filled two-thirds full with acetone, the solvent in the chlorophyll extract. This solvent cuvette is known as a blank. Align the painted mark on the spectrophotometer tube with the mark on the sample holder.
5. Close the sample holder lid. Turn knob (5) to the left or right to adjust the meter to zero absorbance, the last mark on the lower right of the scale.
6. Remove the blank cuvette. In the same manner as the blank cuvette, insert a spectrophotometer cuvette filled two-thirds full with chlorophyll extract into the sample holder and close the sample holder lid.
7. Read the light absorbance from the bottom scale of the meter.

**Table 8–1**   Absorbance Readings of Chlorophyll Extract

| Wavelength (nm) | Absorbance | Wavelength (nm) | Absorbance |
|---|---|---|---|
| 400 | | 550 | |
| 410 | | 560 | |
| 420 | | 570 | |
| 430 | | 580 | |
| 440 | | 590 | |
| 450 | | 600 | |
| 460 | | 610 | |
| 470 | | 620 | |
| 480 | | 630 | |
| 490 | | 640 | |
| 500 | | 650 | |
| 510 | | 660 | |
| 520 | | 670 | |
| 530 | | 680 | |
| 540 | | 690 | |

8. Remove the spectrophotometer cuvette and close the sample holder lid. Turn knob (3) to the next desired wavelength and adjust the instrument as you did in Steps 1 through 6. Repeat Steps 3 through 6 each time you select a different wavelength. Be sure to keep all spectrophotometer cuvettes free of smudges by wiping them with lintless tissue paper.

Once you have completed the above steps, you are prepared to determine the absorption spectrum for an extract of chlorophyll.

1. Begin with a wavelength of 400 nm (nanometers) and proceed upward, increasing the wavelength 10 nm after each reading up to an end point of 690 nm.
2. Record your absorbance readings in Table 8–1.
3. Plot the absorbance readings on a graph. Make sure you appropriately define the axes. The peaks and valleys in the final form of the graph are the absorption spectrum of chlorophyll for the various wavelengths, or colors, of visible white light.

## POSTLAB QUESTIONS

1. Why was the spectrophotometer adjusted with a blank tube containing acetone?

2. Review the graph created in Part 3. At what wavelength(s) and color(s) of visible light would the greatest rate of photosynthesis occur?

3. Review the graph created in Part 3. At what wavelength and color of visible light is most of the light reflected?

## LAB 8.C
## SEPARATION OF PLANT PIGMENTS BY CHROMATOGRAPHY

### MATERIALS REQUIRED

| | |
|---|---|
| filter paper rectangle | chromatography |
| pencil | solvent |
| small metric ruler | stapler |
| chlorophyll extract | calculator |
| capillary pipets | |
| chromatography jar with lid | |

### PROCEDURE

Plant leaves contain several pigments that function in photosynthesis. Two types of green pigments, chlorophyll a and b found within the chloroplasts of plant cells, give plants their characteristic green color. Chloroplasts also contain smaller amounts of carotene and xanthophyll, pigments that range in color from yellow to orange. Carotene and xanthophyll become evident in the brilliant leaves of autumn. These various pigments absorb light energy of several different wavelengths that drive the biochemical reactions of photosynthesis.

The individual pigments within a chlorophyll extract can be separated by a process called chromatography. Chromatography involves taking the mixture to be separated and dabbing it as droplets onto a matrix, or ground substance, that absorbs the mixture. The matrix is then partially placed into a liquid solvent, which subsequently migrates upward through the matrix as it is absorbed. The mixture becomes dissolved into the solvent as it soaks through the matrix. The individual components of the mixture usually vary in how well they dissolve in the solvent and cling to the matrix. As a result, the most soluble substances easily migrate along with the solvent as it soaks upward through the matrix. The less soluble substances and those that tend to cling to the matrix fall behind. When the process is finished, the chromatogram reveals the individual components of the original mixture as spots or bands along the length of the matrix. In this exercise, the mixture is an extract of chlorophyll consisting of several different pigments. The matrix of the chromatogram is filter paper. The solvent is a solution of acetone and petroleum ether.

---

**C A U T I O N**   Acetone and petroleum ether (chromatography solvent) are highly flammable! Make sure there are no open flames in the room during the entire laboratory period.

---

1. Obtain a small piece of filter paper that has been specially cut for chromatography use. Keep the filter paper as clean as possible by handling it with forceps.
2. Place the filter paper on a clean surface. Draw a pencil line 1.5 cm above the longer side of the paper as in Figure 8–6a. Do not use a pen—the ink will dissolve in the chromatography solvent!
3. Mark a small pencil dot at the center of the line. Mark two other small dots that are centered be-

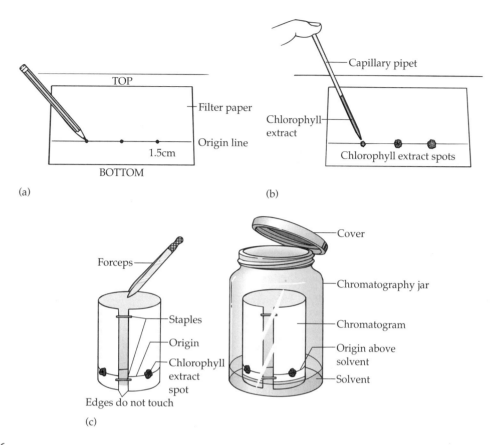

**FIGURE 8–6**

(a) Marking chromatography paper with pencil. (b) Spotting chromatography paper with chlorophyll extract. (c) Stapling of the chromatography paper and its placement into a jar with solvent.

tween the first dot and the edges of the paper as in Figure 8–6a. This is the origin line of your chromatogram.

4. Spot all three pencil dots with chlorophyll extract using a capillary pipet as in Figure 8–6b. Keep your finger over the upper end of the pipet and lightly touch the opposite end to the paper keeping the spots as small as possible. The chlorophyll extract will dry quickly. Apply more chlorophyll as soon as the first spot is dry. Spot the dots at least 10 times each. This doesn't take long, and darker chlorophyll spots will give you better results.

5. Obtain a chromatography jar and lid. Pour enough chromatography solvent into the jar to barely cover the bottom with liquid. Quickly place the lid on the jar and the solvent bottle — the liquid evaporates rapidly. The solvent will saturate the air inside the jar as you prepare the next step.

6. Handling the paper only by its edges, carefully curve the filter paper into a cylinder with the pencil

line on the outside. Staple the ends together as in Figure 8–6c. The staples should bridge the edges of the paper — the edges should not overlap.

7. Open the jar and, using forceps, place the filter paper cylinder inside. The cylinder should stand in the middle of the jar without touching the glass, as in Figure 8–6c. Quickly replace the lid.

8. Watch the chromatography solvent as it rises up the filter paper. The chlorophyll pigments dissolve in the solvent and travel with it. Allow the solvent to rise for several minutes until it reaches about an inch below the top of the filter paper.

9. Open the jar and remove the chromatogram with forceps. Quickly use a pencil to mark the wet solvent line near the top of the filter paper. This is called the solvent front, and it will be important in later calculations. Close the jar and examine your chromatogram, which will dry almost immediately.

**C A U T I O N** Discard the solvent in the container designated for solvent disposal. *Do not* pour solvent down the drain.

You will probably see four bands or areas of color resulting from each spot of chlorophyll extract on your chromatogram. From top to bottom these are carotene (yellow, on or near the solvent front), xanthophyll (yellow, approximately midway between the solvent front and origin line), chlorophyll a (blue-green, in the lower portion of the chromatogram), and chlorophyll b (yellow-green, nearest the origin line).

In addition to color, the various pigments on the chromatogram usually migrate a distance that is characteristic of each pigment. This distance is known as the $R_f$ value of the pigment. The $R_f$ value is expressed as a number easily determined by the following formula.

$$R_f = \frac{\text{distance between color band and origin line}}{\text{distance between solvent front and origin line}}$$

Using a small metric ruler, determine the $R_f$ values for each of the pigments on your chromatogram. When measuring the distance between color band and origin line, place a dot in what you judge as the center of the color band and measure from that point. Record the $R_f$ values in Table 8–2.

**Table 8–2** Chromatography of Chlorophyll Extract

| Pigment | Distance between Pigment and Origin (cm) | Distance between Solvent Front and Origin (cm) | $R_f$ Value |
|---|---|---|---|
| Carotene | | | |
| Xanthophyll | | | |
| Chlorophyll a | | | |
| Chlorophyll b | | | |

## POSTLAB QUESTIONS

1. What are $R_f$ values? (Do not give the formula as the answer.)

2. What does it mean for a pigment to have a specific $R_f$ value (for example, a pigment with an $R_f$ value of .66)?

3. Which pigment is most soluble? Least soluble?

4. If you are given the following $R_f$ values (pigment A = .78, pigment B = .31, pigment C = .48, and pigment D = .92), where would these pigments be located on a chromatogram relative to each other?

5. How do you think differences in molecular weight of two substances would affect their $R_f$ values?

## FOR FURTHER READING

Brock, T., et al. 1988. *Biology of Microorganisms.* Englewood Cliffs, N.J.: Prentice-Hall.

Clayton, Roderick K. 1980. *Photosynthesis: Physical Mechanisms and Chemical Patterns.* New York: Cambridge University Press.

Darnell, J., et al. 1990. *Molecular Cell Biology.* New York: Scientific American.

Evans, J. R., ed. 1988. *Ecology of Photosynthesis in Sun and Shade.* Melbourne: CSIRO.

Gregory, R. P. F. 1989. *Biochemistry of Photosynthesis.* New York: Wiley.

Gregory, R. P. F. 1989. *Photosynthesis.* New York: Chapman and Hall.

Hall, D., and K. Rao. 1987. *Photosynthesis.* Baltimore: Arnold.

Lehninger, A. 1982. *Principles of Biochemistry.* New York: Worth.

Nakatani, H. Y. 1988. "Photosynthesis." *Carolina Biology Reader.* Burlington, N.C.: Carolina Biological Supply.

Prescott, D. M. 1988. *Cells.* Boston: Jones and Bartlett.

Rowan, K. S. 1989. *Photosynthetic Pigments of Algae.* New York: Cambridge University Press.

Salisbury, F. B., and C. W. Ross. 1985. *Plant Physiology.* Belmont, Calif.: Wadsworth.

Simpson, G. M. 1990. *Seed Dormancy in Grasses.* New York: Press Syndicate of the University of Cambridge.

Stolz, John F., ed. 1991. *Structure of Photosynthetic Prokaryotes.* Boca Raton, Fla.: CRC Press.

Walker, David. 1987. *The Use of Oxygen Electrode and Fluorescence Probes in Simple Measurements of Photosynthesis.* Sheffield, United Kingdom: Research Institute for Photosynthesis.

Wiessner, W., D. G. Robinson, and R. C. Starr, eds. 1990. *Cell Walls and Surfaces, Reproduction, Photosynthesis.* New York: Springer-Verlag.

# PART IV

❑

# The Continuity of Life:
# Cell Division and Genetics

# Mitosis

## OBJECTIVES

1. Describe the events occurring in each stage of mitosis emphasizing the movement of chromosomes.

2. Identify the stages in the cell cycle and the time the cell spends in each stage.

3. Describe and compare the structure of a late prophase chromosome with an anaphase chromosome.

## INTRODUCTION

When a cell divides, the genetic information within the nucleus is distributed equally between the two resulting cells and each functions normally. The processes that are responsible for that effect are mitosis and cytokinesis.

**Mitosis** produces two new nuclei with the same number of chromosomes as in the original nucleus. By this process, genetic information is inherited throughout generations of cells. Mitosis causes growth of an organism by producing an increase in cell number and replaces **somatic cells** (cells of the body) that die.

Not all somatic cells divide after they become fully differentiated. Those that do divide have a typical life history, called the **cell cycle** (Fig. 9–1). The cycle begins from the time a cell is formed by division and ends when it too divides. The cell cycle has four stages: (1) The $G_1$ (first gap) period is the time from the beginning of a new cell to when it begins to replicate its DNA. This is a time of growth and normal cellular activity. (2) The S (synthesis) period is the time when DNA is replicated. (3) The $G_2$ (second gap) period is a second period of cellular growth and lasts from the end of DNA synthesis until the next cell division. These first three periods are collectively referred to as **interphase,** the total time between cell divisions. (4)

Mitosis (M) is the final stage, when the nucleus and cytoplasm divide and form two new cells.

Mitosis is a continuous, gradual process, but for purposes of study, the process may be divided into four stages: prophase, metaphase, anaphase, and telophase. In **prophase,** the chromatin (DNA) condenses into distinct chromosomes. Because of replication during interphase (S period), each chromosome consists of two **sister chromatids** held together by a common **centromere** (there is some evidence to suggest that the centromere may actually be divided at this stage). A **kinetochore,** which functions in chromosome separation, forms near the centromere. The **mitotic spindle** appears as a framework of microtubules that will be involved in chromosomal movement later in mitosis. Animal cells have **centrioles,** organelles that organize the mitotic spindle during mitosis, whereas centrioles are generally lacking in plant cells. The nuclear membrane disappears at the end of prophase.

**Metaphase** follows prophase. The mitotic spindle is now complete. The microtubules of the spindle and those associated with the kinetochore align the chromatids at the equator of the cell.

**Anaphase** is next. The centromeres holding sister chromatids together split and the chromatids sepa-

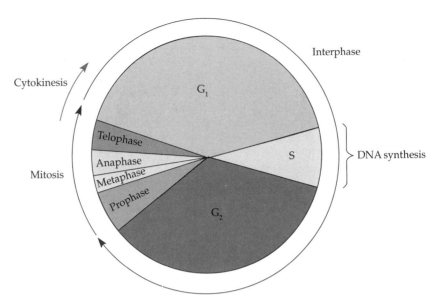

**FIGURE 9-1**

The cell cycle involves four stages: (1) the $G_1$ (first gap) stage, (2) the S (synthesis) period, (3) the $G_2$ (second gap) period, and (4) the mitotic stage.

rate, becoming two new chromosomes. The chromosomes are divided equally when the spindle and kinetochore microtubules pull them toward opposite ends of the cell. It is at this point that substances (such as **colchicine**), which interfere with microtubule function or formation, prevent cell mitosis. Without microtubules pulling apart the chromosomes, the nucleus cannot divide in two. However, the centromeres still split allowing sister chromatids to separate, which results in a doubling of chromosome number in the cell.

**Telophase** is the last stage of mitosis. The chromosomes are now in two groups at each end of the mitotic spindle. The chromosomes uncoil and new nuclear membranes form around them, which define the two new nuclei. In most cells, **cytokinesis** (division of the cytoplasm) becomes obvious in telophase. Cytokinesis gives each new cell half of the original cytoplasm.

The process of cytokinesis differs in plant and animal cells. In animal cells, cytoplasmic division occurs by formation of a **cleavage furrow,** a cleft that continually deepens and pinches apart the new cells. The rigid cell wall of plant cells requires a different method of cytokinesis. Vesicles from the Golgi complexes move along the spindle microtubules to the cell equator where they fuse, which forms a flat **cell plate.** The cell plate grows by the addition of more vesicles and eventually the cell is partitioned in two. Each new cell builds a cell wall on its side of the cell plate, and the process is completed.

The following laboratory studies have been selected to provide you with firsthand experience regarding the important biological phenomenon of mitosis.

## PRELAB QUESTIONS

1. What is the significance of mitosis?

2. What major event occurs during the S (synthesis) period of the cell cycle?

3. During what phase of mitosis are the chromatids aligned at the equator of the cell?

4. What is cytokinesis and during what stage of mitosis does cytokinesis occur?

# LAB 9.A
# CHROMOSOME MOVEMENT IN MITOSIS

## MATERIALS REQUIRED (per pair of students)

120 pop-it beads (60 each of two colors)

4 magnetic centromeres

2 centrioles

4 pieces of string, 75 cm long

transparent tape

small plastic bag

meter stick (optional)

## PROCEDURE

In this exercise, you will simulate the gradual chromosome movements of mitosis by using "chromosomes" made of beads and magnetic "centromeres." Before beginning, be sure to have on hand the items listed above so that the exercise runs smoothly.

1. Obtain 60 beads each of 2 colors and 4 magnetic centromeres. Assemble a single chromosome by snapping together 30 beads of 1 color into a magnetic centromere. The centromere should be in the middle with an equal number of beads at each end as in Figure 9–2.

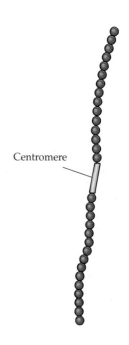

**FIGURE 9–2**

A single chromosome assembled with 30 beads and a magnetic centromere.   *(Redrawn after Chromosome Simulation Set Illustrations by Carolina Biological Supply, © 1976.)*

2. Assemble a second bead strand identical to the one you just made and stick together their magnetic centromeres as in Figure 9–3. This represents a replicated, or doubled, chromosome as it appears before mitosis begins. Each strand of the replicated chromosome is called a chromatid—two chromatids equal *one* replicated chromosome.

3. Assemble two chromatids of the second color and stick their centromeres together to make another replicated chromosome as in Figure 9–3. You will now follow these two replicated chromosomes through the stages of mitosis.

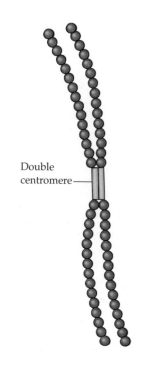

**FIGURE 9–3**

A replicated chromosome assembled from two magnetic centromeres and 60 beads.   *(Redrawn after Chromosome Simulation Set Illustrations by Carolina Biological Supply, © 1976.)*

4. *Interphase:* Place the two chromosomes in a small plastic bag, which represents the nuclear membrane, and lay it on your desk. Obtain two centrioles and position them near the nucleus at right angles to one another as in Figure 9–4. This simulates interphase, a time of normal cell activity when the chromosomes are replicated before the beginning of mitosis. Interphase is *not* one of the phases of actual cell division.

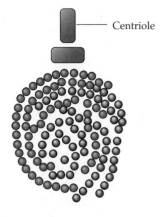

**FIGURE 9–4**

A representation of two chromosomes in interphase, the stage in which chromosomes are replicated before mitosis.   *(Redrawn after Chromosome Simulation Set Illustrations by Carolina Biological Supply, © 1976.)*

5. *Prophase:* Separate the two centrioles. About 50 cm from each side of the nucleus, tape them down so that they are pointing toward the nucleus as in Figure 9–5. Dump the chromosomes out of the bag between the two centrioles. This represents the breakdown of the nuclear membrane during prophase. The chromatids of each chromosome should be intertwined, as in Figure 9–5, to represent their condition in an actual cell during prophase.

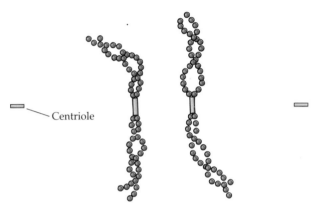

**FIGURE 9–5**

A representation of prophase, the stage in which condensation of chromosomes occur. *(Redrawn after Chromosome Simulation Set Illustrations by Carolina Biological Supply, © 1976.)*

6. Tie a small loop at one end of each of the four strings. Draw one string tightly around each centromere of a chromosome as in Figure 9–6. Thread the string of each centromere through opposite centrioles; that is, one to the left, one to the right.

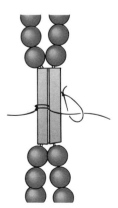

**FIGURE 9–6**

Strings represent spindle fibers that connect the centromeres and the centrioles. *(Redrawn after Chromosome Simulation Set Illustrations by Carolina Biological Supply, © 1976.)*

Slip the remaining two strings around the centromeres of the other chromosome in the same way as above, and again thread them through opposite centrioles. The strings represent the spindle fibers that connect the centromeres and the centrioles. Spindle fibers are microtubules that actually move the chromosomes during mitosis.

7. *Metaphase:* Position the chromosomes along the "cell equator" to align them between the two centrioles. Pull gently on the strings at the centrioles to straighten them as shown in Figure 9–7.

8. *Anaphase:* Gently pull on the strings to separate the centromeres. When the centromeres of a replicated chromosome separate, the chromatids are then called chromosomes (Figure 9–8). Continue pulling on the strings until the chromosomes reach the centrioles.

9. *Telophase:* Remove the strings from the centromeres. Roll up each pair of chromosomes and pile them near their centriole as in Figure 9–9. A nuclear membrane would now form around the chromosomes and division of the cytoplasm (cytokinesis) would occur, which completes cell division. You now have two new cells, each identical to the original parent cell with the same number and color of chromosomes.

## POSTLAB QUESTIONS

1. What condition appears to be required for the movement of chromosomes during mitosis?

2. What event determines when a chromatid becomes a chromosome?

3. A certain cell has eight chromosomes before cell division has started. How many chromosomes will there be in one of the daughter cells as a result of mitotic division?

4. In which of the phases of mitosis is the nuclear membrane visible?

5. Identify the mitotic phase in which the nuclear membrane is reformed.

6. Identify the mitotic phase in which the number of chromosomes increases with no increase in the total amount of nuclear material.

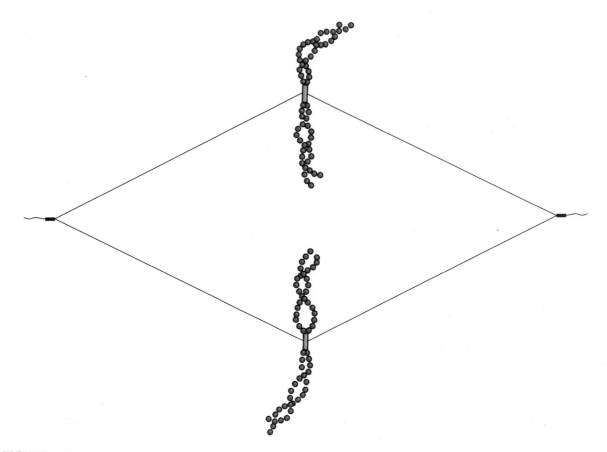

**FIGURE 9–7**

A representation of metaphase, the stage in which the chromosomes align themselves be-
tween the centrioles.   *(Redrawn after Chromosome Simulation Set Illustrations by Carolina Biological Supply, ©*
*1976.)*

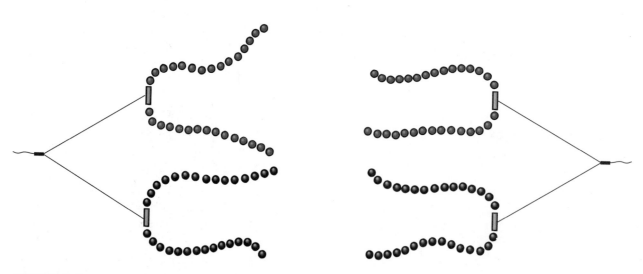

**FIGURE 9–8**

A representation of anaphase, the stage in which the centromeres separate.   *(Redrawn after*
*Chromosome Simulation Set Illustrations by Carolina Biological Supply, © 1976.)*

## FIGURE 9–9

A representation of telophase, the stage in which cell division is completed. *(Redrawn after Chromosome Simulation Set Illustrations by Carolina Biological Supply, © 1976.)*

7. Draw a late prophase chromosome and an anaphase chromosome.

## LAB 9.B
## MITOSIS IN PLANTS

### MATERIALS REQUIRED

onion root tip mitosis, prepared slide

compound microscope

### PROCEDURE

Mitosis followed by cytokinesis (cell division) is how organisms grow in size, by an increase in cell number, and repair damaged tissue, by replacing worn and dead cells. In plants, cell division occurs mainly in specific areas of embryonic cells called **meristems** located at the growing tips of roots and stems. The cells of a plant meristem are unspecialized cells that actively divide, thereby increasing the length of a root or stem as a plant grows. The older cells produced by the meristem then develop into specialized cells that perform a specific function within the plant, such as photosynthesis or sap transportation. Plant meristems are ideal places to see the various phases of mitosis that you created in the previous exercise with model chromosomes.

In the following exercise, you will examine the dividing cells in the meristem of an onion root tip. The chromosomes are darkly stained to make them easily visible within the cells. If your instructor directs you to do so, make sketches of the various mitotic stages studied.

1. Obtain a prepared slide of onion root tip mitosis. Your slide will probably contain two to three lengthwise sections of onion root. Observe one of the sections under low power. Examine the area indicated in Figure 9–10 where the cells are actively dividing. The older cells beyond the tip region (which would be toward the onion bulb) have begun to specialize into functional root cells and have stopped dividing. As you look through the meristematic region, you should see cells in all stages of mitosis and others in interphase.

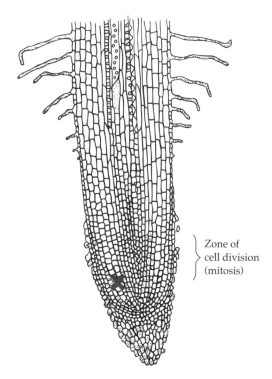

Zone of cell division (mitosis)

## FIGURE 9–10

Meristematic region of the onion root tip. Look at the position marked "X" to find actively dividing cells.

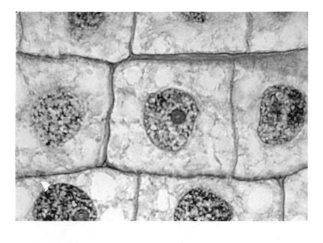

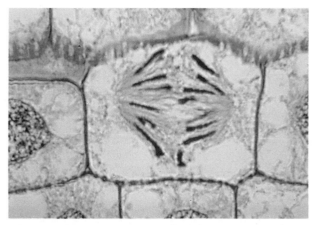

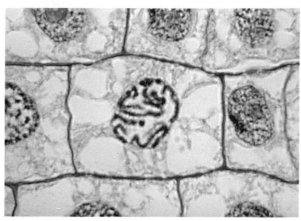

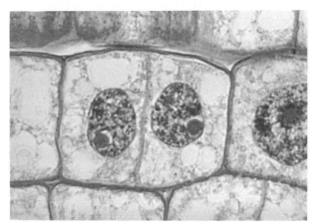

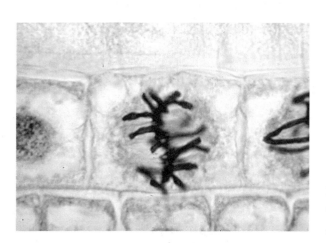

**FIGURE 9–11**

Light micrographs of mitosis in onion root tip cells.   *(Carolina Biological Supply Co.)*

2. Switch to high power. Using the micrographs of Figure 9–11 as a guide, look through the meristematic area and find a cell that matches those pictured for interphase and the four phases of mitosis. Begin with interphase and continue in sequence through prophase, metaphase, anaphase, and telophase to reinforce the order of the mitotic phases in your memory. As you observe the actual phases of mitosis on the onion slide, try to associate the arrangements and movements of the model chromosomes in the previous exercise with those you see on the slide. If you can't find a particular mitotic phase, switch to another root section on the slide and search there.

## POSTLAB QUESTIONS

1. Suppose a plant cell undergoes mitosis, but no cell plate is formed. What would be the result?

2. If a corn cell that normally has 20 chromosomes is dividing mitotically, how many chromosomes would there be during metaphase? During anaphase?

3. A biologist observing mitosis in corn roots (normally 20 chromosomes) located one daughter cell in late telophase that had only 19 chromosomes. How might this abnormality have occurred?

## LAB 9.C
## MITOSIS IN ANIMALS

### MATERIALS REQUIRED

whitefish mitosis, prepared slide

compound microscope

### PROCEDURE

Mitosis in animal cells is similar to what you observed in the onion slide with some minor differences. Remember, animal cells have centrioles and cytoplasmic division occurs by a cleavage furrow and not by a cell plate as in plants.

Obtain a prepared slide of whitefish mitosis. Note that it contains several sections of a fertilized whitefish egg that has begun early embryonic development by mitosis. As you did with the onion slide, look through the sections with low and high power to find cells in interphase and the phases of prophase, metaphase, anaphase, and telophase. The general appearance of the chromosomes illustrated in onion root tip mitosis in Figure 9–11 may be used as a guide in identifying the mitotic stages in the early whitefish embryo. If you can't find a particular phase, switch to another egg section and look for it there. Note the absence of a cell wall as you observe the whitefish cells. You will not see centrioles in the cells because

they are too small to be observed with a compound microscope.

## POSTLAB QUESTIONS

1. In cardiac (heart) muscle cells, there can be many nuclei in a single cell. How can this occur?

2. While using a compound microscope, you observe a cell that is undergoing mitosis. You see double-stranded chromosomes and a partial nuclear membrane. What phase is this cell in?

3. You are examining a slide showing cells undergoing mitosis. You see a cell in which a centriole is at each end of the cell and single-stranded chromosomes are about halfway between the midregion of the cell and each centriole. What stage is this cell in?

4. Why would it be important for daughter cells to receive a portion of the cytoplasm during cell reproduction?

## LAB 9.D
## TIMING THE PHASES OF MITOSIS

### MATERIALS REQUIRED

onion root tip mitosis, prepared slide

compound microscope

### PROCEDURE

The cell cycle begins when a cell is formed by mitosis and ends when the cell itself divides. The cell cycle of an onion cell is approximately 20 hours. By counting a number of cells on a slide in the various stages of mitosis and using simple math, it is possible to estimate the duration in time of the mitotic phases.

1. Obtain a prepared slide of onion root tip mitosis. Observe the meristem region of the root with high power as you did before. In one field of view of your microscope, count all cells that you see in interphase and record the number in the appropriate space in Table 9–1.

2. Without moving the slide, next count the number of cells in each of the four phases of mitosis and record the numbers in Table 9 – 1.

3. Randomly choose another field of view on the slide in the meristem region and again count the number of cells in interphase and the four mitotic phases until you have counted a total of at least 100 cells. The more cells you count in additional fields of view, the more accurate will be your results.

4. Calculate the time of each mitotic phase by using the following formula.

$$\text{Time} = \frac{\text{number of cells in each phase}}{\text{total number of cells counted}} \times 20 \text{ hours}$$

Record the time duration of the mitotic phases in Table 9 – 1.

**Table 9 – 1** Time Duration of Interphase and Mitotic Phases in Onion Root Cells

| Phase | Number of Cells | Duration of Phase |
|---|---|---|
| Interphase | | |
| Prophase | | |
| Metaphase | | |
| Anaphase | | |
| Telophase | | |

Total number of cells counted _____ .

## POSTLAB QUESTIONS

1. What phase of the cell cycle was the longest in duration? Why?

2. It is known that onion root tip cells take about 20 hours to complete the mitotic cycle. If 51 cells out of 100 were found to be in interphase, then how much time does each cell spend in interphase?

## FOR FURTHER READING

Becker, W. M. 1986. *The World of the Cell.* Menlo Park, Calif.: Benjamin/Cummings.

Baserga, R. 1985. *The Biology of Cell Reproduction.* Cambridge, Mass.: Harvard University Press.

Brooks, Robert, ed. 1989. *Symposium: The Cell Cycle.* Cambridge, England: British Society for Cell Biology.

Burgess, Jeremy. 1985. *An Introduction to Plant Cell Development.* New York: Cambridge University Press.

Burns, M. 1983. *Cells.* Philadelphia: Saunders.

Buvat, Roger. 1989. *Ontogeny, Cell Differentiation, and Structure of Vascular Plants.* New York: Springer-Verlag.

Conrad, Gary W., and T. E. Schroeder, eds. 1990. *Cytokinesis: Mechanisms of Furrow Formation During Cell Division.* New York: New York Academy of Sciences.

John, P., ed. 1981. *The Cell Cycle.* Cambridge, England: Cambridge University Press.

Fisher, Paul B., ed. 1990. *Mechanisms of Differentiation.* Boca Raton, Fla.: CRC Press.

Prescott, D. M. 1988. *Cells: Principles of Molecular Structure and Function.* Boston: Jones and Bartlett.

Roberts, Lorin W. 1988. *Vascular Differentiation and Plant Growth Regulators.* New York: Springer-Verlag.

Smith-Klein, C., and V. Kish. 1988. *Principles of Cell Biology.* New York: Harper & Row.

Summer, A. T. 1990. *Chromosome Banding.* Boston: Unwin Hyman.

Whitfield, James F. 1990. *Calcium, Cell Cycles, and Cancer.* Boca Raton, Fla.: CRC Press.

# TOPIC 10

❏

# Meiosis

## OBJECTIVES

1. Describe the events occurring in each stage of meiosis with an emphasis on the behavior of chromosomes.

2. Summarize the significance of meiosis in sexual reproduction.

3. Cite the similarities and differences between spermatogenesis and oogenesis.

## INTRODUCTION

**Meiosis** is the important mechanism that maintains the constant number of chromosomes of a given species from generation to generation. The process of meiosis results in the production of four nuclei each with half the number of chromosomes of the original nucleus. In animals, meiosis results in the formation of reproductive cells known as gametes (eggs and sperm). In plants, meiosis results in the formation of spores that germinate into haploid plants that eventually produce eggs and sperms. Fertilization, the union of an egg and a sperm, results in the restoration of the full chromosome number.

Were it not for meiosis, the chromosome number of a species would double upon fertilization of the egg by the sperm cell. A detailed account of meiosis is given in your text. In the account given here, we shall concern ourselves only with the salient aspects of the process and with the significance of those aspects.

Chromosomes of somatic cells (generally speaking, any plant or animal cells other than germinal cells, but particularly cells that are terminally differentiated) occur in pairs called **homologous chromosomes.** They are identical in appearance and carry the same corresponding genes. One chromosome of each homologous pair is inherited from each parent by means

of the sperm and egg cells. The somatic cells of most organisms have pairs of homologous chromosomes; these cells are **diploid.** Cells produced by meiosis have unpaired chromosomes because the chromosome number has been reduced by half; such cells are **haploid.**

The overall movements and arrangements of chromosomes within a cell undergoing meiosis are similar to mitosis in that prophase, metaphase, anaphase, and telophase stages are recognizable. However, meiosis involves two distinct cell divisions referred to as Meiosis I and Meiosis II. Each meiotic division exhibits the four phases of cell division distinguished as prophase I, prophase II, or metaphase I, etc. The two meiotic divisions result in haploid cells derived from diploid cells that started the process. Before the first division, homologous chromosomes (each consisting of a pair of identical sister chromatids, the pair of sister chromatids being the replication products of the original chromosome) align side-by-side, a process called **synapsis.** During synapsis, segments of nonsister chromatids may become exchanged **(crossing over),** which rearranges genes between the nonsister chromatids. This produces chromosomal configurations and, consequently, gene combinations that may never have existed before and results in genetic variety in the resulting gametes. Further genetic variations between gametes occur as a result of the random alignment of synapsed chromosomes along the equatorial plane during metaphase of the first meiotic division. Different lineup configurations result in different gene combinations in the gametes produced. Genetic variety in offspring is advantageous since, because of it, the chances of survival for the young in a changing environment may be increased.

There are basic similarities in the meiotic divisions of **spermatogenesis** (formation of sperm cells) and **oogenesis** (formation of egg cells). However, there is one significant difference. Every germ cell (cells that give rise to gametes) that enters spermatogenesis produces four functional sperm cells. In oogenesis, every germ cell produces only one functional egg cell due to unequal cytokinesis throughout meiosis. The unequal cytokinesis supplies the egg with a food source and cytoplasmic components for the early embryo. The remaining cells of oogenesis occur as tiny **polar bodies** made of extra chromosomes and very little cytoplasm.

The following laboratory exercises illustrate the important aspects of meiosis.

### PRELAB QUESTIONS

1. What is the significance of meiosis?

2. How does the chromosome number in cells differ before and after they undergo meiosis?

3. What is synapsis?

4. What is the difference between spermatogenesis and oogenesis?

## LAB 10.A
## CHROMOSOME MOVEMENT IN MEIOSIS

### MATERIALS REQUIRED (per pair of students)
120 pop-it beads (60 each of two colors)

4 magnetic centromeres

4 centrioles

4 pieces of string, 75 cm long

transparent tape

small plastic bag

meter stick (optional)

### PROCEDURE
The following exercise will demonstrate how haploid cells are produced from diploid cells during meiosis. Be sure to have all required materials on hand so that the exercise runs smoothly.

1. Obtain 60 pop-it beads of one color. Assemble a single chromosome by snapping together 30 beads into a magnetic "centromere," as shown in Figure 10–1, with the centromere in the center.

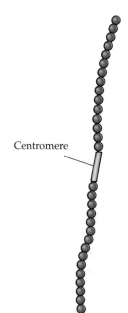

Centromere

**FIGURE 10–1**

A single chromosome.

2. Assemble a second identical structure and stick together the magnetic centromeres (Fig. 10–2). This represents a replicated, or doubled, chromosome consisting of two chromatids as seen before meiosis begins.

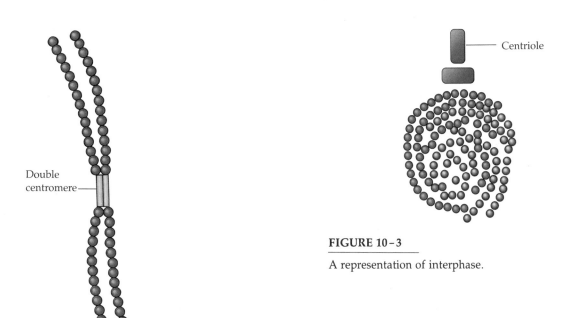

FIGURE 10-2

A replicated chromosome.

FIGURE 10-3

A representation of interphase.

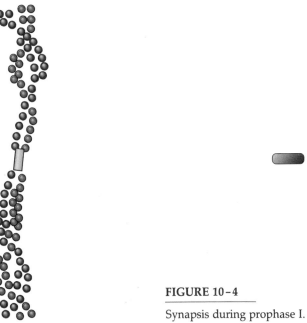

Centriole

FIGURE 10-4

Synapsis during prophase I.

3. Obtain 60 pop-it beads of another color. Assemble another replicated chromosome as you did in Steps 1 and 2. You now have two homologous chromosomes in a diploid nucleus. One member of the chromosome pair originated from the female parent, the other from the male parent.

4. *Interphase:* Place the two chromosomes into the plastic bag, which represents the nuclear membrane, and lay this "nucleus" on your workspace. Position two centrioles above the nucleus at right angles to one another (Fig. 10-3). This arrangement simulates interphase, a time of cell activity

when the chromosomes are replicated before meiosis begins.

**Meiosis I (First Meiotic Division)**

5. *Prophase I:* This phase occurs entirely inside the nuclear membrane before it disappears. However, you must remove the chromosomes from the plastic bag so you can manipulate them. Place the chromosomes together and wrap them around one another (Fig. 10-4). This represents synapsis, a

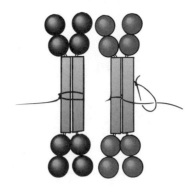

**FIGURE 10-5**

Placement of string around chromosomes.

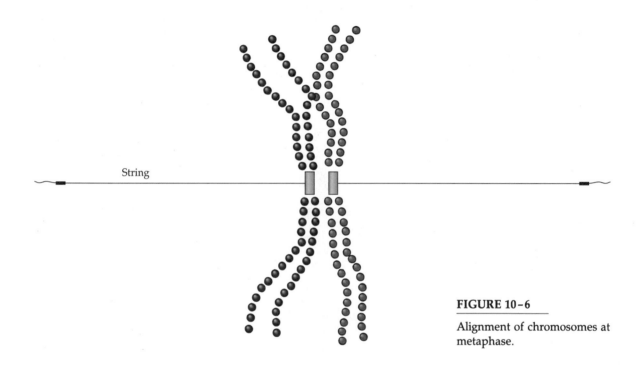

String

**FIGURE 10-6**

Alignment of chromosomes at metaphase.

close pairing of chromosomes that occurs during prophase I. Place the centrioles about 50 cm on each side of the nucleus and tape them down (Fig. 10–4).

6. Tie a loop in one end of two of the strings. Loop a string around the double centromere of each chromosome (Fig. 10–5). Rewrap the chromosomes around one another after attaching the strings and thread the strings through the centrioles (Fig. 10–6). The strings represent spindle fibers that move the chromosomes during cell division.

7. *Metaphase I:* Arrange the chromosomes at right angles to the centrioles and midway between them as shown in Figure 10–6. This simulates the chromosomes aligning along the cell equator, which is characteristic of metaphase.

8. *Anaphase I:* Gently pull on the strings to separate the paired chromosomes. Continue pulling the strings until the chromosomes reach the centrioles (Fig. 10–7).

9. *Telophase I:* Remove the strings and pile each chromosome near its centriole. Obtain two additional centrioles and place them at right angles to the original ones (Fig. 10–8). Development of a nuclear membrane and cytokinesis (division of the cytoplasm) occurs in this phase resulting in two new haploid cells. The new cells are haploid because they have one-half the number of chromosomes as the original cell (one instead of two). This reduction by half in chromosome number is the most significant event of meiosis I. Note that the chromosomes of each haploid cell are still replicated at this point. Proceed to meiosis II below to see how they become singular.

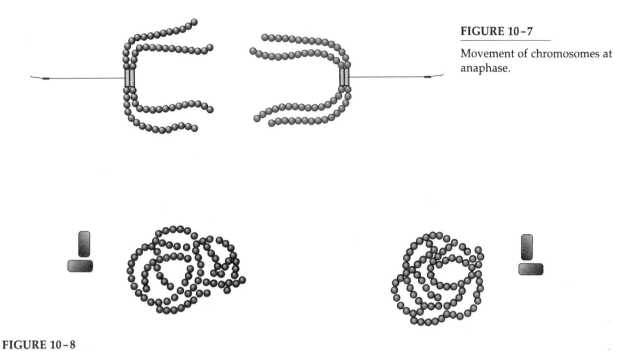

**FIGURE 10-7**

Movement of chromosomes at anaphase.

**FIGURE 10-8**

A representation of telophase.

## Meiosis II (Second Meiotic Division)

Before meiosis II begins, some cells have a brief interphase period also called interkinesis. It is important to note that no chromosome replication occurs during this interphase as it does before mitosis and meiosis I; the chromosomes are already replicated before meiosis II begins.

1. *Prophase II:* Place the two centrioles of each new cell from telophase I approximately 50 cm from the nucleus as shown in Figure 10-9 and tape them down.
2. Tie a loop in one end of all four strings. Loop a string around each of the chromosome centromeres (Fig. 10-10). Thread the strings (spindle fibers) through the centrioles (as in Fig. 10-9).
3. *Metaphase II:* Arrange the chromosomes at right angles to the centrioles and midway between them (as in Fig. 10-9).
4. *Anaphase II:* Pull gently on the strings to separate the chromatids of the replicated chromosomes. Continue pulling the strings until the chromatids, now chromosomes, reach the centrioles (Fig. 10-11).

5. *Telophase II:* Remove the strings and pile each chromosome near its centriole (Fig. 10-12). Development of a nuclear membrane and cytokinesis occurs in this phase resulting in a total of four new haploid cells. (Save this phase for Lab 10.B that follows.)

## POSTLAB QUESTIONS

1. With respect to the number of chromosomes, compare interphase of mitosis to interkinesis (between division I and II) of meiosis.

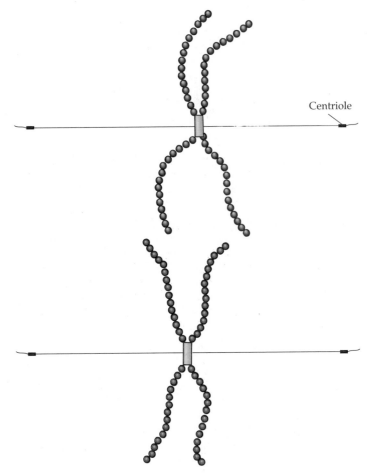

**FIGURE 10-9**

Placement of centrioles in prophase II.

Centriole

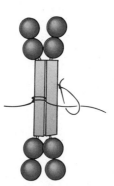

**FIGURE 10-10**

Placement of string around chromosome.

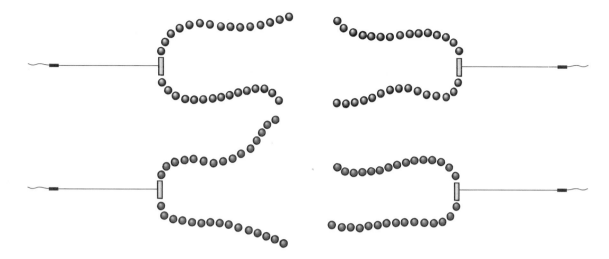

**FIGURE 10-11**

Movement of chromosomes during anaphase II.

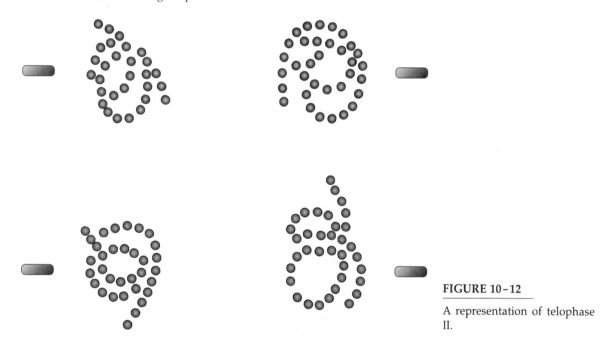

**FIGURE 10-12**

A representation of telophase II.

2. Which cell division process (meiosis I, meiosis II, mitosis, or some combination) corresponds to the following statements?
   a. The resulting cell contains one chromosome of each pair.

   b. There is pairing of homologous chromosomes.

   c. Each chromosome duplicates itself.

   d. Each daughter cell contains the same kind and number of chromosomes as the parent cell.

   e. Haploid cells are formed from diploid cells.

   f. The nuclear membrane disintegrates.

3. During what phase of meiosis does division of centromeres occur?

4. During what phase of meiosis do double-stranded chromosomes move to opposite poles?

5. At the end of telophase I of meiosis, how many cells are produced and how does the chromosome number of these cells compare to the original cell? During telophase II?

6. How many chromosomes of paternal origin (that is, originating from your father) are normally found in the nucleus of an epidermal (skin) cell in your body? In humans, the diploid number of chromosomes is 46.

## LAB 10.B
## CROSSING OVER IN MEIOSIS

### MATERIALS REQUIRED (per pair of students)
4 single-stranded chromosomes, two of each color, from telophase II of previous exercise

### PROCEDURE
Crossing over is an exchange of chromosome segments between homologous chromosomes during meiosis. This frequently occurs when homologous chromosomes closely pair during synapsis as shown in Figure 10–13. When chromosome segments are traded during synapsis, the genes contained on the segments are also traded. This results in genetic diversity among the gametes and is ultimately produced at the end of meiosis. The following exercise will simulate synapsis and crossing over—two very important meiotic events.

1. Obtain the four single chromosomes from telophase II of the previous exercise and reposition (wrap) the chromosomes as they were in prophase I (Fig. 10–13). This simulates synapsis.
2. Remove a segment of beads from one of the chromatids at one of the points where they intersect, such as A, B, or C in Figure 10–13.
3. Remove a segment of equal length at the same point on the neighboring chromatid of a different color. Switch places with the chromatid segments and attach them in their proper place. You have simulated crossing over. Some of the genes, represented by beads, of the two chromosomes were exchanged by crossing over during synapsis.
4. Unwind the chromosomes, note their composition, and answer the questions below.

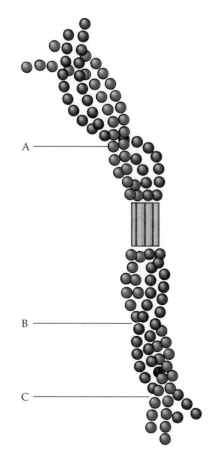

A ——
B ——
C ——

**FIGURE 10–13**

Crossing over.

### POSTLAB QUESTIONS

1. If you were to take these chromosomes through the same steps of meiosis as you did above, would the resulting gametes be identical as those you obtained earlier?

2. During what phase of meiosis does synapsis and crossing over occur?

3. If crossing over occurs between genes A(a) and B(b) of these homologous chromosomes, what would be the genetic composition of the resulting cells of meiosis?

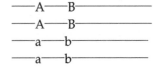

## LAB 10.C
## SITES OF MEIOSIS IN ANIMALS

### MATERIALS REQUIRED

cross section of mammalian testis, prepared slide

cross section of mammalian ovary, prepared slide

*Ascaris* eggs with polar bodies

compound microscope

### PROCEDURE

Now that you understand the mechanics of meiosis, it is important to know specifically where within organisms that meiosis occurs. In animals, including humans, meiosis is limited to particular cells within the gonads, or sex glands, that give rise to gametes. The male gonad is the testis. It is composed of numerous, tightly coiled seminiferous tubules. The interior walls of the tubules are lined with specialized diploid cells that divide by meiosis and eventually give rise to mature haploid sperm cells, called male gametes. Specifically, meiosis that produces sperm cells is called spermatogenesis.

1. Using the low power of your microscope, observe a prepared slide of a cross section of a mammalian testis, such as a rat. The numerous seminiferous tubules where sperm are produced appear as small circles throughout the testis. The structure of a human testis would appear much the same.

2. Switch to high power and closely observe one of the seminiferous tubules. The larger cells nearer the outer edge of the tubule are diploid spermatogonia that develop into haploid sperm cells by meiosis. The cells within the tubule wall closer to the lumen (opening) of the tubule are in various phases of meiosis. Each diploid spermatogonium cell ultimately produces four haploid sperm cells in the same manner as you produced four haploid cells using the simulated chromosomes.

3. Closely examine the lumen of this or other tubules and look for extremely fine threads that are the tails of maturing sperm cells. The heads of the sperm cells are embedded in the tubule wall. Compare what you see on your slide with the electron micrograph in Figure 10–14.

The gonad in female animals is the ovary. In mammals, including humans, small spherical pockets called follicles are scattered throughout the ovary. Each follicle originally contains one specialized diploid cell, an oogonium, that will give rise to a single

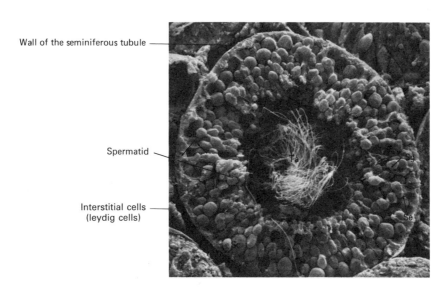

Wall of the seminiferous tubule

Spermatid

Interstitial cells (leydig cells)

### FIGURE 10–14

Electron micrograph of a seminiferous tubule cross section. The threadlike structures in the center of the tubule are tails of mature sperm. The rounded cells throughout the tubule wall are in various stages of meiosis (spermatogenesis). Some spermatogonia enlarge and undergo meiosis. These cells are called primary spermatocytes. Spermatids are the result of the second meiotic division in spermatogenesis. Spermatids differentiate into sperm. Interstitial cells lie between the seminiferous tubules in the testes.   (R.G. Kessel and R.H. Kardon, *Tissues and Organs: A Text-Atlas of Scanning Electron Microscopy.* San Francisco, W.H. Freeman & Co., 1979.)

Follicle cells   Ovary surface

**FIGURE 10-15**

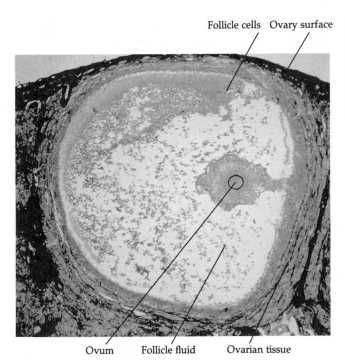

Ovum      Follicle fluid     Ovarian tissue

Light micrograph of a mature follicle in a mammalian ovary. Note the large ovum surrounded by fluid within the follicle.

haploid egg cell by meiosis. As you may realize, this is different from the spermatogonium which produces four sperm cells through meiosis. We will explore this difference later in the exercise. Specifically, meiosis that produces an egg cell is called oogenesis.

1. Obtain a prepared slide of a mammalian ovary cross section. Observe its general structure using the scanning objective or low power of your microscope. You will notice follicles of various sizes which correspond with various stages of follicle development.

2. Search for a large, obvious follicle similar to the micrograph in Figure 10–15. This is a mature follicle that should contain one large cell within its cavity. The large cell in a well-developed follicle such as this is nearly a mature egg cell, or ovum, and is called a female gamete. The large space within the follicle was filled with fluid before the slide was made. The ovum is released when the mature follicle ruptures and breaks through the ovary wall. Switch to high power and observe the developing ovum.

As stated earlier, a diploid spermatogonium (the cell that begins meiosis in males) gives rise to four haploid sperm cells, whereas the equivalent cell in a female, the oogonium, produces only one egg cell at the end of meiosis. Why is there such a difference in the number of gametes produced in males than in females? The answer lies in the amount of cytoplasm (cell content) found within a sperm and an egg cell. Consider the tiny sperm cell—essentially a microscopic cell nucleus with very little cytoplasm and a tail for movement. However, egg cells typically have abundant cytoplasm that becomes the cytoplasm of the early embryo should the egg become fertilized. The abundant cytoplasm in egg cells results from an unequal division of cytoplasm during oogenesis (meiosis). During the meiotic divisions, a majority of the cytoplasm continually goes to just one haploid cell that eventually becomes the ovum. The other three resulting cells, called polar bodies, are also haploid, but are small with very little cytoplasm. These polar bodies eventually degenerate. Consequently, each oogonium produces only one egg cell through meiosis.

1. Obtain a prepared slide containing the eggs of *Ascaris,* a roundworm that is an intestinal parasite of mammals, including humans. The numerous eggs on the slide are in various stages of meiosis. Examine the eggs under low and then high power of your microscope.

2. Some of the eggs will exhibit polar bodies produced by meiosis as described above. A polar body appears as a tiny, elongated dot on the inner side of the egg shell as pictured in Figure 10–16. Search the slide and locate an egg containing a polar body.

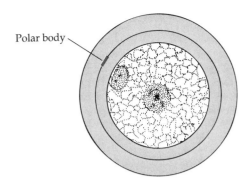

**FIGURE 10–16**

Diagram of an *Ascaris* egg with a polar body. Some eggs may have two polar bodies if they have completed the second meiotic division.

## POSTLAB QUESTIONS

1. How does the chromosome number of somatic cells compare to the chromosome number of gametes?

2. In a hypothetical animal in which the diploid chromosome number is 16, what is the chromosome number of each ovum?

3. If a mature egg cell of a hypothetical animal has 12 chromosomes, how many chromosomes would be found in a somatic cell of this animal?

4. Spermatogonia of *Schistosoma mansoni,* a blood parasite that infects humans, contain 10 chromosomes. How many chromosomes will be present in cells during prophase II of meiosis?

5. From one diploid nucleus in spermatogenesis, how many haploid sperm nuclei are formed?

6. During oogenesis, 50 oogonia could potentially give rise to how many mature ova?

## LAB 10.D
## SITES OF MEIOSIS IN PLANTS

### MATERIALS REQUIRED

cross section of lily anther, prepared slide

cross section of lily ovary, prepared slide

### PROCEDURE

As stated previously, meiosis usually results in some kind of reproductive cells—gametes in animals and spores in plants. In higher plants that produce flowers, the flower holds the structures where meiosis occurs in this particular group of plants. Specifically, the anther (Fig. 10–17) is the site in the flower where meiosis occurs, producing haploid male spores that give rise to pollen grains containing sperm cells. Note that the sperm cells do not come directly from meiosis as in animals. Sperm cells in plants are derived from

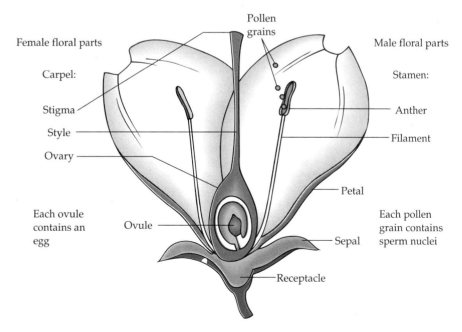

**FIGURE 10–17**

Flower structure showing male and female floral parts.

haploid **spores** produced by meiosis. You will investigate plant reproduction in detail during a future laboratory exercise. Presently, we are concerned only with the location of meiosis in flowering plants.

1. Examine a prepared slide of a cross section of a lily anther under low and high power of your microscope. The numerous small spheres you see within the anther are spores (cells) in various stages of meiosis during which they develop into pollen grains.

    Another site of meiosis in flowering plants is the ovule, or undeveloped seed, contained within the ovary of the flower (Fig. 10–17).

2. Examine a prepared slide of a cross section of a lily ovary under low power. Search for the ovules. They occur in pairs within the three large chambers of the ovary. Select an ovule and examine its interior under high power. The central area of the ovule contains female spores (cells) in various stages of meiosis that will give rise to an egg cell and other cell types associated with reproduction. Again, as with plant sperm, the egg cell in the ovule is not directly produced from meiosis, but derives from haploid spores that are the products of meiosis. This is one of the major differences between plants and animals.

## POSTLAB QUESTIONS

1. In animals and plants, the zygote will have ____ as much genetic material as the gametes.
2. What are the direct products of meiosis in plants? In animals?

## FOR FURTHER READING

Adolph, Kenneth W., ed. 1991. *Advanced Techniques in Chromosome Research.* New York: Decker.

Baserga, R. 1985. *The Biology of Cell Reproduction.* Cambridge, Mass.: Harvard University Press.

Bell, G. 1982. *The Masterpiece of Nature: The Evolution of Genetics of Sexuality.* Berkeley, Calif.: University of California Press.

Bernard, John. 1990. *Meiosis.* New York: Cambridge University Press.

Burns, G. W., and P. J. Bottino. 1989. *The Science of Genetics.* New York: Macmillan Publishing Company.

Christidis, Les. 1990. *B. Aves.* Berlin: Gebr under Borntraeger.

Conrad, Gary W., and T. E. Schroeder, eds. 1990. *Cytokinesis: Mechanisms of Furrow Formation during Cell Division.* New York: New York Academy of Sciences.

Davies, Kay E., and Shirley M. Tilghman, eds. 1990. *Genetic and Physical Mapping.* Cold Spring Harbor, N.Y.: Cold Spring Harbor Press.

Gillies, Christopher B., ed. 1989. *Fertility and Chromosome Pairing: Recent Studies in Plants and Animals.* Boca Raton, Fla.: CRC Press.

Haseltine, Florence P., and Neal L. First, eds. 1988. *Meiotic Inhibition: Molecular Control of Meiosis. Proceedings of A Symposium Held at the National Institutes of Health, Bethesda, Maryland, January 1987.* New York: A. R. Liss.

Moens, Peter B., ed. 1987. *Meiosis.* Orlando Fla.: Academic Press.

Strickberger, M. 1985. *Genetics.* New York: Macmillan Publishing Company.

Summer, Adrian T. 1990. *Chromosome Banding.* Boston: Unwin Hyman.

Verma, Ram S. 1990. *The Genome.* New York: VCH Publishers.

❏

# Mendelian Genetics

## OBJECTIVES

1. Define the basic terms relating to genetic inheritance, such as alleles, chromosome, dominance, gene, genotype, heterozygous, homologous, homozygous, phenotype, and recessiveness.

2. Explain how segregation and independent assortment affect the inheritance of genetic traits.

3. Describe the inheritance of sex-linked genes.

4. Solve simple problems in genetics involving monohybrid crosses, dihybrid crosses, and sex linkage.

---

## INTRODUCTION

Broadly speaking, **genetics** is the study of heredity, inherited characteristics that are passed from parents to offspring. The discipline of genetics includes the spectrum of transmission genetics, molecular or biochemical genetics, cytogenetics, and population genetics.

The basic principles of transmission genetics come from the research of Gregor Mendel, a 19th century monk. Mendel studied inheritance in the garden pea by crossing many pea plants with differing traits and by analyzing the traits found in the resulting offspring. Mendel noted that genetic factors that determine traits are inherited as separate particles.

Many, many genetic discoveries were made after Mendel's pioneering research; the science of genetics flourished. Today, we refer to Mendel's "inherited particles" as genes. Genes appear in alternate forms called **alleles**. The genes are carried on chromosomes that occur in like pairs in diploid eukaryotic cells. The two chromosomes of a pair are termed **homologous chromosomes.** Since chromosomes are paired, so too are the genes carried on the chromosomes.

An individual organism produced by sexual reproduction receives one chromosome of each pair from each parent. Consequently, the individual receives one gene (allele) of a gene pair from each parent. An individual with two same alleles for a trait is termed **homozygous** for that trait. An individual with two different alleles for a trait is called **heterozygous.** In many cases, only one allele of a gene pair is expressed as an observable trait in the organism and masks the presence of the other allele. The expressed allele is termed **dominant,** and the masked allele is termed **recessive.**

Because of dominance and other factors, the **genotype** (the specific genes or alleles) that an individual has cannot be determined with certainty by observation alone. Individuals that are homozygous or heterozygous for a dominant gene may have the same **phenotype** (expression of genes that can be observed). However, if an individual exhibits the recessive phenotype, its genotype is known to be homozygous recessive because the presence of only one dominant gene would mask the recessive trait. It is important to distinguish between an individual's genotype (the genes or alleles that the individual has) versus the phenotype (the appearance of the individual or the type of trait shown).

A genetic cross in which only one trait of the parents is considered is called a **monohybrid cross.** A monohybrid cross can be conveniently diagramed by a **Punnett square,** named after the individual who devised the method for illustrating genetic crosses. A Punnett square for a monohybrid cross consists of four boxes similar to a window with four panes. The genes present in the gametes of one parent are written above the top two squares, and those from the other parent are written down the side of the two left squares. The offspring genotypes are determined by combining genes from the top of the column with those from the left-hand row and recording them in the "window pane" or cell formed by the intersection of that column and that row. The completed diagram also reveals the ratio in which they occur.

In a **dihybrid cross,** two different traits are considered in the same cross. A dihybrid cross can also be diagramed by a Punnett square, but 16 squares (four across and four down) are required. The procedure for determining the offspring is the same as in a monohybrid cross, except that each gamete contains two genes, one for each trait being considered. When analyzing a cross, the original parents are referred to as the **parental** ($P_1$), generation. The offspring of the cross are called the **first filial** ($F_1$) generation. When members of the $F_1$ generation reproduce, the offspring are the **second filial** ($F_2$) generation.

Mendel's most important contributions to genetics can be stated in two laws that arose from his observations of many genetic crosses: Mendel's law of segregation and Mendel's law of independent assortment. Mendel noted—using modern terminology—that paired genes separate from one another in the formation of gametes. This observation became **Mendel's law of segregation** and reflects the events of meiosis. The members of each pair of homologous chromosomes (and the genes they carry) separate into different haploid nuclei when gametes are formed. The genes pair again at fertilization when gametes combine at random forming new gene pairs.

When Mendel observed the inheritance pattern of two different traits in a dihybrid cross, he noticed—again, using modern terminology—that different gene pairs assorted into gametes independently of the other gene pair. This is known as **Mendel's law of independent assortment** and is again a result of chromosome movement in meiosis. The alleles of one locus become distributed into the resulting cells independently of the distribution of the alleles at a second locus on another chromosome.

However, independent assortment does not always occur because it applies only to genes carried on different chromosomes (or, actually, to genes widely separated from one another on the same chromosome). Were it not for **crossing over** (review meiosis) all alleles on the same chromosome **(linkage groups)** would stay together during meiosis and occur within the same haploid nucleus. Crossing over can result in new chromosome configurations (new combinations of genes) and may lead (if those new gene combinations interact) to the establishment of advantageous genetic diversity in the offspring, diversity which could, for example, help the offspring survive in a changed environment.

In general, the two sexes of a given organism have identical pairs of chromosomes, except in the case of the sex chromosomes which, in fact, determine the sex of the individual. There are several different modes of chromosomal determination of sex, but the most common mode is for the female to have a pair of identical sex chromosomes and for the male to have one of those chromosomes. The homologous sex chromosome in the male is different in size and/or shape. Thus, designating chromosomes, females are XX and males are XY. Genes located on the X-chromosome are called **sex-linked genes.** *Drosophila melanogaster,* the common fruit fly, is useful for demonstrating the inheritance of a sex-linked trait. For example, eye color in fruit flies is sex linked. Sex-linked characteristics are observed much more in males than females. Many sex-linked genes are recessive. A female carrying a recessive sex-linked gene on one of her X-chromosomes has a good chance of having the dominant allele on her other X-chromosome; the dominant allele, of course, will be expressed in her physical features (phenotype). However, if a male has a recessive sex-linked gene on his single X-chromosome, the recessive trait will be expressed in his phenotype because there is no possibility for the presence of a dominant allele to mask the effect of the recessive gene. The Y-chromosome has no role in the inheritance of sex-linked traits because it carries no allele for the trait; sex linkage is restricted to the X-chromosome. A recessive sex-linked trait can appear in the phenotype of a female only if she has the recessive gene on both of her X-chromosomes.

The experiments of this chapter will give you some direct and, we hope, fascinating experiences with the basic principles and concepts of genetics.

**PRELAB QUESTIONS**

1. What term refers to an individual's appearance with respect to a certain inherited trait?

2. An organism possessing two different alleles for a given characteristic is said to be _____ for that trait.

3. What is Mendel's law of segregation?

4. What do we call the type of cross between two individuals in which only one heritable trait is being studied?

5. What is the difference in meaning between the terms "allele" and "gene"?

## LAB 11.A

## A MONOHYBRID CROSS IN GENETIC CORN

### MATERIALS REQUIRED (on display)

genetic corn mounts, monohybrid cross

### PROCEDURE

A monohybrid cross is a mating between two organisms where *one* genetic trait is of interest. Each parent organism carries two alleles of a particular gene. For example, Figure 11–1a concerns the alleles for fur color in guinea pigs: B = black, b = brown. Capital letters traditionally represent dominant alleles, and lowercase letters indicate recessive alleles. Each parent guinea pig carries one allele (gene) for black fur and one allele (gene) for brown fur. Because black is dominant over brown, both guinea pigs have black fur (their phenotype) even though they carry one of each allele (their genotype). The alleles become separated into individual gametes during meiosis. The possible gametes produced by the parents are placed along the top and left side of the Punnett square (Fig. 11–1a). The possible genotypes of the offspring can be predicted within the four boxes by bringing together gametes in a down and across fashion. Figure 11–1b shows the predicted offspring genotypes of this monohybrid cross concerning fur color in guinea pigs.

The proportions of offspring can be expressed as a phenotypic (phenotype) ratio based on fur color—3 black to 1 brown, or 3 : 1. The proportions of offspring can also be expressed as a genotypic (genotype) ratio based on the alleles they carry for fur color—1 BB, 2 Bb, and 1 bb, or 1 : 2 : 1. Notice that two of the offspring carry one of each allele for fur color (Bb); this is known as a heterozygous genotype. The other offspring have two of the same alleles (BB or bb), which are called homozygous genotypes. A monohybrid cross between two heterozygous parents, such as the parent guinea pigs, typically results in a 3 : 1 pheno-

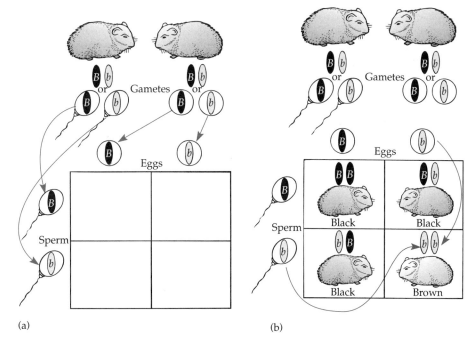

(a)                    (b)

**FIGURE 11–1**

Gametes produced by the parents are placed along the top and down the side of the Punnett square as in (a). Predicted offspring genotypes are shown in (b).

typic ratio among the offspring. From a genetic standpoint, a monohybrid cross between two heterozygous parents typically results in a 1 : 2 : 1 genotypic ratio among the offspring, just as you see in the guinea pigs.

By making similar crosses with his now famous pea plants and observing phenotypic ratios in the offspring, Gregor Mendel correctly reasoned that genetic factors (alleles) occur in pairs and become separated when gametes are formed. The factors (genes) randomly pair again at fertilization resulting in the typical ratios he noticed in his monohybrid crosses. Mendel remarkably predicted the existence and behavior of genes through his pea crosses even though chromosomes and genes were not described until after his death.

You can see the results of a monohybrid cross similar to that of the guinea pigs by observing a cross between corn of two colors.

1. Observe the display mount demonstrating a monohybrid cross of corn. The trait of interest here is kernel color having two alleles, R (purple dominant) and r (yellow recessive). Each kernel on the corn cob represents one new individual formed by the union of an egg and a sperm cell in the corn plant. The original parent plants, or $P_1$ generation, are represented by the top two corn sections. Note that the original parents are homozygous purple (RR) and homozygous yellow (rr).
2. Observe the two middle corn sections. These represent the first generation of offspring, or $F_1$ generation, from the original homozygous parents. Record the color(s) of the kernels. ____ Complete the Punnett square, below left, to determine the genotype(s) of the $F_1$ generation.
3. Observe the bottom corn section representing a cross between two $F_1$ individuals, producing the $F_2$ generation. Record the color(s) of the kernels. ____ Complete the Punnett square, below right, to determine the genotype(s) of the $F_2$ generation.

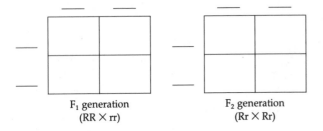

F$_1$ generation (RR × rr)          F$_2$ generation (Rr × Rr)

## POSTLAB QUESTIONS

1. What was the phenotype and genotype of the $F_1$ generation?

2. What was the genotypic ratio of the $F_2$ generation?

3. What was the expected phenotypic ratio of the $F_2$ generation? Does this agree with your observations?

4. Perform the following cross: Aa × Aa. Assume that A is dominant to a. What proportion of the offspring would you expect to be homozygous dominant? Homozygous recessive? Heterozygous?

5. A tall plant is crossed with a dwarf plant. The offspring are all tall. If we let T represent the allele for tall and let t represent the allele for dwarfism, what are the probable genotypes of the parents?

# LAB 11.B
# A DIHYBRID CROSS IN GENETIC CORN

## MATERIALS REQUIRED (on display)
genetic corn mounts, dihybrid cross

## PROCEDURE
An extension of the monohybrid cross is the dihybrid cross in which *two* traits are of interest. A dihybrid cross can also be represented with a Punnett square, but the square must contain 16 boxes to accommodate all possible genotypes of offspring that could be produced when considering two genetic traits. For example, suppose two organisms are crossed having the genotypes AaDd × AaDd. Each sperm or egg produced by these individuals will contain one allele for each trait. Because of independent assortment, the gamete possibilities for each parent are AD, aD, Ad, and ad. The four gamete types are lined across the top and left side of the Punnett square as before giving a total of 16 boxes (Fig. 11–2). The gametes are combined in the same down and across method as you did in the monohybrid cross. If you carefully determined the various phenotypes of the offspring, you would find a 9 : 3 : 3 : 1 phenotypic ratio that is typical of a dihybrid cross between two heterozygous parents.

|     | AD   | aD   | Ad   | ad   |
|-----|------|------|------|------|
| AD  | AADD | AaDD | AADd | AaDd |
| aD  | AaDD | aaDD | AaDd | aaDd |
| Ad  | AADd | AaDd | AAdd | Aadd |
| ad  | AaDd | aaDd | Aadd | aadd |

**FIGURE 11-2**

A Punnett square illustrating a dihybrid cross.

1. Observe the display mount demonstrating a dihybrid cross of corn. Here there are two traits of interest — kernel color and seed texture. The kernel color alleles are the same as those in the monohybrid cross, R (purple dominant) and r (yellow recessive). The seed texture alleles are Su (smooth dominant) and su (wrinkled recessive). The allele for kernel color and the allele for seed texture are located on different chromosomes. The original $P_1$ (parent) generation is again at the top of the display. Note that one parent is homozygous dominant for both alleles, and the other is homozygous recessive.

2. Observe the two middle corn sections. These represent the $F_1$ (first generation) offspring from the homozygous parents. Record the color(s) and texture(s) of the kernels.

Draw a Punnett square and determine the genotype(s) of the $F_1$ generation.

3. Observe the corn section at the bottom of the display mount representing the $F_2$ (second generation) offspring from a cross between two heterozygous $F_1$ individuals. Again, each kernel represents a new individual produced by an egg and sperm carrying various combinations of the alleles for kernel color and texture. Record the color(s) and texture(s) of the kernels.

Draw a Punnett square to determine the genotype(s) of the $F_2$ generation.

**POSTLAB QUESTIONS**

1. What was the genotype and phenotype of the $F_1$ generation?

2. What was the phenotypic ratio of the $F_2$ generation?

3. Which law of Mendel's explains the combinations of phenotypes present on the corn cob?

4. How many genetically different types of gametes would be produced by a parent with the genotype TtWw?

5. In four-o'clock flowers, bushy (B) is dominant over creeping (b). Red (R) is dominant over white (r). From a BbRr × BbRr cross, how many are expected to be creeping white?

6. A tall, red-flowered plant is crossed with a dwarf, white-flowered plant. Tall (T) is dominant to short (t) and red (R) is dominant to white (r). All the $F_1$ offspring are tall and red flowered. What were the parents' genotypes?

## LAB 11.C

### SEX LINKAGE IN *DROSOPHILA*, THE FRUIT FLY

**MATERIALS REQUIRED (per pair of students)**

container of preserved fruit flies

vial of live fruit flies

hand lens or stereomicroscope

**PROCEDURE**

The common fruit fly, scientifically named *Drosophila melanogaster,* is one of the most studied organisms on Earth because many of its characteristics make it a perfect subject for genetic studies. Fruit flies are small and easy to culture in large numbers; one culture bottle can hold hundreds of individuals. The fruit fly life cycle is very short (two weeks) making it possible to produce many generations in the same time it would

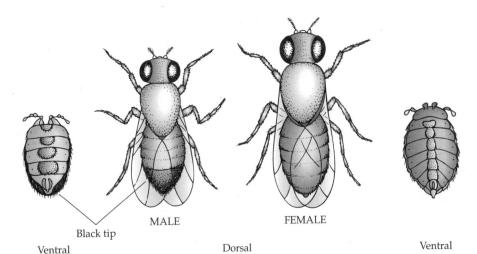

**FIGURE 11–3**

Male and female fruit flies. Note the black tip of the male's abdomen.

Black tip

MALE          FEMALE

Ventral              Dorsal                          Ventral

take to produce a single generation if one were studying inheritance in, say, sheep or dogs. Fruit flies also have a small number of chromosomes (eight), which simplifies the study of genes and chromosomes themselves. The study of fruit flies has greatly contributed to what is known about genetics in all organisms. For example, *Drosophila* is useful for demonstrating the inheritance of a sex-linked trait.

In order to proceed with this exercise, it is necessary that you learn how to distinguish between live male and female fruit flies. This is easier than it may seem —males are slightly smaller than females and the male's abdomen ends in a black tip. The female's abdomen lacks the obvious black tip and has only dark bands across it (see Fig. 11–3).

In addition to sexing the flies, you will also be looking at differences in eye color, a sex-linked trait in *Drosophila*. Wild type flies (those normally found in nature) have a red eye color, which is dominant to all other eye colors. A number of eye color mutations (genetic changes) are known in fruit flies; here, we consider white eyes as a recessive sex-linked mutation. You will be observing the offspring from a cross between a wild type male fly and a white-eyed female fly.

1.  Obtain a container of preserved fruit flies and place it on the stage of your stereomicroscope or work area if using a hand lens. Determine the sex of the individual flies by carefully observing them and comparing them to Figure 11–3. Also note the red and white eye color. Ask your instructor for assistance if you have any difficulty in distinguishing between female and male flies. Proceed to Step 2 when you are skilled at sexing the flies.

2.  Obtain a vial of live fruit flies and lay it horizontally on your stereomicroscope stage or work space. Note the presence or absence of female and male flies with red or white eyes by simply placing a check mark in the appropriate space of Table 11–1. The flies will be active, but with a little patience you can determine the eye color among the flies of both sexes. Do not remove the plug from the vial!

**Table 11–1**   Offspring from Wild Type Male × White-eyed Female *Drosophila*

| | |
|---|---|
| Female, white eyes | _____ |
| Male, white eyes | _____ |
| Female, red eyes | _____ |
| Male, red eyes | _____ |

A genetic cross involving a sex-linked trait can be represented by a Punnett square in much the same way as a monohybrid cross. The only difference is the X and Y sex chromosomes that must be included as part of the gametes.

Draw a Punnett square diagram to represent the above cross in *Drosophila* (wild type male × white-eyed female). Use a capital R to represent the allele for red eyes (dominant), and a lowercase r for white eyes (recessive). Place the proper allele associated with the X-chromosomes in the cells of the square indicating sex linkage. The Y-chromosome should have a dash in the area for the alleles to indicate that the Y-chromosome does not carry an allele for the sex-linked trait and therefore has no role in its inheritance.

## POSTLAB QUESTIONS

1. What was the phenotypic and genotypic ratio of the above cross?

2. What would be the phenotypic and genotypic ratio if a white-eyed male fruitfly were crossed with a homozygous red-eyed female fruitfly? (Remember to set up a Punnett square).

3. Red-green color blindness is a sex-linked recessive trait in humans. A color-blind man mates with a woman with normal vision whose father was color-blind. If only female offspring are considered, what is the probability that their daughters will be color-blind?

4. Czar Nicholas II of Russia and his wife, Empress Alexandra, had five children. The youngest was their only son, Alexis. The Empress was a carrier of the sex-linked gene causing hemophilia, and Alexis had the disease. With respect to the four sisters of Alexis, answer the following questions.

(a) What is the probability that none of them were carriers of the hemophilia gene? (b) What is the probability that all of them were carriers? (c) What is the probability that at least one of them was a carrier?

## FOR FURTHER READING

Adolph, Kenneth W. 1991. *Advanced Techniques in Chromosome Research.* New York: Decker.

Ayala, F. J., and J. A. Kiger, Jr. 1984. *Modern Genetics.* Menlo Park, Calif.: Benjamin/Cummings Publishing Company, Inc.

Burns, G. W., and P. J. Bottino. 1989. *The Science of Genetics.* New York: Macmillan Publishing Company.

Hawkins, John D. 1991. *Gene Structure and Expression.* New York: Cambridge University Press.

King, Max. 1990. *Amphibia.* Berlin: Gebru der Borntraeger.

King, Robert C. 1990. *A Dictionary of Genetics.* New York: Oxford University Press.

Klug, William S., and Michael R. Cummings. 1991. *Concepts of Genetics.* New York: Maxwell Macmillan International.

Latchman, David S. 1990. *Gene Regulation: A Eukaryotic Perspective.* Winchester, Mass.: Unwin Hyman.

Mange, A. P., and E. J. Mange. 1980. *Genetics: Human Aspects.* Philadelphia: W. B. Saunders Company.

Mendel, G. 1866. "Experiments on Plant Hybridization." Translated and reprinted in C. Stern and E. Sherwood (eds.). *The Origins of Genetics: A Mendel Source Book.* 1966. San Francisco, W. H. Freeman and Co.

Strickberger, M. 1985. *Genetics.* New York: Macmillan.

Sturtevant, A. 1965. *A History of Genetics.* New York: Harper & Row.

Suzuki, D. T., et al. 1989. *Introduction to Genetic Analysis.* New York: W. H. Freeman.

TOPIC 12

❏

# DNA and Gene Function and Regulation

OBJECTIVES

1. Describe or diagram the basic chemical structure of a nucleic acid strand.

2. Distinguish chemically between DNA and the types of RNA.

3. Given the base sequence of one strand of DNA, predict the sequence of a complementary strand of DNA.

4. Summarize the process in DNA replication.

5. Define and describe the process of transcription.

6. Summarize the sequence of events that occur in translation.

7. Describe the function of transfer RNA, messenger RNA, and ribosomal RNA.

8. Draw a diagram illustrating an operon and describe how it functions.

## INTRODUCTION

**Deoxyribonucleic acid** (DNA) is the molecule that contains genetic information. A gene is actually a specific segment of the long DNA molecule. Thousands of genes occur along the length of the DNA molecule. The DNA molecule is composed of subunits called nucleotides. Each nucleotide is made up of 3 parts: a nitrogenous base (adenine, thymine, guanine, or cytosine), a five-carbon sugar (deoxyribose), and a phosphate group. Nucleotides are linked together between the sugar and phosphate group by covalent bonds. This forms a long strand. The bases protrude to one side of the sugar-phosphate "backbone." Two of these nucleotide strands are held together in a spiral **double helix** by hydrogen bonds between the nitrogenous bases. Inside of the double helix, the bases always pair in a specific order: adenine with thymine and guanine with cytosine. This is why the amount of adenine in DNA always equals the amount of thymine and the amount of guanine always equals the amount of cytosine.

The scientists who worked out the details of the structure of DNA also realized that the structure im-

mediately suggested a model for **replication** (duplication) of the molecule. This property of replication means that each new cell receives a copy of the original genetic information. Because bases in the two strands of a DNA molecule pair only in a specific way, the nucleotide sequence of each strand automatically supplies the information for the production of its partner strand.

When replication begins, the double helix unwinds as the hydrogen bonds between the nitrogenous bases break. The separate strands of the DNA molecule now have exposed bases to which new bases may join by new hydrogen bonds. The nucleotides are joined by **DNA polymerase,** an enzyme. The two original strands each receive a new partner strand and two copies of the DNA molecule are thus formed.

**Ribonucleic acid** (RNA), which is another nucleic acid, has some similarities to DNA but also some important differences:

1. RNA usually consists of a single strand of nucleotides. DNA is double stranded.

2. Ribose is the sugar in RNA, whereas deoxyribose is the sugar in DNA.

3. RNA and DNA differ by one nitrogenous base in the composition of their nucleotides. RNA contains uracil instead of thymine, which is found in DNA.

DNA carries genetic information as a sequence of nucleotides that ultimately specifies the sequence of amino acids in a polypeptide. Twenty common amino acids compose the diverse proteins of organisms. Therefore, various combinations of the 4 nucleotide bases of DNA must code for the 20 amino acids. The nucleotides arranged in groups of any number less than 3 do not provide enough combinations for 20 amino acids. The nucleotides in groups of 3 (triplets) provide 64 possible combinations, more than enough for 20 amino acids. Some of the "extra" base triplets code for the same amino acid and certain triplets have special functions in the process of protein synthesis.

Protein synthesis begins by the production of **messenger RNA** (mRNA) from the nucleotide sequence of DNA, a process called **transcription.** By transcription, genetic information is coded from DNA to mRNA. The enzyme **RNA polymerase** starts RNA synthesis by binding to a DNA strand serving as a template. The enzyme moves along the DNA molecule and joins the complementary nucleotides. Guanine pairs with cytosine, as in DNA replication, but uracil pairs with adenine in RNA. The enzyme stops when it reaches a termination signal.

In the cytoplasm, mRNA binds to the ribosomes, which are composed of protein and ribosomal RNA and are the actual sites of protein synthesis. **Transfer RNA molecules** (tRNA) carry amino acids to the ribosomes. Amino acids are bonded to the tRNA molecule by the aminoacyl attachment site of the tRNA molecule. At the ribosome, **codons** (base triplets) of mRNA pair with the **anticodons** (base triplets) of tRNA. As a result, the amino acids become properly arranged linearly and are bonded together by enzymes. The tRNA, now freed from its amino acid, leaves the ribosome and may pick up another amino acid. The process is repeated as the ribosome moves along the mRNA molecule. This process results in many amino acids being bonded as a polypeptide, which eventually becomes a functional protein. In this way, the genetic information of DNA is expressed as part of a living cell.

Given the nucleotide sequence of one strand of a portion of DNA molecule, it is possible to determine the nucleotide sequence of the corresponding mRNA (or even the nucleotide sequence of the other DNA strand) because the bases pair only in a specific manner in transcription (or in DNA replication). The mRNA codons that correspond to specific amino acids are known. The amino acid sequence of a polypeptide that would be produced by an mRNA strand can be established by "reading" the codon sequence of the mRNA molecule and determining the complementary amino acids. Because of all of the above, a substantial amount of information can be obtained about the synthesis of a particular protein when only part of the process is known.

In prokaryotic bacterial cells, the synthesis of specific enzymes is controlled by the presence or absence of particular food molecules in the environment. Bacterial genes are arranged in groups called **operons.** An operon contains **structural genes** that code for proteins, **regulatory genes** that control transcription of structural genes, an **operator** that is a binding site for a repressor protein, and a **promoter** site where RNA polymerase binds and starts transcription. The regulatory genes control protein synthesis by producing a **repressor** protein that binds to the operator and prevents RNA polymerase from binding to the promotor site. This is the situation when a structural gene is "off." Food molecules function as **inducer** molecules that bind with the repressor protein and remove it from the operator. This results in RNA polymerase binding with the promotor, in transcription of mRNA, and in synthesis of the enzyme for metabolizing the food molecules. The structural genes are now "on."

This account of DNA, RNA production, and protein synthesis is known as "the central dogma" of molecular biology. We know, however, that RNA can also act as an enzyme and that it can copy itself. This discovery led to the Nobel Prize for Chemistry in 1989 for Thomas Cech and Sidney Altman, the researchers involved. When answering the questions in this section, however, base your answers only on "the central dogma."

## PRELAB QUESTIONS

1. What are the three components of a DNA nucleotide?

2. In a DNA molecule, what nitrogenous base always pairs with cytosine?

3. What process results in the production of two exact copies of the DNA molecule?

4. What sugar is found in RNA?

5. What is transcription?

6. Which RNA molecule carries amino acids to the ribosomes?

7. What process results in many amino acids being bonded as a polypeptide?

8. When the repressor protein binds to the operator and prevents RNA polymerase from binding to the promotor site, is the operon on or off?

## LAB 12.A
## STRUCTURE OF THE DNA MOLECULE

### MATERIALS REQUIRED

1 DNA model kit per group of students

### PROCEDURE

In this exercise, you will examine the subunits that compose the DNA molecule and will use them to create a model that simulates the structure of this nucleic acid. Your instructor will divide the class into smaller groups, each group having its own DNA model pieces.

Observe the various colored beads in your kit. These beads represent the building blocks you will use to construct a DNA molecule. The subunits mentioned earlier that compose DNA are called nucleotides. A nucleotide contains the following three parts:

1. a five-carbon sugar molecule called deoxyribose represented by the white beads;
2. a phosphate group represented by the red beads;
3. a nitrogenous (nitrogen-containing) base represented by the yellow, orange, blue, and green beads.

The nitrogenous base within a nucleotide may be one of four possible kinds — adenine (A) as an orange bead, thymine (T) as a yellow bead, cytosine (C) as a blue bead, and guanine (G) as a green bead.

1. Closely examine a deoxyribose sugar (white) bead, which forms the central portion of a nucleotide. Hold the bead so that the peg is pointing upward. The peg and various holes in the bead

represent carbon atoms of the sugar molecule to which the phosphate group and nitrogenous base will bond. The peg represents the 3' (pronounced "three prime") carbon position on the sugar molecule. The opening at the bottom of the bead represents the 5' carbon atom. The hole at the left side of the bead is the 1' carbon position (see Fig. 12–1).

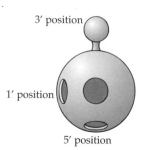

**FIGURE 12–1**

Carbon positions on the deoxyribose sugar (white) bead.

2. Begin construction of a nucleotide by taking a red phosphate group and snapping it into the 5' position of the white sugar bead. This attachment represents a chemical bond. Complete the nucleotide by snapping any one of the nitrogenous bases into the 1' position of the sugar bead. Congratulations! You have created your first nucleotide. Many of these nucleotides bonded together in a certain way produce a DNA molecule.

3. Your next step in constructing DNA is to create more nucleotides. Assemble 7 nucleotides containing adenine (orange beads) and place them in a pile. Assemble 7 more nucleotides containing thymine (yellow beads) and pile them together. Again assemble 7 additional nucleotides with guanine (green beads) and 7 more with cytosine (blue beads), all sorted into piles by the nitrogenous base (color) they contain.

4. Begin to build a DNA molecule by snapping together 14 of the nucleotides into a single strand. Do this by snapping the phosphate group (red bead) of 1 nucleotide onto the 3' peg of the deoxyribose sugar (white bead) of another nucleotide. Attach the nucleotides in any mix of colors, but always be sure to snap the phosphate group (red) onto the 3' peg of the deoxyribose (white). When completed, you will have one-half of the double stranded DNA molecule.

5. You will use the remaining nucleotides to build

the other complementary strand of the DNA molecule. The sequence of nucleotides in the strand you just completed will determine the sequence of nucleotides in its complementary strand. This is because the nitrogenous bases of the 2 DNA strands pair together only in a specific way — adenine always pairs with thymine (orange with yellow) and guanine always pairs with cytosine (green with blue).

Place the nucleotide strand horizontally on the table with the 3′ ends (pegs) of the white sugar beads pointing to the right. Arrange all the nitrogenous bases so that they are pointing downward. Starting with the base on the left end of the strand, determine which nitrogenous base would pair with this base according to the pairing scheme — adenine with thymine and guanine with cytosine. The colors of the beads can serve as a guide — related colors, such as orange and yellow, must always pair together.

6. Choose the appropriate nucleotide from your stock piles and place it on the table at the end of the strand so that the 2 bases are touching. Do not snap together any of the nucleotides at this time. Move to the next nitrogenous base on the nucleotide strand, determine its complementary base, and pair it with another nucleotide from your stock piles. Continue making nucleotide pairs until all bases on the nucleotide strand have a mate.

7. Pause for a moment and make sure that the bases of all nucleotides are properly paired. Correct any errors you find. Before connecting the nucleotides, be aware that the red phosphate and white sugar beads must be snapped together in a specific manner. The 3′ ends (pegs) of the white sugar beads must all point in the *opposite direction* as those in the original strand. This arrangement of the bead nucleotides simulates the **antiparallel** nature of the strands in a DNA molecule. Position the new nucleotides so that the alternating phosphate groups and sugars are antiparallel to the original strand.

8. After you have arranged the new nucleotides in their proper alignment, form a second nucleotide strand by snapping together the alternating phosphate groups and sugar molecules. Always be sure to snap the red phosphate group onto the 3′ end (peg) of the white sugar molecule.

9. Next, use the clear pieces to link the paired bases

with one another. These clear pieces represent hydrogen bonds that form between complementary bases in a DNA molecule. Hydrogen bonds hold together the 2 nucleotide strands and stabilize the molecule. Your completed DNA molecule should resemble Figure 12–2. If it does not, ask your instructor for assistance.

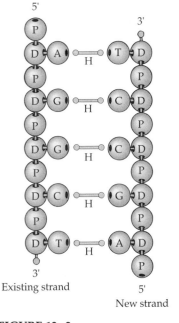

D: deoxyribose sugar
P: phosphate group
A: adenine
T: thymine
C: cytosine
G: guanine
H: hydrogen bond

Existing strand    New strand

**FIGURE 12–2**

A simulated segment of double stranded DNA.

10. Carefully pick up the 2 ends of your completed DNA molecule. Gently twist the strands about 2 turns to simulate the double helix form of the DNA molecule. Keep in mind that your model represents only a very short segment of an actual DNA molecule. If you built an entire DNA molecule using this scale, your model would be many miles long. Save this model for Lab 12.B.

## POSTLAB QUESTIONS

1. In a DNA molecule, would you expect the amount of adenine to equal the amount of thymine? Why or why not?

2. If one side of a DNA molecule was composed of the

bases A-G-T-A-C-G-C-C-T-A, what would be the bases of the complementary strand?

3. Using a ladder analogy to explain the structure of the DNA molecule, what would constitute the rungs of the ladder versus the sides of the ladder?

4. What is meant by the term "antiparallel" in regard to the strands of a DNA molecule?

## LAB 12.B
## REPLICATION IN THE DNA MOLECULE

### MATERIALS REQUIRED

1 DNA model kit per group of students as above

### PROCEDURE

Using the DNA molecule assembled in Lab 12.A, you will follow the outlined replication process to produce two DNA molecules from your original model.

1. Place your DNA model on the table top and position it as shown in Figure 12–3. The 5′ end of the upper strand and the 3′ end of the lower strand should be on the left. The 3′ end of the upper strand and the 5′ end of the lower strand should be on the right.
2. Beginning on the right side of the molecule, break apart the hydrogen bonds between the first 8 pairs of nitrogenous bases and separate the 2 sin-

gle strands to form a "Y" as in Figure 12–4.
3. During replication, the sequence of bases in each separated nucleotide strand of the original DNA molecule serves as a template, or pattern, for the production of a new nucleotide strand. The addition of nucleotides to a newly forming strand always proceeds in a 5′ to 3′ direction on the *new DNA strand.*

Begin replication of the upper strand by pairing the last base on the free end of the strand with an appropriate nucleotide. Place the new nucleotide so that the paired bases touch. Next, position the new nucleotide so that it is running antiparallel to the original strand. (That is, the pegs of the white sugar beads in the original and new nucleotides are pointing in opposite directions.) When the new nucleotide is properly positioned, attach it to the original strand with a hydrogen bond between the paired bases. Continue adding nucleotides to the upper strand in this way, one at a time, until you reach the separation point of the 2 original strands (see Fig. 12–4). Ask your instructor for help if you are having any difficulty.

This replicating strand proceeding *toward* the separation point is called the leading strand. In an actual DNA molecule, the leading strand replicates continuously as the original molecule continues to separate.

4. On the lower strand, replication proceeds *away from* the separation point of the original strands, but the new DNA strand still grows in a 5′ to 3′ direction as before (see Fig. 12–4). This lower strand is called the lagging strand because repli-

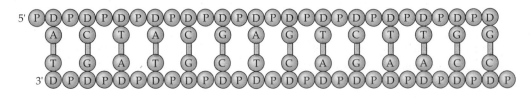

D: deoxyribose sugar
P: phosphate group
A: adenine
T: thymine
C: cytosine
G: guanine

**FIGURE 12–3**

Proper positioning of DNA segment before replication.

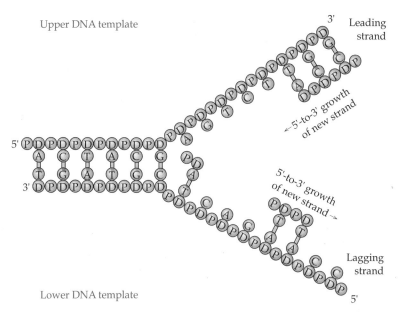

Upper DNA template

3' Leading strand

5'-to-3' growth of new strand

5'-to-3' growth of new strand

Lagging strand

Lower DNA template

5'

**FIGURE 12-4**

Separation of the two DNA strands and initiation of replication.

cation occurs in short, discontinuous segments to keep pace with the separating strands of the original molecule.

To simulate replication of the lagging strand, count back to the fourth nucleotide from the free end of the lower strand and attach the appropriate nucleotide with a hydrogen bond between the paired bases. Be sure that this nucleotide is positioned antiparallel to the original lower strand. Continue to attach new complementary nucleotides to the right, one at a time, until you complete a new DNA segment that reaches the end of the lower strand.

5. Next, count back to the eighth nucleotide from the right end of the lower strand and attach the appropriate nucleotide with a hydrogen bond between the paired bases. Be sure that this nucleotide is positioned antiparallel to the original lower strand. Continue to attach new complementary nucleotides to the right, one at a time, until you complete a new DNA segment that reaches the first completed segment in Step 4. This procedure simulates the discontinuous replication that occurs on the lagging strand.

6. At this point, you have replicated approximately half of the original DNA molecule. Closely examine each new half and check to see that they are identical. If not, determine where the error occurred and make corrections.

7. Continue replication by unsnapping the hydrogen bonds between the remaining 6 nucleotide pairs in the original molecule. Resume replication of the upper leading strand by continuously adding the appropriate nucleotides and hydrogen bonds until you reach the left end of the original upper strand.

8. Resume replication of the lower lagging strand by counting back 3 nucleotides to the left from where replication last occurred. Attach the appropriate nucleotide with a hydrogen bond and continue replication to the right until the last completed new segment is reached.

9. Complete replication of the lower lagging strand by moving to the last nucleotide on the left end and adding the complementary nucleotide with a hydrogen bond. Continue adding new nucleotides to the right until replication is complete.

10. Behold! You now should have two complete DNA copies, each identical to your original molecule. Each copy should be composed of antiparallel strands of nucleotides that are paired correctly, adenine with thymine and cytosine with guanine. If your two copies are not identical, determine where the error occurred and correct it.

## POSTLAB QUESTIONS

1. Replication of DNA is bidirectional. Prove or disprove the validity of this statement.

2. What is the result of DNA replication?

# LAB 12.C
# PROTEIN SYNTHESIS

## MATERIALS REQUIRED

1 RNA and protein synthesis kit per group of students

## PROCEDURE

In this exercise, you will observe protein synthesis using a model kit that simulates the process. Again you will work in groups as you did before, this time using the protein synthesis kit.

**Transcription,** the formation of messenger RNA from DNA, is the first step in protein synthesis. You will first make messenger RNA, or mRNA, from the model pieces. As before, you will begin by assembling nucleotides for making mRNA. Notice the pink beads in the kit. They represent ribose, the five-carbon sugar found in RNA. The green, blue, and orange beads again represent the nitrogenous bases guanine, cytosine, and adenine, respectively, just as they did in the DNA kit. The purple beads represent uracil, another nitrogenous base found in RNA instead of thymine (yellow bead), which is found in DNA. The red beads still represent phosphate groups as they did before.

1. Hold a pink ribose sugar bead with its peg pointing upward. The peg represents the 3′ carbon atom as in the DNA kit. The hole at the bottom of the bead is the 5′ carbon position and the hole at the left is the 1′ carbon atom. Start an RNA nucleotide by snapping a red phosphate bead into the 5′ position of the pink ribose bead. Complete the nucleotide by snapping any one of the nitrogenous bases into the 1′ position of the ribose bead. Assemble a total of 24 nucleotides by making 6 with each of the 4 kinds (colors) of nitrogenous bases. Place the nucleotides in 4 piles according to their bases.

2. Spread out the long paper strip that is the DNA segment, or gene, to be transcribed into mRNA. Next, place the RNA polymerase "molecule" into the far left of the slit between the DNA strands with the active site visible as shown in Figure 12–5. The attachment of RNA polymerase to the DNA molecule marks the beginning of transcription. RNA polymerase first binds to DNA at a promoter region shown in gray on the left end of the DNA molecule. When transcription begins, RNA polymerase moves along the DNA molecule separating the two strands as it goes. The enzyme also "reads" the sequence of bases on just 1 of the DNA

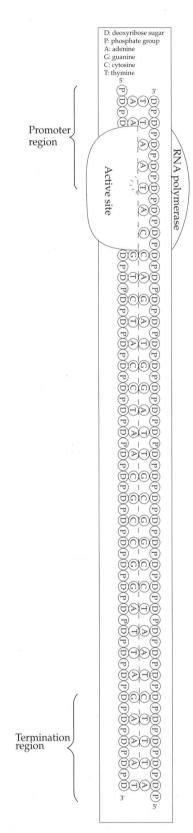

**FIGURE 12–5**

Attachment of RNA polymerase to the DNA molecule.

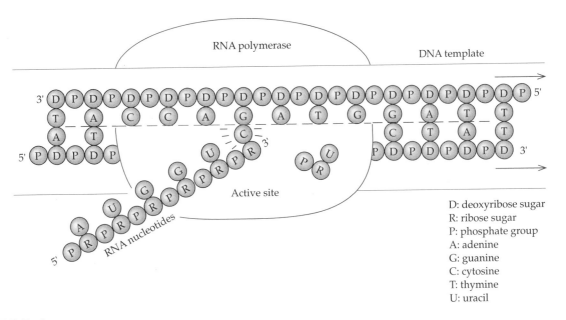

**FIGURE 12-6**

Synthesis of an mRNA strand.

strands as it moves and uses them as a template, or pattern, for bonding free nucleotides into a single mRNA strand.

3. The upper strand of this DNA molecule will be transcribed into mRNA by RNA polymerase. Check to be sure that the active site is positioned at the first DNA base (T) just after the gray promoter region. Take an RNA nucleotide with an adenine base from your stock pile and pair it with the T in the active site by touching the 2 bases together. As in DNA, adenine pairs with thymine in transcription. Cytosine and guanine also pair in RNA synthesis, which is the same as in DNA. An RNA strand grows in a 5' to 3' direction just as does a DNA strand. Position the RNA nucleotide so that it is antiparallel to the upper DNA strand. That is, the 3' peg of the RNA nucleotide should point to the right since the 3' end of the DNA strand is to the left.

4. Hold the RNA nucleotide stationary and move the RNA polymerase forward until the next DNA base (A) is in the active site. Take a nucleotide containing uracil and pair it with adenine (A) by touching the bases together. *Notice that adenine pairs with uracil in RNA,* an important point to remember! Position the nucleotide so that it is antiparallel to the upper DNA strand. When properly positioned, connect the first and second RNA nucleotides by snapping the phosphate group (red) of the newest

nucleotide onto the 3' end (peg) of the previous ribose sugar (pink).

5. Again hold the RNA nucleotides stationary and move the RNA polymerase forward until the next DNA base is in the active site. Pair the base with the appropriate RNA nucleotide and position it properly. Connect the new nucleotide with the growing mRNA strand. Continue these steps, one base at a time, along the entire length of the DNA molecule (see Fig. 12-6). Notice how the DNA strands separate when RNA polymerase moves forward and then rejoin after it passes, which is what actually occurs during transcription.

6. When the RNA polymerase reaches the gray termination region at the right end of the DNA strand, remove the enzyme from the DNA molecule. This simulates the release of both DNA and the mRNA strand by RNA polymerase when it arrives at a termination region. You now have a single strand of messenger RNA that is complementary to the DNA molecule. Transcription is now complete — the genetic information of DNA has been transcribed into messenger RNA.

Now that you have a messenger RNA strand by transcription of DNA, the next step in protein synthesis is translation. The sequence of bases in mRNA determine the sequence of amino acids in a protein during translation. Just as Spanish can be translated

**Table 12-1**  Messenger RNA Codons and Their Corresponding Amino Acids

| First Letter | Second Letter | | | | Third Letter |
|---|---|---|---|---|---|
| | U | C | A | G | |
| U | UUU — Phe<br>UUC<br>UUA — Leu<br>UUG | UCU<br>UCC — Ser<br>UCA<br>UCG | UAU — Tyr<br>UAC<br>UAA — term.<br>UAG | UGU — Cys<br>UGC<br>UGA — term.<br>UGG — Trp | U<br>C<br>A<br>G |
| C | CUU<br>CUC — Leu<br>CUA<br>CUG | CCU<br>CCC — Pro<br>CCA<br>CCG | CAU — His<br>CAC<br>CAA — Gln<br>CAG | CGU<br>CGC — Arg<br>CGA<br>CGG | U<br>C<br>A<br>G |
| A | AUU<br>AUC — Ile<br>AUA<br>AUG — Met | ACU<br>ACC — Thr<br>ACA<br>ACG | AAU — Asn<br>AAC<br>AAA — Lys<br>AAG | AGU — Ser<br>AGC<br>AGA — Arg<br>AGG | U<br>C<br>A<br>G |
| G | GUU<br>GUC — Val<br>GUA<br>GUG | GCU<br>GCC — Ala<br>GCA<br>GCG | GAU — Asp<br>GAC<br>GAA — Glu<br>GAG | GGU<br>GGC — Gly<br>GGA<br>GGG | U<br>C<br>A<br>G |

Ala: alanine
Arg: arginine
Asn: asparagine
Asp: aspartic acid
Cys: cysteine

Gln: glutamine
Glu: glutamic acid
Gly: glycine
His: histidine
Ile:  isoleucine

Leu: leucine
Lys: lysine
Met: methionine
Phe: phenylalanine
Pro: proline

Ser: serine
Thr: threonine
Trp: tryptophan
Tyr: tyrosine
Val: valine

into French, translation in protein synthesis changes the genetic code into an actual protein structure.

1. Position the mRNA strand horizontally on your desk so that the 5′ end is on the left and the 3′ end is on the right. Arrange all of the nitrogenous bases so that they are pointing upward.

2. Genetic information in nucleic acids is encoded by the nitrogenous bases in groups of 3 along the length of the molecule. Each group of 3 bases is appropriately called a triplet. In mRNA, a triplet is referred to as a codon to distinguish it from triplets found in other kinds of nucleic acids. Look at the 5′ end of the mRNA strand and notice that the first triplet, or codon, consists of adenine, uracil, and guanine. This codon can be abbreviated as AUG. On a piece of paper, list the 8 codons of the mRNA strand in order, beginning with AUG at

the 5′ end and proceeding to the 3′ end of the molecule.

3. Next, determine the kinds of amino acids that correspond to the codons on the list by using Table 12 – 1. For example, consider the first codon AUG. Find A in the column marked "First Letter." Next find the U column in the wide middle area marked "Second Letter" and follow down the U column to the A region. Then find G in the last column marked "Third Letter." From this G, follow directly across to the left and you will find the word "Met," which stands for methionine, the first amino acid in the protein coded for by your mRNA strand. Write "methionine" next to AUG in your list. Continue to find the corresponding amino acids for the next 6 mRNA codons. The last codon on the 3′ end of the mRNA strand (UAG) is a ter-

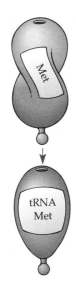

**FIGURE 12-7**

Attachment of an amino acid to a transfer RNA molecule.

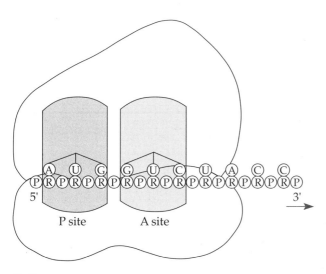

R: ribose sugar
P: phosphate group
A: adenine
G: guanine
C: cytosine
U: uracil

**FIGURE 12-8**

Association of the ribosome with mRNA.

spond to the codons. The white oval beads in your kit represent transfer RNA molecules, or tRNA, that carry amino acids to the site of protein synthesis in the cytoplasm. Notice they are also labelled with the same amino acid names from your list. Transfer RNA molecules carry amino acids by temporarily bonding with them. Transfer RNA molecules are also specific for the kind of amino acid they carry. Simulate this specificity by snapping an amino acid into the hole of a matching tRNA (Fig. 12-7). Continue until all tRNA molecules are carrying a matching amino acid.

5. Make sure the mRNA strand is positioned horizontally with the 5' end to the left and bases pointed upward. Take the ribosome and slide it under the mRNA so that the first codon on the left lies in the P site and the second codon lies in the A site (Fig. 12-8).

6. Pick up the tRNA carrying the amino acid corresponding to the first codon in the ribosome P site (methionine). Snap the peg of the tRNA into the middle base of this codon (U) (see Fig. 12-9). This simulates the temporary bonding of tRNA with mRNA. The tRNA molecules have a triplet of bases called an anticodon that pairs with the codon on the mRNA. In reality, the tRNA anticodon bonds with all 3 bases of the mRNA codon, but this is not possible with the structure of the beads.

minator signal and does not code for an amino acid.

4. The twisted white beads in the kit represent amino acids. Notice they are labelled with the same amino acid names on your list that corre-

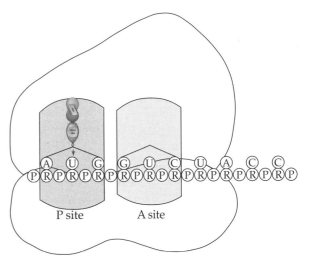

**FIGURE 12-9**

Bonding of the first tRNA-amino acid complex to mRNA in the P site.

7. Pick up the tRNA carrying the amino acid that corresponds to the second codon in the ribosome A site and snap it into the middle base of the codon. Both P and A sites should now be occupied as in Figure 12–10. Now a bond forms between the amino acid in the P site and the amino acid in the A site. Simulate this by pulling out the P site amino acid from its tRNA and attaching it to the amino acid in the A site (Fig. 12–11). You now have the beginning of a growing chain of amino acids that will eventually form a protein molecule.

8. Move the ribosome ahead to the next codon and position it in the A site. The growing amino acid chain should now be in the P site and the tRNA lacking its amino acid should be to the left of the ribosome. This free tRNA is expelled from the mRNA strand and may return with another amino acid of its kind (see Fig. 12–12). Simulate this by removing the free tRNA and placing it aside.

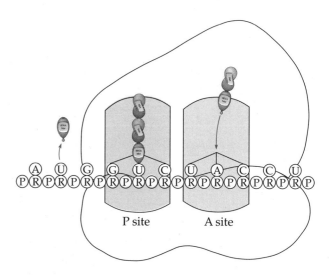

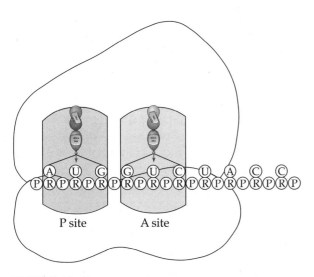

**FIGURE 12–10**

Bonding of the second tRNA-amino acid complex to mRNA in the A site.

**FIGURE 12–12**

Forward movement of the ribosome to the next codon and subsequent release of "empty" tRNA from the P site.

9. Bring in the next tRNA-amino acid complex that corresponds to the codon in the A site and attach it to the middle base (Fig. 12–12). Remove the short amino acid chain from the tRNA in the P site and snap it into the amino acid of the A site. This simulates the bonding of a new amino acid to the growing chain.

10. Move the ribosome ahead to the next codon and repeat Steps 8 and 9 until you reach the end of the mRNA strand. When the ribosome arrives at the end of the strand, the completed protein chain of 7 amino acids should be attached to the mRNA in the P site and the last codon should be in the vacant A site. The last codon UAG signals for termination of the translation process and the eventual release of the protein molecule. The anticodons of tRNAs do not bond with termination codons.

11. Next, a release factor binds to the A site and causes the transfer of the completed protein from

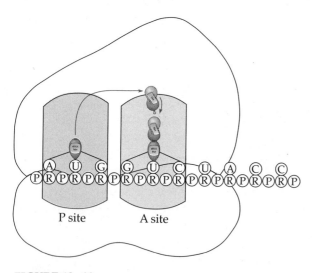

**FIGURE 12–11**

Transfer of the amino acid(s) to the A site, which lengthens the incomplete protein.

the P site to the A site. Then the protein is released from the ribosome. Simulate this by removing the protein from its tRNA in the P site, place it briefly in the A site, and then remove it from the ribosome. Remove the tRNA in the P site and separate the ribosome from the mRNA strand. Translation is now complete.

## POSTLAB QUESTIONS

1. What two molecules will you find in RNA but not in DNA?

2. Radioactive thymine is introduced into a suspension of cells. Two hours later, all excess thymine is removed. Only the nuclei of the cells show any radioactivity. What does this indicate?

3. In the formation of messenger RNA, which molecule acts as a template?

4. Anticodons on tRNA correspond to particular amino acids. There are 64 possible different triplets, but only 20 known amino acids. Give an explanation for this discrepancy.

5. If the DNA triplet for a given amino acid is ATG, what is the corresponding three base sequence on the codon? The anticodon?

## LAB 12.D
## THE LACTOSE OPERON

### MATERIALS REQUIRED (per pair of students)

6 culture tubes of *E. coli*, as follows:

a. normal, grown in maltose

b. normal, grown in lactose

c. mutant, grown in maltose

d. mutant, grown in lactose

e. unknown strain, grown in maltose

f. unknown strain, grown in lactose

1 test tube rack

1 dropper bottle of toluene

1 small flask of ONPG

1 one ml pipet with pipetor

hot water bath or incubator

## PROCEDURE

In this exercise, you will observe a visual demonstration of gene function and regulation concerning the lactose operon in the bacterium *Escherichia coli*. This exercise is based on beta-galactosidase, an enzyme produced by *E. coli* when lactose sugar is present in its surroundings. Beta-galactosidase breaks apart the lactose molecule into its two component sugars, glucose and galactose. These simple sugars are then used by the bacterial cell.

If grown in a culture with both glucose and lactose sugars, *E. coli* will not produce the enzyme beta-galactosidase until all glucose has been used. The enzymes for glucose metabolism are continually produced by most cells of any kind, including *E. coli*. Therefore, if glucose is present with lactose, the bacteria first metabolize glucose for energy. The cells will not produce enzymes for metabolizing another sugar until they are stimulated to do so by the absence of glucose.

Observe Figure 12–13 for an illustration of how the lactose operon functions. This diagram shows what happens when lactose is not present and beta-galactosidase is not produced. The repressor gene (R) continually is transcribed and produces repressor protein. The repressor protein binds to the operator gene (O) which prevents RNA polymerase from binding to the promoter site (P) on the DNA molecule. This in turn prevents transcription of the structural gene (S). The promoter site and operator gene are DNA base sequences that are not translated into proteins, but serve a regulatory function.

Figure 12–14 shows what happens when beta-galactosidase production is turned on, or induced by the presence of lactose. Notice that lactose binds with the repressor proteins which alters their shape. Because of this shape change, the repressor protein cannot bind with the operator gene (O). This allows RNA polymerase to bind with the promoter site (P). Transcription of the structural gene (S) follows and beta-galactosidase is produced.

In the case of one particular strain of *E. coli*, a mutation, or change, has occurred in the operator gene. This mutation alters the function of the lactose operon as illustrated in Figure 12–15. Because of the change in the operator gene (O), the repressor protein cannot

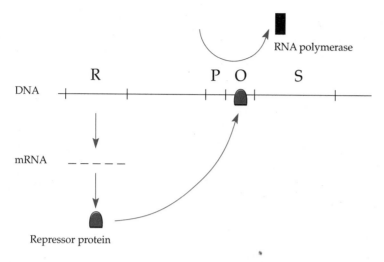

**FIGURE 12–13**

Inhibition of enzyme production in lactose operon by repressor protein; RNA polymerase can't bind to promoter gene.

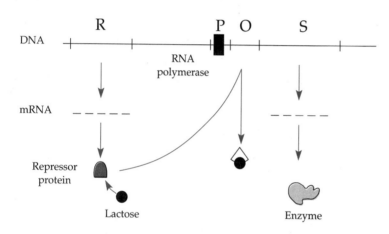

**FIGURE 12–14**

Induction of enzyme production in lactose operon by lactose; lactose prevents repressor from binding to operator gene.

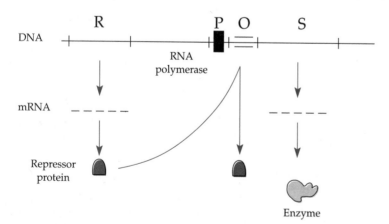

**FIGURE 12–15**

Continual enzyme production in lactose operon; repressor protein can't bind to mutated operator gene.

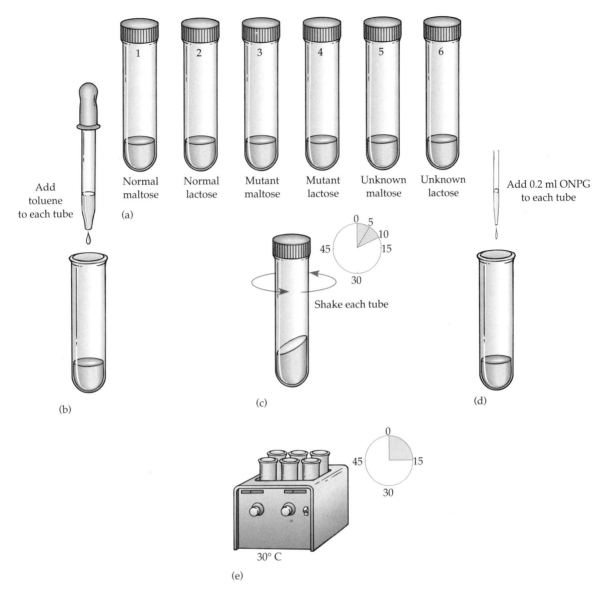

Add toluene to each tube

(a) Normal maltose | Normal lactose | Mutant maltose | Mutant lactose | Unknown maltose | Unknown lactose

Add 0.2 ml ONPG to each tube

(b)

Shake each tube

(c)

(d)

30° C

(e)

**FIGURE 12-16**

Procedure for indicating beta-galactosidase activity in wild type and mutant *E. coli*.

bind with it regardless of whether lactose is present or absent. This permits RNA polymerase always to bind with the promoter site (P). The structural gene (S) is transcribed for the continuous production of beta-galactosidase.

Working with a partner for this exercise, you will be looking for beta-galactosidase activity in the six culture tubes listed under Materials Required that contain normal or mutant *E. coli* grown in the two kinds of sugars. The presence and activity of the enzyme is indicated by a color change in a colorless substance you will add to the culture tubes called ONPG (o-nitrophenyl galactoside). ONPG is split by beta-galactosidase into galactose and o-nitrophenyl, which

is bright yellow in color. The appearance of yellow within a tube is a positive test for beta-galactosidase (see Fig. 12-16).

1. Add 2 drops of toluene to all 6 test tubes. Cover the tubes and shake them frequently over a period of 10 minutes. The toluene breaks down the cell wall of the bacteria, which releases the cell contents into the test tube liquid.

2. Add .2 ml of ONPG to all 6 test tubes and place them in a 30°C water bath or incubator for 15 minutes.

3. After the tubes have incubated, observe them for any results. Record any color change in Table 12-2

**Table 12–2**   Results of Lactose Operon Induction

| E. coli | (a) Normal | (b) Normal | (c) Mutant | (d) Mutant | (e) Unknown | (f) Unknown |
|---------|-----------|-----------|-----------|-----------|-------------|-------------|
| Sugar   | Maltose   | Lactose   | Maltose   | Lactose   | Maltose     | Lactose     |
| Result  |           |           |           |           |             |             |

using a minus sign for no change, a plus sign for some appearance of yellow, or 2 plus signs for obvious yellow color in the tube.

## POSTLAB QUESTIONS

1. Which test tubes show activity of beta-galactosidase?

2. Do your results agree with what you would expect based on what you know of the lactose operon?

3. Based on your results, are you able to determine if the unknown strain grown in maltose is normal or mutant? If so, which strain did this culture contain?

4. Based on your results, are you able to determine if the unknown strain grown in lactose is normal or mutant? If so, which strain did this culture contain?

5. In the lactose operon, how does a repressor protein "turn off" enzyme production?

6. If the lactose operon is "turned off," how can it be activated for synthesis of beta-galactosidase?

7. What advantage does the lactose operon offer to the bacteria?

8. Regarding energy and cell resources, present an argument whether or not the mutant strain has an advantage over normal E. coli.

## FOR FURTHER READING

Adams, R. L. P. 1991. *DNA Replication.* New York: IRL Press.

Cherayil, Joseph D. 1990. *Transfer RNAs and Other Soluble RNAs.* Boca Raton, Fla.: CRC Press.

Cozzarelli, Nicholas R., ed. 1990. *DNA Topology and Its Biological Effects.* Cold Spring Harbor, N.Y.: Cold Spring Harbor Laboratory Press.

Hawkins, John D. 1991. *Gene Structure and Expression.* New York: Cambridge University Press.

Herrmann, H. 1989. *Cell Biology.* New York: Harper & Row.

Judson, H. F. 1979. *The Eighth Day of Creation.* New York: Simon and Schuster.

Langridge, J. 1991. *Molecular Genetics and Comparative Evolution.* New York: Wiley.

Lewin, B. 1985. *Discovery—The Search for DNA's Secrets.* New York: Van Nostrand Reinhold.

Mol, Joseph N. W., and Alexander R. van der Krol, eds. 1991. *Antisense Nucleic Acids and Proteins: Fundamentals and Applications.* New York: M. Dekker.

Rickwood, D., and B. D. Hames. 1990. *Gel Electrophoresis of Nucleic Acids: A Practical Approach.* New York: IRL Press.

Selander, Robert K., Andrew G. Clark, and Thomas S. Whittam, eds. 1991. *Evolution at the Molecular Level.* Sunderland, Mass.: Sineaur Associates.

Sierra, Felipe. 1990. *A Laboratory Guide to Invitro Transcription.* Boston, Mass.: Birkhauser.

Spedding, G., ed. 1990. *Ribosomes and Protein Synthesis.* New York: IRL Press.

Taylor, J., ed. 1965. *Selected Papers on Molecular Genetics.* New York: Academic Press.

Watson, J. 1980. *The Double Helix.* New York: Atheneum.

Watson, J. D., and F. H. C. Crick. 1953. "A Structure for Deoxyribose Nucleic Acid." *Nature* 171:737.

❏

# Genetic Engineering

## O B J E C T I V E S

1. Outline the primary techniques utilized in recombinant DNA experiments.

2. Identify the role of vectors in recombinant DNA experiments and give several specific examples of such vectors.

3. Explain the special measures that have been employed to introduce genes experimentally into plant and animal cells.

## INTRODUCTION

The field of **genetic engineering** is relatively new, but it is a rapidly developing area of scientific endeavor that offers considerable benefits for humankind. Genetic engineering is concerned, generally, with the direct manipulation of the hereditary material — DNA. From genetic engineering and molecular biology have come great possibilities for the curing of genetic diseases, such as diabetes, as well as other diseases, such as cancer, and the improvement of domestic plants and animals for the benefit of the human race.

Research into gene structure has produced many methods of genetic engineering. The discovery of **restriction enzymes** marked the beginning of genetic engineering. Restriction enzymes in bacteria were found that protect the bacteria from invading phages (viruses). The enzymes attack specific nucleotide sequences in the viral DNA and break the molecule into harmless pieces with known nucleotides at the fragment ends. The DNA fragments can be joined with other DNA segments by a different enzyme, making it possible to insert a specific section of DNA into another DNA molecule. This results in **recombinant DNA,** the combined genes of more than one organism. Recombinant DNA has produced commercial products such as human insulin for diabetics from bacteria that carry the gene for production of the protein hormone.

In genetic engineering, a particular gene of interest must first be isolated for it to be manipulated by researchers. An efficient way of isolating a gene is to obtain cells containing the active desired gene, cells which also carry a quantity of messenger RNA transcribed from the particular gene of interest. The mRNA is isolated and its corresponding DNA is produced by the viral enzyme via **reverse transcriptase.** This enzyme synthesizes DNA using RNA as a template, the reverse of the normal process. In this way, a copy of the gene can be obtained and then duplicated for use.

**Gene transplantation,** the moving of genes from one organism into another, is a promising method in genetic engineering. Genes transplanted into higher organisms must be spliced into a carrier that can enter a cell and introduce the new gene into the genome of that cell. **Plasmids** of bacteria are used as carriers. A plasmid is a small, circular molecule of DNA that is separate from the genome of the bacterial cell. Viruses are also used as carriers in gene transplants.

Once a gene has been transplanted into, for example, an individual plant cell, it is possible to grow an entire mature plant from this one cell. This cell is said to be **totipotent,** or all powerful. Genetic totipotency is very useful because many plants can be cloned from the cells of one particular plant and the seeds made available for sale. In this way, plants with genetically engineered characteristics are made available for agricultural use.

Totipotency in animals has yet to be clearly demonstrated. Clearly, however, certain related experiments offer tempting speculations of what might be possible. In **nuclear transplantation** experiments involving frogs, the nucleus of a differentiated body cell is placed in an egg cell from which the nucleus has been removed. The resulting zygote develops normally into a sexually mature adult frog. All genes of the transplanted nucleus occur in all cells of the frog, including the germ cells (cells that give rise to eggs or sperm). Therefore, the transplanted genes will be passed on to the next generation. As in plants, production of adult animals from single cells with desired genetic traits could have widespread effects on animal husbandry.

Genetic engineering may hold some hope for cancer treatment since cancers occur due to changes in the cell's genome. **Oncogenes** (genes with the potential to cause cancer) may become active or mutate thereby causing normal cells to become cancerous. Some cancers are caused by environmental **carcinogens,** cancer causing substances found in the environment. Cancer begins as cells grow and divide abnormally forming clumps, or **tumors.** Some tumors are harmless, but a **malignant** tumor grows uncontrollably and destroys surrounding healthy tissue. The cells of some malignant tumors do not adhere well to one another. They become free and **metastasize,** traveling to other areas of the body where they start new tumors. Researchers are investigating many methods of genetic engineering that may lead to cancer treatments.

In this chapter, we have designed some exercises that show some of the more fundamental aspects of genetic engineering in the broad sense.

## PRELAB QUESTIONS

1. What macromolecule is directly altered or manipulated in genetic engineering?

2. Genetic engineering was made possible by what discovery?

3. What is the function of reverse transcriptase?

4. What process involves the movement of genes from one organism to another?

5. Identify at least three commercial (either actual or potential) applications of genetic engineering.

## LAB 13.A
## GENETIC TRANSFORMATION OF *ESCHERICHIA COLI*

### MATERIALS REQUIRED (per group of students)
petri plate culture of *Escherichia coli*

inoculating loop

test tube rack

2 test tubes of sterile, cold calcium chloride solution

plasmid solution

labelling tape

ice bath

3 sterile 1 ml pipets, 1 kept on ice

nutrient broth

warm water bath, 42°C

2 sterile petri plates with nutrient agar containing ampicillin

### PROCEDURE
Bacteria are commonly used as subjects for genetic engineering because they are relatively simple organisms (compared to higher plants and animals) and they reproduce at a rapid rate. Rapid reproduction is an advantage because under ideal conditions one genetically engineered bacterium cell can produce billions of copies of itself within 24 hours. This means that the numerous altered cells will perform a desired task or produce in great quantity a desired substance for which it was engineered.

Inserting a gene or genes into the DNA of an organism requires a vector, a DNA molecule that carries the desired gene(s) into the recipient organism. In bacteria, a common vector is a plasmid, a small circular molecule of DNA that is separate from the larger circular chromosome of the bacterial cell. Bacteria can

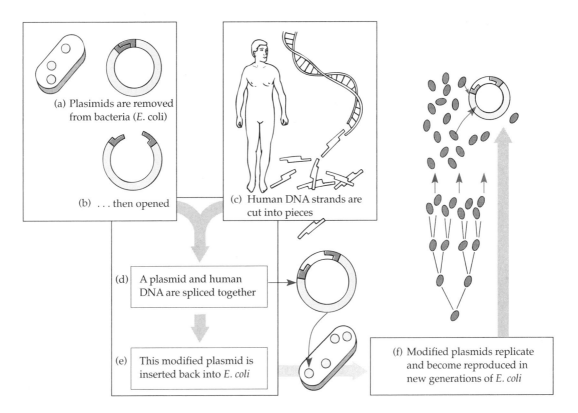

(a) Plasimids are removed from bacteria (*E. coli*)

(b) ... then opened

(c) Human DNA strands are cut into pieces

(d) A plasmid and human DNA are spliced together

(e) This modified plasmid is inserted back into *E. coli*

(f) Modified plasmids replicate and become reproduced in new generations of *E. coli*

**FIGURE 13-1**

Procedure for using plasmids to place a desired gene in bacteria cells. (a) Plasmids are removed from *E. coli* cells. (b) Plasmids are opened with the enzyme restriction endonuclease. (c) Human DNA fragments are produced with restriction endonuclease. (d) DNA fragment with the desired gene is spliced into the plasmid with the enzyme ligase. (e) *E. coli* cell takes in modified plasmid. (f) Transformed cell produces many new generations of cells that replicate the plasmid and produce a human substance.

be artificially induced to take in plasmids from outside their cell wall, although exactly how this occurs is not known. The newly received plasmid, and all the genes it carries, will then be reproduced with each cell division. Consequently, all the new cells will possess the characteristics produced by the gene(s) of the introduced plasmid. This process is known as transformation because the bacterium has been transformed, or changed, to a form that differs from the original organism. Naturally occurring plasmids may be introduced into bacteria (as in this exercise), or a gene from another species may be spliced into a plasmid that is then taken in by bacteria that produce the desired product of the transplanted gene (Fig. 13–1).

In the following exercise, you will transform the bacterium *Escherichia coli,* a beneficial bacterium that grows in the large intestine of humans, into a form that is resistant to the antibiotic ampicillin that kills unaltered *E. coli* cells. The transformation occurs when *E. coli* cells take in plasmids that carry a gene for

ampicillin resistance. The plasmids are obtained from strains of bacteria that naturally carry this type of plasmid.

*Note:* Using good sterile technique is important for the success of this exercise. Closely follow your instructor's directions for sterile procedure when handling bacteria cultures, media, and solutions.

1. Obtain a petri plate of nutrient agar with growing colonies of *E. coli.*
2. Sterilize an inoculating loop by holding it over a flame until it glows briefly. Allow the loop to cool, then use it to scoop up a few colonies from the surface of the agar (see Fig. 13–2a).
3. Remove the cap from a test tube containing 3 ml of sterile, ice cold calcium chloride solution. To maintain test tube sterility, do not put down the test tube cap! Hold it in your hand and do not touch its interior. Gently swirl the inoculating loop through the solution to introduce bacteria

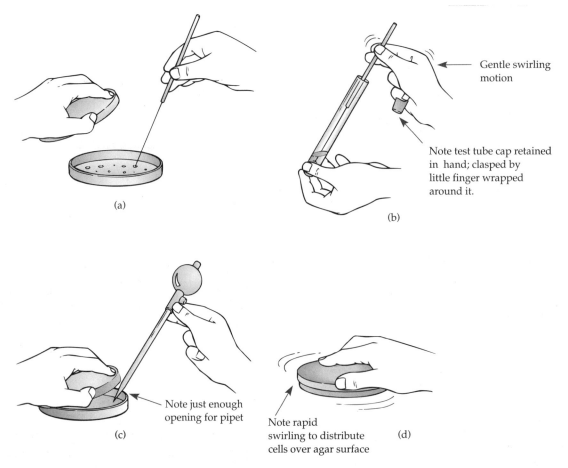

(a)

Gentle swirling motion

Note test tube cap retained in hand; clasped by little finger wrapped around it.

(b)

Note just enough opening for pipet

(c)

Note rapid swirling to distribute cells over agar surface

(d)

**FIGURE 13–2**

Steps in preparing *E. coli* for transformation. (a) Collecting colonies with sterile transfer loop. (b) Introducing cells into calcium chloride solution. (c) Transferring *E. coli* to agar plate containing ampicillin. (d) Swirling plate on table surface to distribute cells.

cells into the liquid, but do not agitate the solution (see Fig. 13–2b). Replace the test tube cap and label as tube 1.

4. Repeat Steps 2 and 3 with another test tube of cold, sterile calcium chloride solution labelled as tube 2. Allow both tubes to sit in the ice bath for 30 minutes.

   *Note:* While the tubes are timing in the ice bath, start Lab 13.B EXTRACTION OF DNA. Return to Step 5 of this exercise when 30 minutes have expired.

5. Add .1 ml of ice cold plasmid solution to tube 1 using a cold sterile pipet. Use another pipet to add .1 ml of nutrient broth to tube 2. Place both tubes in the ice bath for 20 minutes.

6. Remove the tubes from the ice bath and place them in a warm water bath (42°C) for 2 minutes. This change in temperature facilitates uptake of plasmids by *E. coli* cells.

7. Cool both tubes in the ice bath again for 10 minutes.

8. Obtain 2 sterile petri plates containing nutrient agar with ampicillin. Tilt open one of the plate lids exposing just enough room to pipet .5 ml of solution from tube 1 onto the surface of the agar (see Fig. 13–2c). Cover and quickly swirl the plate on the lab bench to spread the cells in solution over the surface of the agar (see Fig. 13–2d). Label this plate "transformed bacteria" and include your names and the date.

9. Repeat Step 8 using the remaining agar plate and solution from tube 2. Label this plate "control" and include your names and the date. Incubate both agar plates upside down for 24 hours at 37°C (human body temperature). After the incubation period, the plates may be stored in the refrigerator until the following week's lab period for observation.

10. Examine the plates for any colonies of bacteria growing on the agar. A colony results from the repeated cell divisions of a single cell. The presence of colonies on the ampicillin-containing agar indicates *E. coli* cells that have been transformed to an ampicillin resistant strain.

## POSTLAB QUESTIONS

1. How do we know that the original bacterial culture of *E. coli* was not ampicillin resistant?

2. Would anything have been added to this experiment by attempting to culture the plasmid solution?

3. What inference would you draw if both test tubes 1 and 2 produced bacterial colonies on the agar plates at the conclusion of the experiment?

4. What is the possible effect of plasmid DNA concentration on transformation rate?

5. In bacterial transformation experiments in which eukaryotic DNA is used, the eukaryotic DNA hardly ever becomes inserted into the bacterial chromosome. Why?

## LAB 13.B
## EXTRACTION OF DNA FROM PLANT CELLS

### MATERIALS REQUIRED (per group of students)

weighing scale

centrifuge

2 centrifuge tubes

2 test tubes

test tube rack

glass stirring rod

chloroform/isoamyl solution

small graduated cylinder

fresh, young bean leaves

mortar and pestle

warm water bath, 60° C

dropper

pipet and pipetor

tube of isolation buffer

ice cold isopropanol

### PROCEDURE

Before the desirable genes of an organism can be introduced into the DNA of a recipient organism, the DNA of the donor organism must first be extracted, or isolated, from the donor's cells. In this exercise, you will extract the DNA from the leaves of a bean plant (see Fig. 13–3).

1. Grind 1.5 g of leaf tissue in 7.5 ml of warm isolation buffer using a mortar and pestle until a smooth paste is formed (see Fig. 13–3a). Pour the leaf paste into a clean test tube. Rinse the mortar with an additional .5 ml of buffer solution and add the liquid to the test tube containing the leaf paste.

2. Incubate the test tube of leaf paste in a warm water bath at 60° C for 30 minutes (see Fig. 13–3b). Shake the tube from side to side every 10 minutes to mix.

3. Place the test tube on ice for 5 minutes to cool (see Fig. 13–3c). After cooling, divide the leaf paste equally into 2 centrifuge tubes (see Fig. 13–3d).

4. Working in a ventilation hood, add an equal volume of chloroform/isoamyl alcohol solution to the leaf paste in each centrifuge tube. Mix thoroughly by plunging with a glass rod (see Fig. 13–3e).

5. Spin the tubes in a centrifuge at high speed for 5 minutes to separate liquids and solids (see Fig. 13–3f).

6. Use a dropper to transfer the upper liquid layer (contains dissolved DNA) into a clean test tube (see Fig. 13–3g). *Note:* Discard the solid lower layer (and all waste of this exercise) into the designated container.

7. Add 5 ml of ice cold isopropanol to the DNA solution and stir gently with a glass rod (see Fig. 13–3h and i). Within a few moments, DNA will appear as white strands throughout the liquid in the test tube.

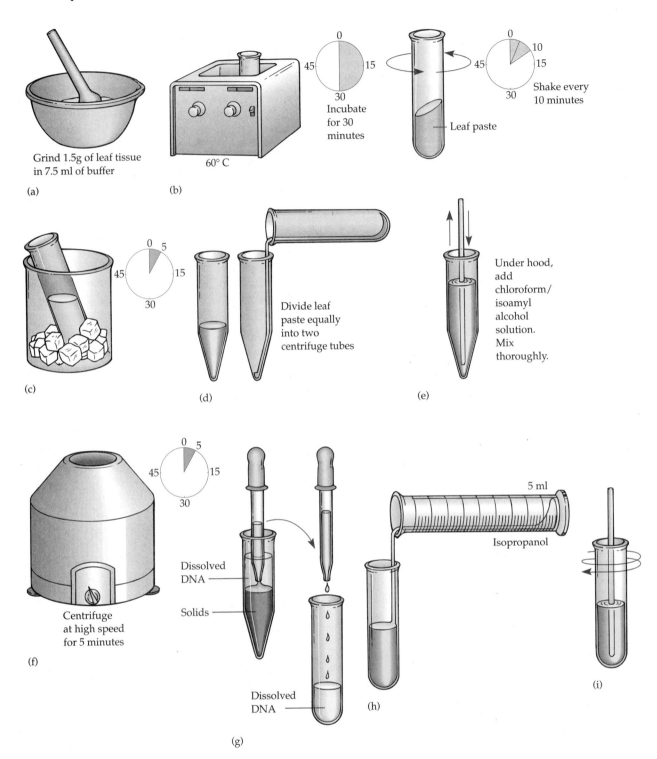

Grind 1.5g of leaf tissue
in 7.5 ml of buffer

(a)

60° C

Incubate
for 30
minutes

(b)

Leaf paste

Shake every
10 minutes

Divide leaf
paste equally
into two
centrifuge tubes

(c)

(d)

Under hood,
add
chloroform/
isoamyl
alcohol
solution.
Mix
thoroughly.

(e)

Centrifuge
at high speed
for 5 minutes

(f)

Dissolved
DNA

Solids

Dissolved
DNA

5 ml

Isopropanol

(g)

(h)

(i)

**FIGURE 13–3**

Procedure for extracting DNA from plant tissue.

This procedure of DNA extraction is easy compared to other methods, but it does not yield pure DNA — it also contains proteins associated with DNA and RNA (ribonucleic acid) as well. Purification methods would be necessary if this DNA were to be used for genetic engineering purposes.

## POSTLAB QUESTIONS

1. Why did the directions for this exercise call for fresh, young bean leaves to be used?

2. What types of plant material commonly available from an ordinary grocery store might be suitable alternatives to the young bean leaves called for in this experiment?

3. DNA could be extracted from what type of cells in human blood?

4. In laboratory investigations such as the one described in this exercise, why is it important that all waste materials be properly disposed of rather than simply poured down the drain?

5. Would it be productive to attempt to isolate DNA from bacterial ribosome preparations? If not, why not?

6. Would it be possible to isolate the base uracil from preparations of bacterial DNA?

7. Students performing this laboratory exercise were not required to wear latex gloves; however, in other laboratory work of this nature, why would wearing such gloves be important?

## FOR FURTHER READING

Anderson, John W., and John Beardall. 1991. *Molecular Activities of Plant Cells: An Introduction to Plant Biochemistry.* Oxford, England: Blackwell Scientific Publications.

Bothwell, Alfred et al., eds. 1990. *Methods for Cloning and Analysis of Eukaryotic Genes.* Boston: Jones and Bartlet Publishers.

Brock, Thomas D. 1990. *The Emergence of Bacterial Genetics.* Cold Spring Harbor, N.Y.: Cold Spring Harbor Laboratory Press.

Brown, T. A. 1990. *Gene Cloning: An Introduction.* New York: Chapman and Hall.

Cozzarelli, Nicholas R., and James C. Wang, eds. 1990. *DNA Topology and Its Biological Effects.* Cold Spring Harbor, N.Y.: Cold Spring Harbor Laboratory Press.

Dix, Philip J., ed. 1990. *Plant Line Selection: Procedures and Applications.* New York: VCH.

Doyle, J. 1985. *Altered Harvest: Genetics, and the Fate of the World's Food Supply.* New York: Viking.

Emery, A. E. H. 1984. *An Introduction to Recombinant DNA.* New York: John Wiley and Sons.

Fogarty, William M., and Catherine T. Kelly, eds. 1990. *Microbial Enzymes and Biotechnology.* New York: Elsevier Applied Science.

Grange, J. M. et al., eds. 1991. *Genetic Manipulation: Techniques and Applications.* Boston: Blackwell Scientific Publications.

Hames, B. D., and D. M. Glover, eds. 1990. *Gene Rearrangements.* Washington, D.C.: IRL Press.

Kimura Motoo, and Naoyuki Takahata, eds. 1991. *New Aspects of the Genetics of Molecular Evolution.* New York: Springer-Verlag.

Levin, Morris A., and Harlee S. Strauss. 1991. *Risk Assessment in Genetic Engineering.* New York: McGraw-Hill.

Lewin, B. 1990. *Genes IV.* Oxford: Oxford University Press.

Maniatis, T. et al. 1989. *Molecular Cloning: A Laboratory Manual.* Cold Spring Harbor, N.Y.: Cold Spring Harbor Laboratory.

Parker, Peter J., and Matilda Katan, eds. 1990. *Molecular Biology of Oncogenes and Cell Control Mechanisms.* New York: E. Horwood.

Primrose, S. 1991. *Modern Biotechnology.* Boston: Blackwell.

Rickwood, D., and B. D. Hames, eds. 1990. *Gel Electrophoresis of Nucleic Acids: A Practical Approach.* New York: IRL Press.

Sandell, Linda J., and Charles D. Boyd. 1990. *Extracellular Matrix Genes.* San Diego, Calif.: Academic Press.

Seetharam, Ramnath, and Satish K. Sharma, eds. 1991. *Purification and Analysis of Recombinant Proteins.* New York: M. Dekker.

Setlow, Jane K., and Alexander Hollaender, eds. 1979. New York: Plenum Press.

Singer, Maxine, and Paul Berg. 1991. *Genes and Genomes: A Changing Perspective.* Mill Valley, Calif.: University Science Books.

Watson, J. 1987. *Molecular Biology of the Gene.* Menlo Park, Calif.: Benjamin/Cumings.

Watson, J. et al. 1983. *Recombinant DNA: A Short Course.* New York: W. H. Freeman.

# PART V

□

# The Diversity of Life

❑

# Monera and Protista

1. Describe the distinguishing characteristics of the kingdom Monera.

2. Briefly characterize the eubacteria and cyanobacteria.

3. Characterize the common features of the kingdom Protista.

4. Briefly characterize representative protozoa phyla.

5. Briefly characterize representative algae groups.

6. Briefly characterize the slime molds and water molds.

7. Summarize the impact of monerans and protists upon humans.

---

## INTRODUCTION

The kingdom Monera contains the prokaryotic bacteria. These organisms live in many different habitats and have diverse metabolic capabilities. The archaebacteria are a primitive group thought to resemble the earliest cells that first evolved on Earth. They are biochemically different from other bacteria and live in harsh environments such as hot springs and extremely salty water.

In contrast to the archaebacteria, the two other bacteria groups are more similar to one another in cell structure and biochemistry. The cyanobacteria (blue-green bacteria) are **autotrophs,** that is they make their own food from simple inorganic compounds. Photosynthetic pigments contained in their cytoplasmic membranes function in water-splitting photosynthesis similar to that of eukaryotic algae and plants. Cyanobacteria were the first organisms to carry out photosynthesis. As millions of years passed, they produced enough oxygen to change the environmental atmosphere and to set the stage for the oxygen-using life forms that were to follow. Cyanobacteria may grow in filaments or colonies. Some species cause problems in polluted bodies of water when they re-

produce rapidly and produce toxins and odors that foul the water. Other autotrophic bacteria produce their food by chemosynthesis, which uses energy from inorganic chemical reactions instead of light as in the cyanobacteria.

The eubacteria, those commonly referred to as "bacteria," are mostly **heterotrophs**—they use food made by other organisms. More specifically, many bacteria are **saprobes**—their food is dead organic material. Bacteria absorb nutrients from the organic matter by secreting enzymes that break down the food in the immediate vicinity of the bacterial cell. Bacteria are important **decomposer organisms** that release materials bound in dead organisms back into the environment where recycling by autotrophs is possible.

In addition to being food producers and decomposers, bacteria have other important roles in ecology. Some have the capability of **nitrogen fixation,** which is the reduction of atmospheric nitrogen gas into ammonia. Next, chemosynthetic **nitrifying bacteria** oxidize the ammonia to nitrites and then to nitrates that green plants can use for producing proteins. In this way, bacteria link gaseous nitrogen to the living world.

Bacteria vary in their oxygen utilization. **Aerobes** use oxygen in respiration. **Obligate anaerobes** live in the complete absence of oxygen and obtain energy by fermentation or other processes. **Facultative anaerobes** have the ability to live with or without the presence of oxygen.

**Symbiosis** is an intimate relationship between members of different species. Those organisms involved in symbiosis are called symbionts. There are three types of symbiosis:

1. **Parasitism**—one species, the parasite, benefits at the expense of the host.
2. **Commensalism**—one species benefits and the other is not harmed or benefited.
3. **Mutualism**—both species benefit from the association.

Symbiotic bacteria that normally live on or within an animal are called its **flora.** The flora are important to the animal because they can benefit the host by eliminating the disease-causing bacteria, making vitamins that are absorbed by the host, or producing enzymes that digest a cellulose diet. Many examples of mutualism between plants, animals, and their flora have been discovered.

Prokaryotes usually reproduce by binary fission, which results in new individuals that are genetically identical. Mutation produces genetic change in prokaryotes. Mutation does not occur frequently, but bacteria reproduce at a rapid rate, presenting ample opportunity for mutations to arise. This high reproductive capability combined with the sources of genetic variation have produced the great diversity of prokaryotes found everywhere.

Some bacteria form thick-walled, dormant **spores** that are resistant to heat and drying. As a result of this spore-forming ability, the bacteria can survive otherwise lethal environments. When conditions improve or when the spore is transported to a better habitat, an active cell emerges from the spore.

The effect of bacteria upon humans is both positive and negative. On the negative side, some bacteria are **pathogens,** disease-causing organisms that create suffering and cost much in medical expenses. Bacteria also cause diseases in agricultural plants and animals. Food spoilage is due to bacteria as well. However, many more bacteria species are beneficial, or at least harmless. Bacteria are very important to the ecology and to agriculture as decomposers and symbionts of many plants. Bacteria are used to produce medicines and antibiotics to fight the diseases caused by pathogens. Certain bacteria are essential to the production of many dairy products.

The kingdom Protista contain the eukaryotic unicellular algae and protozoa. They live wherever there is water in all types of habitats. Most are free-living heterotrophs or autotrophs. Some are serious parasites. Protists usually live as solitary cells, although some form colonies of independent cells.

There are three means of locomotion in the protists that are also important in classification. Some move by a **flagellum** (plural flagella), a long, thread-like organelle that turns with corkscrew-like motions that push or pull the protist through its liquid environment. **Cilia** are shorter than flagella and generally more numerous. Cilia propel a protist by coordinated rowing motions. Some protists move by **pseudopodia** that are cytoplasmic extensions of the cell. The remainder of the cytoplasm flows into the extension moving the cell forward. Some protists have no means of locomotion and move passively via environmental forces.

**Phytoplankton** are the photosynthetic algae and cyanobacteria floating near the surfaces of oceans, lakes, and ponds. Phytoplankton are ecologically important as food to many aquatic organisms and produce much of the atmospheric oxygen as a by-product of photosynthesis.

The kingdom Protista is usually divided into several phyla. Phylum Euglenophyta contains mostly freshwater, photosynthetic members that resemble the common genus *Euglena.* These protists usually have two flagella, but one is highly modified into an organelle that detects light aiding in photosynthesis. Euglenophytes lack a cell wall.

Phylum Bacillariophyta are the diatoms and their relatives found in both fresh and salt water. The diatoms have intricate cell walls that contain silica. Diatoms are important as phytoplankton and the hard cell walls that have accumulated over geologic time have many commercial uses. Diatoms lack flagella, but other members have one or more flagella.

Phylum Chlorophyta are the green algae that show a variety of form, reproductive methods, and habitats. Most green algae live in fresh water or moist places. These photosynthetic organisms appear to have evolved from ancestors that also gave rise to land plants.

Phylum Rhodophyta consists mainly of multicellular, marine red algae. The cell walls of red algae are often embedded with mucilaginous polysaccharides. One prominent mucilaginous polysaccharide is agar, which is used as a growth medium for culturing microorganisms. Some members have the ability to deposit calcium into their cell walls. These algae are very important in building coral reefs.

Phylum Phaeophyta consists entirely of brown algae that are almost exclusively marine. Many brown algae contain in their cell walls a polysaccharide called algin. Algin is used as a thickening agent in such things as ice cream, candies, and cosmetics.

Members of phylum Dinoflagellata have a cell wall, two flagella, and are photosynthetic. Most are marine and important as phytoplankton. *Gonyaulax* is one of a few dinoflagellates responsible for the "red tide," a population explosion that can result in nerve toxin accumulating in shellfish that consume the dinoflagellates.

The previous phyla are mostly photosynthetic. The remaining phyla are mainly heterotrophs. Members of phylum Sarcomastigophora lack a cell wall and have one or more flagella. They are freshwater organisms that may be free living, symbionts, or parasites. Some serious tropical diseases are caused by this group, such as African sleeping sickness produced by *Trypanosoma*.

Phylum Apicomplexa contains all parasitic members. Some cause serious diseases in humans. Many have complex life cycles that may involve two different host species. *Plasmodium*, which includes the protozoa that cause malaria, is a member of this phylum.

The ciliates are contained in the phylum Ciliophora, named for the cilia that move them. They have no cell wall and occur in both fresh and salt water. Ciliates have two nuclei of different sizes and functions. *Paramecium* is a common ciliate.

The final phyla are fungus-like protists: phylum Myxomycota (plasmodial slime molds), phylum Acrasiomycota (cellular slime molds), and the phylum Oomycota (water molds). Despite their names, the slime molds and water molds are not true molds belonging to the kingdom Fungi. These "molds" are not true fungi because of cellular differences that set them apart from typical fungi. Nor are the slime molds and water molds related to one another. The kingdom Protista contains many such unusual organisms that do not fit neatly into any one of the four other kingdoms.

As can be seen from the preceding discussion, there are many fascinating characteristics of microbes. The following exercises deal with some of those characteristics.

## PRELAB QUESTIONS

1. What groups of organisms comprise the kingdom Monera?

2. Identify one ecological advantage of bacteria.

3. Identify the three means of locomotion in the protists.

4. What phylum do diatoms belong to?

5. Identify the distinguishing feature of organisms in the phylum Ciliophora.

## LAB 14.A

### KINGDOM MONERA: EUBACTERIA

#### MATERIALS REQUIRED

plain yogurt

toothpicks

slide and coverslip

dropper bottle of water

compound microscope

demonstration oil-immersion microscope with bacterial types, prepared slide

6 sterile petri plates per class

3 permanent markers per class

#### PROCEDURE

People commonly think of bacteria as harmful disease agents. Actually, there are many more beneficial bacteria than there are harmful bacteria. For example, bacteria are used in human industries, such as in the production of dairy products.

1. Obtain a clean slide and toothpick. Scoop up a tiny dab of yogurt with the toothpick and spread it over a small area on the slide. Add a drop of water to the yogurt smear, mix well with the toothpick, then place a coverslip over the mixture.

2. Carefully observe the yogurt smear with high power in an area where the yogurt is thinnest. In the box below, draw the bacteria observed in the yogurt smear. Try adjusting the light intensity of your microscope if you have difficulty seeing the

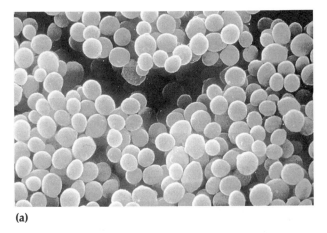

(a)

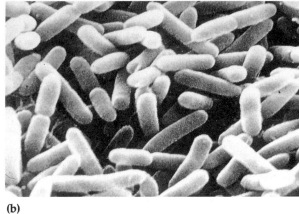

(b)

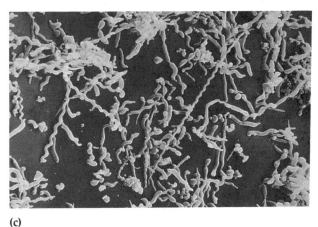

(c)

**FIGURE 14-1**

Electron micrographs illustrating the three basic shapes of bacteria—(a) coccus (spherical), (b) bacillus (rod shaped), (c) spirillum (spiral).

bacteria. The bacteria secrete digestive enzymes into the milk that chemically change the milk protein and sugars. These chemical changes result in the characteristic texture and flavor of yogurt.

3. Observe the prepared slide on demonstration that shows the three main shapes of bacteria—coccus (spherical), bacillus (rod shaped), and spirillum (spiral) (Fig. 14-1). These bacteria are magnified by a special objective on the microscope that uses oil for increased magnification (1000 ×). Use only the fine focus knob when viewing the slide. Locate at least 1 bacterial cell of all 3 shapes. These cell shapes are used in the classification of bacteria and commonly occur in their names, such as in *Staphylococcus aureus* or *Bacillus subtilis*.

4. One of the remarkable traits of bacteria is their presence—they are everywhere! The instructor will distribute 6 sterile petri plates of nutrient agar throughout the class. Nutrient agar is a gel-like substance extracted from marine algae (seaweed) and blended with beef broth and other nutrients for culturing microorganisms. Take the petri plates and contaminate them in any way you like—lightly use your fingertip to mark your initials across the agar, cough into the plate, expose it for a few minutes to the open air, or kiss the agar surface.

5. Cover the petri plate after contaminating it. Take a marker and write on the plate how it was contaminated. Place the petri plates in a dark warm place and examine them the next lab period. Colonies

(spots) of bacterial growth will appear on the plates. Each colony begins as a single bacterium cell that divides numerous times and eventually forms a colony large enough to be visible with the unaided eye. Look for differences in the shape, color, and texture of the colonies that are characteristic of different bacteria species. You may be surprised at what you find when you examine the plates next week!

## POSTLAB QUESTIONS

1. What shape were the yogurt bacteria?

2. The antibiotic penicillin inhibits the production of certain substances required in the composition of the bacterial cell wall. The bacterial cell wall is subsequently weakened and the bacteria are destroyed. Why are human cells not destroyed?

## LAB 14.B
### KINGDOM MONERA: CYANOBACTERIA

#### MATERIALS REQUIRED

*Oscillatoria* culture

*Anabaena* culture

slides and coverslips

droppers

#### PROCEDURE

The cyanobacteria are similar to the bacteria in that they have no membrane-bound nucleus or cell organelles, making them both prokaryotes. One major difference in the cyanobacteria is the presence of photosynthetic pigments within their cells. The cyanobacteria make their own food by photosynthesis whereas heterotrophic bacteria, lacking such pigments, absorb nutrients from their surrounding environment. You will observe two cyanobacteria representatives in this exercise.

1. Obtain a clean slide and coverslip. Use a dropper to pick up the fine, dark green strands of *Oscillatoria* from the sides and bottom of the culture jar. Place a few drops of the culture water on the slide and add a coverslip.

2. Observe *Oscillatoria* under low and high power. The long filaments of *Oscillatoria* are formed by many cells stacked like coins. Draw several filaments of *Oscillatoria*. Carefully watch an individual strand of *Oscillatoria* under high power for a few moments. How do you think *Oscillatoria* got its name?

3. Clean your slide and make a wet mount of the *Anabaena* culture. Observe the cells under both low and high power. *Anabaena* cells also grow in filaments, here resembling a string of beads (Fig. 14–2). *Anabaena* and other species of cyanobacteria are important ecologically because of their ability to fix nitrogen from the atmosphere into compounds that can be used by plants. Nitrogen is an essential nutrient for plant growth. The rice paddies of Southeast Asia consistently produce grain for years without the addition of chemical fertilizers because of nitrogen-fixing cyanobacteria that grow in the water of the fields. Note the enlarged cells called heterocysts in the *Anabaena* filaments. Nitrogen fixation occurs in these cells.

## POSTLAB QUESTIONS

1. To what extent would filamentous cyanobacteria be considered multicellular?

2. Which of the following words or phrases describes the heterotrophic bacteria, the cyanobacteria, both, or neither?
   a. lack a nuclear envelope
   b. photosynthesize similarly to green plants

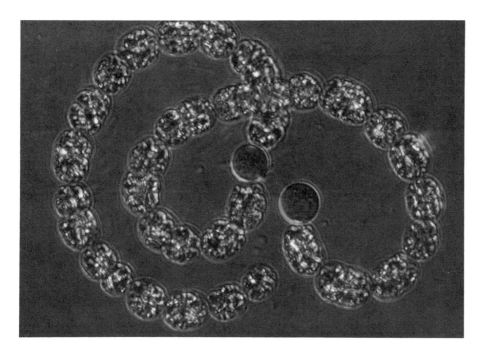

**FIGURE 14-2**

*Anabaena*, a nitrogen-fixing cyanobacterium.

c. possess a cell wall

d. eukaryotic cells

e. symbiotic relationships

f. saprobes

## LAB 14.C
## KINGDOM PROTISTA: PROTOZOA— ANIMAL-LIKE PROTISTS

### MATERIALS REQUIRED

live cultures of the following:

a. *Amoeba*

b. *Paramecium*

demonstration microscope with *Trypanosoma*, prepared slide

clean slides and coverslips

droppers

compound microscope

prepared slide of *Plasmodium*—demonstration

### PROCEDURE

The protozoa are a diverse group of unicellular, animal-like microorganisms that exhibit a tremendous variation of cell shapes, means of locomotion, and ways of life. Protozoa are found in every kind of environment—marine, fresh water, and in the soil. Some are parasitic, others are predators, and others are decomposers. As with the bacteria, protozoa cause some of the worst human diseases and also play important beneficial roles as decomposers in the ecosystem. The protozoa, and all other organisms beyond the bacteria and cyanobacteria, are eukaryotes—the cells contain an organized, membrane-bound nucleus as well as other organelles with membranes such as vacuoles, mitochondria, and so on.

*Phylum Sarcomastigophora*—The protozoa of this phylum are diverse, but they typically have one of two means of locomotion. Some protozoa of this phylum move by pseudopodia (singular pseudopodium, "false foot"). Others have at least one flagellum (plural flagella), a long whip-like appendage that beats in rhythmic waves and pulls the protozoan forward through its watery environment.

1. Obtain a clean slide and coverslip. Use a dropper to pick up an amoeba from the bottom of the culture jar. Do this by placing the jar on a dark surface and look for the amoebae (plural) on the jar bottom appearing as tiny specks of white lint. Using the dropper, draw up a few amoebae and place them on the slide in a few drops of water. Add the coverslip and watch the amoebae under low and high power. Observe the pseudopodia as they form and note how the amoeba moves forward. Note the interior of the amoeba. Do not use bright light to

observe an amoeba; excessive heat from the light may cause it to become inactive. Clean your slide and return it to the proper place when finished.

2. Observe the demonstration slide of *Trypanosoma* (Fig. 14–3), a parasitic, flagellated protozoan, which includes the species that causes African sleeping sickness. *Trypanosoma* can be seen as the darker-stained, snake-like shapes among the round red blood cells on the slide. This protozoan is transmitted from one host to another by the bite of the tsetse fly as it feeds on mammalian blood. Wild hoofed animals, cattle, and humans suffer from the disease in large parts of Africa. *Trypanosoma* attacks the central nervous system in humans, eventually causing death. *Trypanosoma* swims by one flagellum and an undulating membrane along the length of the organism.

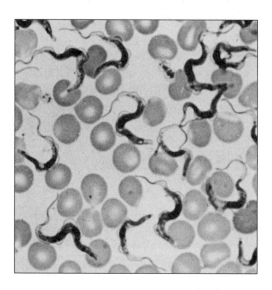

**FIGURE 14–3**

*Trypanosoma* (dark, curved shapes) among red blood cells of a rat.

*Phylum Ciliophora* — Protozoa of this group are some of the most highly modified protists. As the name of the phylum implies, all ciliates have some portion of the cell surface covered with cilia (singular cilium), which are short hair-like appendages. The cilia beat rapidly in regular, rhythmic patterns that propel the cell through its watery environment. In some species that attach themselves to an object, water currents created by the beating cilia bring food particles to the cell. Some ciliates live in very specialized environments such as the digestive tracts of termites and some large grazing mammals where they aid in the digestion of tough plant matter.

3. Obtain a clean slide and coverslip. Look closely at the *Paramecium* culture against a dark background and you may be able to see the protozoa as tiny white specks cruising through the water. Use a dropper to draw up some debris from the bottom of the culture jar. Place a few drops of the water onto your slide, then add one drop of methyl cellulose solution. This is a thick liquid that slows the rapid swimming of *Paramecium* so they can be observed easily under the microscope. Add the coverslip and observe under low and high power.

4. Draw an individual *Paramecium* under high power. Look for the beating cilia by lowering the light of the field and observing the outer edge of the protozoan. Although visible only along the outer boundary of the cell, cilia cover the entire surface of *Paramecium*. Also note the interior of *Paramecium*; look for any organelles or modifications within its cytoplasm.

*Phylum Apicomplexa* — This large group of protozoa are all parasites. Protozoa of this phylum usually have a resistant spore stage in their life cycle and have long been called sporozoans. Unlike the protozoa of the previous phyla, the sporozoans have no organelles of locomotion but are capable of movement by cellular flexing. Malaria, one of the most serious diseases in warmer regions of the world, is caused by the sporozoan *Plasmodium* that is transmitted by the bite of a mosquito. The *Plasmodium* life cycle is complex, involving the cells of the mosquito's digestive tract and human red blood cells in the development of the various life stages (Fig. 14–4).

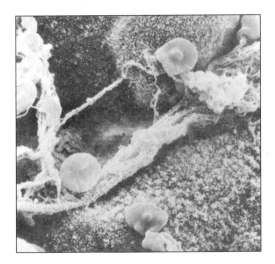

**FIGURE 14-4**

Scanning electron micrograph of malarial parasites *(Plasmodium)* within red blood cells (circular objects). The presence of the parasites within the red blood cells is indicated by the small bumps on the surface of the cells.

5. Examine the prepared slide on demonstration of *Plasmodium* in the red blood cells from a malaria victim. The sporozoan may be difficult to see in the blood cells. If needed, ask your instructor for assistance.

## POSTLAB QUESTIONS

1. Was the interior of the amoeba homogenous or was there an indication of various organelles?

2. Describe the process of amoeboid movement. Does the cytoplasm remain stationary as the amoeba moves or does it shift as movement proceeds?

3. Did the *Paramecium* cytoplasm appear homogenous or show evidence of specialization?

4. Which of the following phrases describe the phylum Ciliophora, the phylum Sarcomastigophora, the phylum Apicomplexa, or none of these phyla?
   a. unicellular organisms
   b. lack membrane-bound organelles
   c. body covered with cilia
   d. two nuclei
   e. move by pseudopodia
   f. all parasitic

## LAB 14.D
## KINGDOM PROTISTA: ALGAE— PLANT-LIKE PROTISTS

### MATERIALS REQUIRED
living cultures of the following:
a. *Euglena*
b. *Spirogyra*
c. *Volvox*

diatoms, prepared slide

dinoflagellates, prepared slide

preserved red algae

preserved brown algae

slides and coverslips

concavity slides

compound microscope

### PROCEDURE
The term ''algae'' (singular alga) is a broad, general term for a group of plant-like protists including both microscopic unicellular forms and giant multicellular seaweed as big as a tree. Algae are mostly aquatic although some do grow on land in moist places. Many kinds of algae are essential to marine and freshwater ecosystems because they are the producers that begin the food chain and release oxygen into the water as a by-product of photosynthesis. Many food and industrial products come from algae, which are known to most people only as a green growth in their aquarium.

*Phylum Euglenophyta*—The euglenoids are peculiar because they have both plant and animal characteristics. In the past, when organisms were considered simply as either plant or animal, the euglenoids were a source of argument. The euglenoids are unicellular and flagellated as represented by *Euglena.*

1. Obtain a clean slide and coverslip. Observe the green color of the culture jar produced by the numerous *Euglena,* which contain green chloroplasts within the cells. Use a dropper to place a few drops of *Euglena* culture onto the slide. There is no need to sample the bottom of the culture because *Euglena* actively swims throughout the water. Add the coverslip and examine under high and low power. Note the swimming movements of *Euglena.* To see the flagellum, turn down the light of your microscope and look closely for its movement at the forward-moving end of the cell.

Draw an individual *Euglena* indicating the flagellum.

Describe the movement of the cell through the water.

When finished, clean your slide and use it for viewing other cultures later in this exercise.

*Phylum Bacillariophyta* — This group of protists is best represented by the diatoms. Diatoms are remarkable because of their intricate cell walls, which are composed of silica, the basic component of glass! Diatomaceous earth, the accumulated cell walls of billions of fossil diatoms, is used as an abrasive in silver polish and toothpaste and as a material for filtering beer.

2. Obtain a prepared slide of diatoms and observe under low and high power. Note the variety of cell shapes and patterns in the cell wall. The cell wall consists of two halves that fit together like a petri plate and a lid. Draw a few representatives of this group.

*Phylum Dinoflagellata* — The dinoflagellates are odd-looking, unicellular organisms found mostly in marine habitats although there are some freshwater species. Most dinoflagellates are photosynthetic and contribute greatly to aquatic food chains. The dinoflagellates are characterized by two flagella occurring in two grooves within the cell wall that propel the organisms through water. Some marine dinoflagellates occasionally reproduce in such tremendous numbers that the cells cause a reddish tinge in the water. This is known as red tide. A few species secrete a toxin that kills many fish in the area of the red tide.

3. Examine a prepared slide of dinoflagellates under low and high power. The dinoflagellate cell wall consists of interlocking plates of cellulose, the same compound that composes the cell wall of higher plants.

*Phylum Chlorophyta* — You are familiar with Chlorophyta, the green algae, as a tinge of green color growing on a clay flower pot, the base of a tree, or perhaps your aquarium. This group is called "green algae" because the chlorophylls found in them are the principal photosynthetic pigment. Green algae are very common in ponds, lakes, and streams as unicellular forms or as large mats of filamentous alga that float at the surface.

4. Use forceps to obtain a bit of the green filaments in the culture jar of *Spirogyra,* a common green alga in ponds. Make a wet mount of the filaments in a few drops of water. Observe under low and high power. The filaments are formed by numerous cells that grow end to end. Note the unusual arrangement of the chloroplasts within *Spirogyra.* Draw a few filaments of *Spirogyra.*

5. Closely observe the culture jar of *Volvox,* a green alga composed of many individual cells living together as a spherical colony. The colonies are visible as tiny green beads in the water. Pick up a few colonies with a dropper and place them in the depression of a concavity slide along with

**FIGURE 14-5**

The red alga *Porphyra* growing on a cultivation net in Japan.

enough water to fill the depression. Add a coverslip and observe the colonies under low and high power.

6. Using high power, note the individual cells embedded in the gel-like matrix that composes the spherical wall of the colony. Each cell of the colony has two flagella that beat in an organized manner, which rolls the colony forward through the water. The depression slide is deep enough to allow *Volvox* to move. Watch for any motion. Some colonies may have miniature daughter colonies within the hollow interior. These will be released through a rupture in the colony wall and grow into a full size colony themselves.

*Phylum Rhodophyta*—From a commercial standpoint, the red algae, so named because of their red photosynthetic pigment, are one of the most important groups of algae. The agar used in the petri plates as part of the first exercise is extracted from marine (ocean-dwelling) red algae. Tons of agar are used annually for culturing microorganisms. Carrageenan, a stabilizer used in foods and toothpaste, is extracted from red algae. You may have seen this ingredient listed on food labels. Some species of red algae are gathered and cultured as a nutritious food in Asia. The red algae are relatively large and plant like in appearance, so you will not require a microscope to observe specimens.

7. Observe the preserved mounts of *Porphyra,* a red alga commercially cultivated as food ("nori") in Japan for centuries (Fig. 14-5).

8. *Chondrus* is also available as a preserved mount. *Chondrus* is a commerical source of carrageenan. Although both of these red algae appear to be very plant like in appearance, their internal structure is far less advanced than the anatomy of a true leaf or stem of a higher plant.

*Phylum Phaeophyta*—The brown algae are some of the most complex and largest of the protists. Kelp, a common "seaweed" of the Pacific coast, grows to a length of a hundred feet or more (Fig. 14-6). Brown algae, like the Rhodophyta, are a commercial source of thickening additives used in foods and cosmetics. Asian countries also use brown algae as a food source. Ecologically, the brown algae are important photosynthesizers in marine ecosystems.

9. Observe the preserved specimens of *Laminaria,* or kelp. Many of the brown algae superficially appear to have leaves, stems, and roots. The "leaves" are simply sheets of cells that lack the complex structure of a true leaf, and the "roots" are holdfasts, structures that anchor the algae in the surf where they commonly grow.

10. *Fucus* is another common brown algae on rocky beaches. Look at the preserved specimens and note the air bladders at the tip of the blades. The bladders provide buoyancy and lift the algae toward light at the water's surface.

**FIGURE 14–6**

A kelp "forest" off the Pacific Coast.

## POSTLAB QUESTIONS

1. Was the "body" of *Euglena* rigid or flexible? Why?

2. How is *Euglena* both animal like and plant like?

3. How did *Spirogyra* get its name?

4. The red pigment in the members of the phylum *Rhodophyta* absorb the blue wavelengths of light that can penetrate deeper water. How would this determine the habitats and body structures of red algae?

5. Identify the plant-like protist phylum or phyla that these phrases describe.
   a. lack a cell wall
   b. mainly multicellular members
   c. photosynthetic
   d. cell wall composed of silica
   e. nuclear envelope
   f. ancestors of higher plants
   g. commercial source of agar

## LAB 14.E
# KINGDOM PROTISTA: SLIME MOLDS AND WATER MOLDS—FUNGUS-LIKE PROTISTS

### MATERIALS REQUIRED

*Physarum* plasmodium culture on agar plate

*Dictyostelium* pseudoplasmodium culture on agar plate, demonstration

live *Saprolegnia* in water, demonstration

### PROCEDURE

*Phylum Myxomycota* — A slime mold could be one of the most unusual organisms you will ever see. A slime mold exists in two distinct life stages — vegetative (or feeding) and reproductive. The vegetative form is called a plasmodium (as distinguished from the genus *Plasmodium* with a capital "P"), a mass of naked cytoplasm containing many nuclei that are not separated by cell walls. This "blob" slowly creeps over the forest floor in moist woods engulfing bacteria and organic debris as food. When the environment becomes dry or food becomes scarce, the plasmodium transforms into numerous stalked sporangia that produce spores resistant to adverse conditions. Each spore releases one amoeba-like or flagellated reproductive cell when favorable conditions return. Two of the reproductive cells fuse and then give rise to a new plasmodium as mitosis occurs, but with no division of the cytoplasm by a cell wall. Hence, the multinucleated plasmodium (Fig. 14–7).

1. Observe the plasmodium of *Physarum* growing on an agar plate. Note the bright yellow color and thinness of the plasmodium with thicker "veins" over its surface. The plasmodium lives by absorbing nutrients and water that have been mixed into the agar.

2. Place the plasmodium agar plate onto a stereomicroscope having a transparent glass stage. Position the light source so that it illuminates from below the stage. Look into the stereomicroscope and adjust the mirror within the stage to achieve a dark background. Increase the magnification and carefully observe the yellow "veins" of the plasmodium. You should see cytoplasmic streaming within the veins that appears as fine sand trickling through the veins. Extensive cytoplasmic streaming results in the plasmodium spreading out and moving to another area. Plasmodia (plural) can

**FIGURE 14-7**

Plasmodium of *Physarum* creeping over a stone on the forest floor.

grow to several feet across under ideal conditions. These giants have been known to startle people who find the alien-like organisms oozing over the ground.

*Phylum Acrasiomycota* — Species of this phylum, called cellular slime molds, are distinctly different from the slime molds discussed above that produce a plasmodium. The vegetative (feeding) stage of this group consists of individual, amoeba-like cells that glide about on moist soil and decaying wood where they feed on other microorganisms. Under conditions of inadequate food or moisture, the individual amoeboid cells gather into a mass of cells called a pseudoplasmodium. The pseudoplasmodium ("pseudo" means false) differs from the plasmodium of phylum Myxomycota because the aggregated cells remain as individuals with separate cell membranes instead of being a continuous mass of numerous nuclei and cytoplasm enclosed within one large membrane as in a "true" plasmodium. The pseudoplasmodium crawls about as one organism for a time and then forms a sporangium that releases spores. In favorable conditions, a vegetative amoeboid cell emerges from each spore and the cycle begins again (Fig. 14-8).

3. Observe the petri plate culture of *Dictyostelium*, a representative cellular slime mold, on display under a stereomicroscope. The culture is a vegetative pseudoplasmodium consisting of numerous cells that respond as one organism.

*Phylum Oomycota* — Many members of this group, as the name "water molds" implies, are common in bodies of fresh water. You may have seen water mold in a pond or aquarium as a white, downy growth on a dead fish or frog. Although the water molds do have more of a mold appearance than do the slime molds, they are not true fungi due to cellular differences such as the composition of the cell wall. A few species of this phylum are terrestrial and serious pests of some crop plants. In the late 1870s, a species of this group nearly wiped out the French wine industry by damaging grape vines. The severe famine in Ireland of the mid-1800s was caused by a water mold that devastated the country's potato crop, an important food source for millions of people.

4. Observe a display culture of *Saprolegnia*, a common water mold that is growing on a sterilized hemp seed, which is its food source. The down-like strands that form the "body" of *Saprolegnia* are called hyphae (singular hypha). The water molds were once considered as true fungi that also are composed of hyphae. Strictly speaking, "hypha" is a fungal term, but it has become closely associated with the structure of fungus-like organisms such as water molds.

## POSTLAB QUESTIONS

1. Explain the difference between a plasmodium and a pseudoplasmodium.

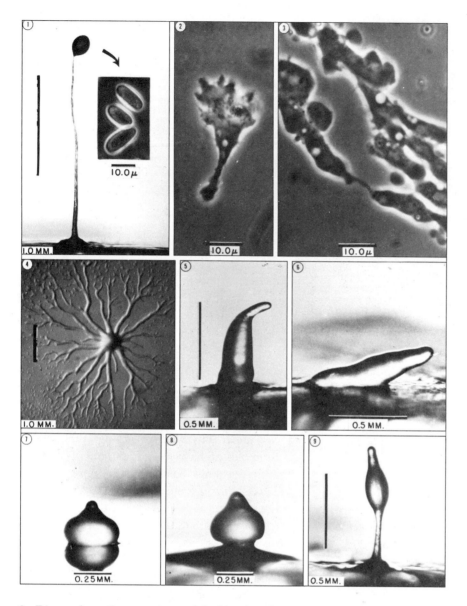

**FIGURE 14–8**

Life cycle of *Dictyostelium discoideum*, a cellular slime mold. (a) Fruiting body releases spores. (b) Independent amoeboid cells emerge from the spores. (c) The cells group together when food or moisture is scarce. (d) A mass of gathering cells. (e) The mass of cells collects into a "slug" that behaves as one organism. (f) The slug moves about for a time. (g, h, i) The slug settles and elongates into a fruiting body that will release spores when mature.

2. Discuss how the organisms of the kingdom Protista are more dissimilar than they are similar. Argue for maintaining an organism in the kingdom Protista versus the kingdom Fungi, Plantae, or Animalia (you choose the organism).

## FOR FURTHER READING

Atlas, R. 1988. *Microbiology.* New York: Macmillan.

Brown, A. D. 1990. *Microbial Water Stress Physiology: Principles and Perspectives.* New York: Wiley.

Cooper, Stephen. 1991. *Bacterial Growth and Division: Biochemistry and Regulation of Prokaryotic and Eukaryotic Division Cycles.* San Diego, Calif.: Academic Press.

Creager, Joan G. et al. 1990. *Microbiology: Principles and Applications.* Englewood Cliffs, N.J.: Prentice Hall.

Dilworth, M. J., and A. R. Glenn. 1991. *Biology and Biochemistry of Nitrogen Fixation.* New York: Elsevier.

Fettner, Ann G. 1990. *Viruses: Agents of Change.* New York: McGraw-Hill.

Frazier, W., and D. Westoff. 1988. *Food Microbiology.* New York: McGraw-Hill.

Hill, L. R., and B. E. Kirsop, eds. 1991. *Bacteria.* New York: Cambridge University Press.

Jones, Dorothy et al., eds. 1990. *Staphylococci.* Oxford, England: Blackwell.

Kingsbury, David W., ed. 1991. *The Paramyxoviruses.* New York: Plenum Press.

Kurstak, Edouard, ed. 1991. *Viruses of Invertebrates.* New York: M. Dekker.

Margulis, L., and K. Schwartz. 1988. *Five Kingdoms: An Il-*

*lustrated Guide to the Phyla*. San Francisco, Calif.: W. H. Freeman.

McFeters, Gordon A., ed. 1990. *Drinking Water Microbiology*. New York: Springer-Verlag.

McKane, L., and J. Kandel. 1985. *Microbiology, Essential and Applications*. New York: McGraw-Hill.

Mozes, Nava et al., eds. 1991. *Microbial Cell Surface Analysis: Structural and Physiochemical Methods*. New York: VCH Publishers.

Poole, Robert K. et al., eds. 1990. *Microbial Growth Dynamics*. New York: IRL Press.

Silver, Simon et al., eds. 1990. *Pseudomonas: Biotransformations, Pathogenesis, and Evolving Biotechnology*. Washington, D.C.: American Society for Microbiology.

Streips, Uldis, and Ronald E. Yasbin, eds. 1991. *Modern Microbial Genetics*. New York: Wiley-Liss.

Tortora, G. J., et al. 1989. *Microbiology*. Menlo Park, Calif.: Benjamin-Cummings Publishing Co.

Tuite, Michael F., and Stephen G. Oliver, eds. 1991. *Saccharomyces*. New York: Plenum Press.

Van der Ploeg, Lex H. T. et al., eds. 1990. *Immune Recognition and Evasion: Molecular Aspects of Host-Parasite Interaction*. San Diego, Calif.: Academic Press.

Vyverman, Wim. 1991. *Diatoms from Papua New Guinea*. Berlin, Germany: J. Cramer.

Wistreich, G., and M. Lechtman. 1984. *Microbiology*. New York: Macmillan Publishing Company.

◻

# Fungi

## OBJECTIVES

1. List the distinguishing characteristics of the kingdom Fungi.

2. Briefly characterize the four major divisions of fungi: Zygomycota, Ascomycota, Basidiomycota, and Deuteromycota.

3. Describe the life cycle of *Rhizopus* emphasizing differences between sexual and asexual stages.

4. Identify the two types of organisms that comprise a lichen.

5. Summarize the impact of fungi upon humans.

---

## INTRODUCTION

The kingdom Fungi contains eukaryotic, mostly multicellular organisms apart from yeast, which are single-celled fungi. All fungi are heterotrophic, being saprobes, parasites, or a partner in mutualistic symbiosis. Fungi are usually terrestrial, and all feed by absorption. They reproduce by **spores** — tiny, usually haploid cells derived from sexual or asexual processes. A spore germinates into a thread-like **hypha** after landing on a food source. The hypha branches into a tangled **mycelium** as it grows. The mycelium secretes digestive enzymes into its food and absorbs nutrients. A mycelium may give rise to **sporangia,** reproductive structures that bear spores. The more complex fungi produce **fruiting bodies** (such as mushrooms) that contain many sporangia.

The hyphae of some fungi contain many nuclei within the same cytoplasm (coenocytic) while hyphae of other fungi are divided into compartments containing one or more nuclei by **septa** (cross walls). Fungal cells have cell walls of cellulose or chitin.

Fungi are placed in their own kingdom because they have qualities that make them neither plant nor animal. The fungi resemble plants superficially. However, they are not autotrophic and the presence of chitin in the cell wall is a major difference. The cell wall certainly indicates that they are not animals. And they lack movement, which is generally associated with animals.

Asexual reproduction is common in fungi. As an example, black bread mold *(Rhizopus)* sends up aerial hyphae from the vegetative hyphae feeding on the food source. Sporangia develop on the tops of the aerial hyphae which produces asexual spores. These are then blown to a new habitat where they germinate.

Some fungal species grow in mutualistic symbiosis such as **mycorrhizae,** fungi associated with plant roots in a symbiotic relationship. The hyphae grow into the soil and absorb minerals that are used by both the fungus and the plant. In return, the fungus receives food from the plant. Some plants are so dependent upon their fungal partner that they cannot grow well without it.

Lichens are another symbiosis between a fungus and a green alga or cyanobacterium. The fungus receives organic compounds from its photosynthetic partner. The autotroph partner possibly receives water and minerals from the fungus. Lichens are important because they are one of the few autotrophs that can grow in harsh environments.

The effect of fungi upon humans is both positive and negative. The pathogenic characteristics of certain fungi are certainly the major, negative aspect of the fungi. Perhaps the most important benefit from the fungi comes from their ecological role as decomposers. Fungi are also an important resource for the production of antibiotics and other medicines. And, many humans would find certain foods considerably less interesting without a widespread fungi—mushrooms.

We have devised the following laboratory exercises to demonstrate some of the more important aspects of the fungi.

## PRELAB QUESTIONS

1. What is the method of nutrition for fungi?

2. Give two reasons why fungi are not placed in the plant kingdom.

3. What two groups of organisms comprise a lichen?

4. Identify one benefit of fungi to humans.

## LAB 15.A
## DIVISION ZYGOMYCOTA: ZYGOSPORE FUNGI

### MATERIALS REQUIRED

*Rhizopus* culture plate with zygospores, on demonstration stereomicroscope

*Rhizopus,* prepared slides

compound microscope

### PROCEDURE

Before dealing specifically with this group of fungi, you should first become familiar with a few basic structures that compose the "body" of a fungus.

1. Begin by observing the petri plate of agar on display containing *Rhizopus,* a black bread mold that

is common when bread does not contain preservatives. Typically, a fungus is composed of fine, thread-like strands called hyphae (hypha, singular). Examine the petri plate under the stereomicroscope and note the numerous hyphae distributed over the surface of the agar. The cells that compose the hyphae of some species (not visible at low magnification) are defined by partitions, or septa. The septa usually have openings that allow a mixing of cytoplasm between neighboring cells. The fungal cells may have one or two nuclei per cell, depending on the stage of the life cycle and species of fungus. Some kinds of hyphae, such as *Rhizopus,* may lack septa entirely (coenocytic hyphae), being essentially fine tubes of free-flowing cytoplasm and nuclei (see Fig. 15–1).

(a) Coenocytic hypha

(b) Hypha with septa; one nucleus per cell

(c) Hypha with septa; two nuclei per cell

**FIGURE 15–1**

Types of fungal hyphae.

2. Note the collective mat of growth formed by the numerous hyphae growing over the agar. This grouping of hyphae is called a mycelium (mycelia, plural). The down-like spot of mold you have seen on spoiled food is a good example of a mycelium. Actually, the mycelium is much more extensive than what is visible on the food surface. The hyphae grow throughout the interior of the food and secrete digestive enzymes that break down the food into smaller molecules that can be absorbed by the hyphae.

The name Zygomycota of this fungi division, of which *Rhizopus* is a member, is derived from zygospore, a reproductive structure produced by all members of this group.

3. Note the dark line of zygospores down the center of the *Rhizopus* petri plate. The zygospores are produced via sexual reproduction when the hyphae of two different mating strains of *Rhizopus* grow together. The two strains cannot correctly be referred to as male and female, but they are different genetically and usually designated as "plus" or "minus"

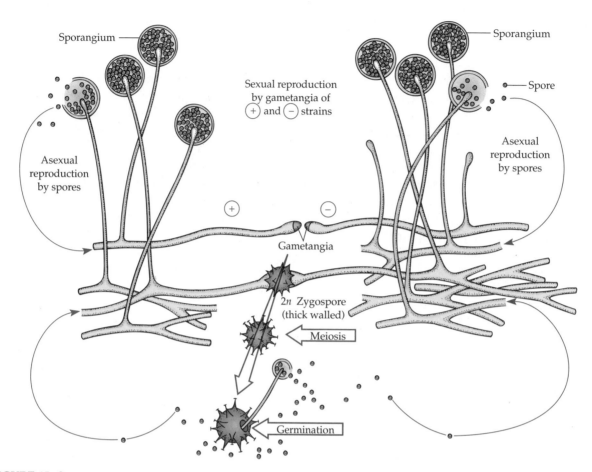

**FIGURE 15–2**

Life cycle of *Rhizopus,* a zygomycete.

strains. In this culture, one side of the plate grows the plus strain, and the other side grows the minus strain. The zygospores form at the center line where the hyphae contact one another. A zygospore is a diploid cell covered by a thick, sculptured cell wall. A zygospore forms by the fusion of two haploid nuclei (a zygote!) when the plus and minus hyphae grow together and merge their cytoplasm.

The diploid cell (zygote) within the protective cell wall undergoes meiosis, then germinates a hypha that develops a sporangium (spore-producing body) that releases many haploid spores into the air. The spores will germinate into hyphae that contain haploid nuclei if they fall onto a food source. Sexual reproduction may then occur again, or the haploid hyphae may asexually produce sporangia (singular of sporangium) that release numerous haploid spores. Both the sexual and asexual life cycles of *Rhizopus* are summarized in Figure 15–2.

4. Examine a prepared slide of *Rhizopus* under your compound microscope. You should be able to identify the structures described above: hyphae, zygospores, and sporangia. You may be able to find a young developing zygospore where two hyphae of different strains have fused.

## POSTLAB QUESTIONS

1. Do the hyphae of zygomycetes contain cross walls (septa)?

2. How would you describe the hyphae of zygomycetes?

3. Is a zygospore the result of sexual or asexual reproduction?

4. Zygomycetes are sometimes called the "conjugation fungi." Why would this be a fitting term for this group?

## LAB 15.B
## DIVISION ASCOMYCOTA: SAC FUNGI

### MATERIALS REQUIRED

yeast suspension

methylene blue

slides and coverslips

*Peziza,* prepared slide

preserved specimens of morels, *Xylaria,* and *Peziza*

### PROCEDURE

This fungal division contains several species that you are probably familiar with because they are an important food or because you have seen them growing along a woodland trail. The ascomycetes are known as the "sac fungi" because of the ascus (asci, plural), a sac-like structure containing spores, produced by sexual reproduction in this group. Again, sexual reproduction in this group involves the fusion of haploid nuclei when hyphae of two different mating types grow together. The spores produced within an ascus are appropriately called ascospores. The asci form within a "fruiting body," an ascocarp, that you would recognize as a fungus growing from the soil or a dead tree branch. The ascocarp consists of tightly compacted hyphae. Many more loosely arranged hyphae compose the extensive mycelium that runs throughout the soil or branch. Some ascomycetes, the yeasts, are unicellular, but still form an ascus with ascospores during sexual reproduction.

1. Obtain a clean slide and place two drops of yeast suspension on it. Add a drop of methylene blue dye to the yeast and place a coverslip over the mixture. Examine the slide under low and high power. Note the numerous individual cells of this fungus. Draw several yeast cells.

The yeast fungi are essential to the alcoholic beverage and baking industries. Yeast produces ethyl alcohol by fermenting fruit juices or grains for the production of wine, beer, and distilled liquors. The carbon dioxide gas produced by yeast after it is mixed into bread and pastry dough makes it rise. This gives baked goods a lighter, fluffier texture.

2. Yeasts commonly reproduce asexually by budding. A small bud grows from a yeast cell, increases in size, and eventually falls away from the parent cell, producing a new, identical individual (see Fig. 15–3). Search through your slide and look for budding cells.

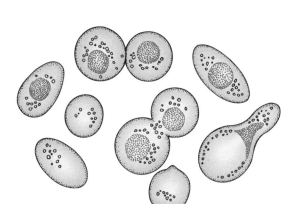

**FIGURE 15–3**

Budding yeast cells.

The multicellular forms of the ascomycetes are far more common than the unicellular yeasts. These fungi become noticeable when the mycelium produces an ascocarp that rises out of the food source in which the hyphae grow.

3. Obtain a prepared slide of *Peziza,* the cup fungus. The slide is made from a thin section of the ascocarp. First use low power to observe the lower area of the ascocarp. Note that it is composed of compact hyphae that were sectioned when the slide was made.

4. Still using low power examine the upper layer of the ascocarp. The vertical, elongated structures lying parallel to one another are the asci, which contain obvious ascospores (Fig. 15–4). Switch to high power and observe the spores within an ascus. Note that no ascus contains more than eight spores. This occurs when one diploid cell within the ascus, produced by the sexual fusion of two

haploid nuclei, divides by meiosis, which produces four cells. Each of these new haploid cells divides once by mitosis resulting in the eight haploid ascospores.

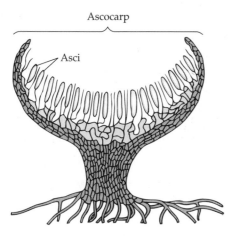

**FIGURE 15–4**

Cross section of an ascocarp that reveals the asci.

5. Observe whole preserved specimens of *Peziza* and you will see why it is called "cup fungus." This fungus is fairly common in wooded areas during the cooler seasons. One species of *Peziza* is bright scarlet in color. Observe *Xylaria*, another woodland ascomycete, and morels *(Morchella),* or sponge mushroom. Edible morels appear in wooded areas during the spring when they are eagerly gathered for their delicious flavor. Draw a representative ascomycete.

**POSTLAB QUESTIONS**

1. Describe the hyphae of *Peziza*.

2. Are ascospores the result of asexual or sexual reproduction?

## LAB 15.C
## DIVISION BASIDIOMYCOTA: CLUB FUNGI

**MATERIALS REQUIRED**

fresh supermarket mushrooms

*Coprinus,* prepared slide

preserved specimens of the following:

a. coral fungus

b. earthstars

c. bird's-nest fungus

d. corn smut

e. bracket fungi

**PROCEDURE**

Everyone is familiar with at least one representative of this fungi division—the edible supermarket mushroom. The commercial mushroom is representative of many basidiomycete species. As you have seen previously, the mushroom is the spore-producing body and is composed of compact hyphae that form a stalk and mushroom cap. The underside of the cap holds numerous delicate gills that are covered with microscopic basidia that produce spores. Each basidium is club shaped, hence the name "club fungi," and the fruiting body (mushroom) is called a basidiocarp. As in other fungi, the basidiocarp grows from a mycelium as a result of sexual reproduction involving the fusion of haploid nuclei in hyphae of different mating types.

1. Examine a fresh mushroom purchased from a grocery. Observe the stalk that supports the cap above ground as the mushroom grows. Gills are located on the underside of the cap. The gills may be partially hidden in younger "button" mushrooms because the cap has not fully expanded. The gills are externally lined with numerous microscopic basidia that produce spores (Fig. 15–5).

2. Obtain a prepared slide of a *Coprinus* mushroom cap cross section. First observe the section under low power. Notice the compacted hyphae that compose the cap and the stalk, which is the circular

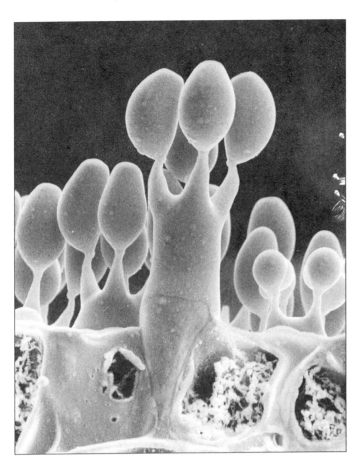

**FIGURE 15-5**

Scanning electron micrograph of basidia, each bearing four basidiospores.

area in the center of the cap. The gills radiate outward from the stalk to the edge of the cap. Switch to high power and carefully observe the outer edge of a gill. Look for small club shaped basidia that extend from the gill wall. Each basidium bears four darkly stained basidiospores, the specific kind of spore produced by basidia in this group of fungi.

3. Examine other preserved specimens of basidiomycete fruiting bodies. These demonstrate the variety of forms within this group of fungi. Many of these fungi are named for objects and other organisms they resemble in appearance. Coral fungus (*Clavaria*), earthstars (*Geaster*), and bird's-nest fungus (*Cyathus*) are all appropriately named. Corn smut (*Ustilago*) is an agricultural pest that destroys growing corn ears. Bracket fungi are commonly seen on both live and dead trees in wooded areas. Although all of these basidiomycetes do not have gills, they still bear basidia along their outer surface or within internal areas, which makes them members of this group of fungi. Draw a representative basidiomycete.

**POSTLAB QUESTIONS**

1. Describe the hyphae of basidiomycetes.

2. Are basidiospores the result of asexual or sexual reproduction?

3. In general terms, compare asci and basidia with respect to the roles of meiosis and mitosis in their development.

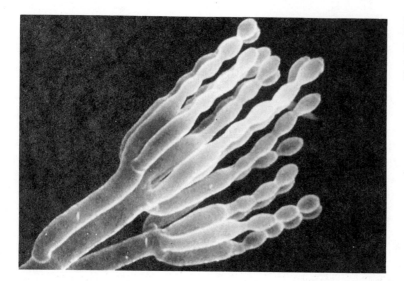

**FIGURE 15-6**

Scanning electron micrograph of *Penicillium* conidiophores bearing conidia (spores).

## LAB 15.D
## DIVISION DEUTEROMYCOTA: IMPERFECT FUNGI

### MATERIALS REQUIRED

*Penicillium* agar plate culture

*Penicillium*, prepared slide

### PROCEDURE

The deuteromycetes contain all fungal species in which no sexual reproduction has been observed. Perhaps some species have not been thoroughly studied in order to discover sexual reproduction or perhaps members of this fungal group produce only by asexual reproduction. Consequently, the deuteromycete division contains varied species that may not be truly related.

Many deuteromycetes are of particular interest because they cause infections such as ringworm, thrush, and athlete's foot. Other species are serious agricultural pests and pose a threat to several crop plants. *Penicillium*, a deuteromycete, produces the antibiotic penicillin. Some *Penicillium* species intentionally cultured within blue cheese are responsible for producing the distinctive flavor and bluish color specks of the cheese. The classification of *Penicillium* is confusing because some species that are known to reproduce sexually are placed in the Division Ascomycota, while other species of this same genus not known to have a sexual phase are classified in Division Deuteromycota.

1. Observe the agar plate culture of *Penicillium* on demonstration under the stereomicroscope. The bluish color of the mycelium is due to the pigmented spores of this fungus.
2. Examine the prepared slide of *Penicillium* under low and high power. Notice that many of the hyphae end in structures that resemble a paint brush. The "handle" of the brush is the conidiophore, which supports the "bristles." These bristles are short chains of asexually produced spores called conidia (conidium, singular) that develop at the end of the conidiophore (Fig. 15–6). The conidia break off and germinate hyphae that begin a new mycelium.

### POSTLAB QUESTIONS

1. Why are these fungi referred to as "imperfect" fungi?

2. *Penicillium roqueforti* (the blue mold that flavors roquefort and blue cheese) is an ascospore-producing fungus. In what division would you place *P. roqueforti*?

## LAB 15.E
## LICHENS: COMPOSITE ORGANISMS

### MATERIALS REQUIRED

lichen, prepared slide

various types of preserved lichens

## PROCEDURE

Lichens are hardy plant-like organisms that grow on rocks, on tree trunks, and in bare soil where true plants would have difficulty surviving. Although plant like in appearance, lichens are actually a combination of a fungus and an alga or a cyanobacterium that grow together as one organism. The fungus is frequently an ascomycete, although some are basidiomycetes. The hyphae of the fungus grow around the photosynthetic algal or cyanobacterium cells and absorb food from them (Fig. 15–7). The fungal hyphae may provide some protection from water loss for the photosynthetic cells. Although the fungus and photosynthetic partner of a lichen are themselves distinct species that can grow independently, the lichen they compose is also given a species name of its own.

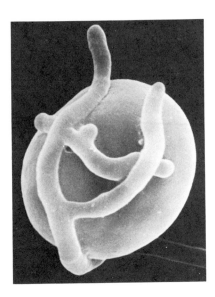

**FIGURE 15-7**

Scanning electron micrograph of fungal hyphae in a lichen associated with an algal cell.

1. Examine the prepared slide of a lichen section under low and high power. Look for the photosynthetic alga or cyanobacterium as a loose layer of cells among the fungal hyphae in the upper portion of the lichen interior. Note the close association between the photosynthetic cells and hyphae.
2. Observe the various kinds of lichens on display. Lichens occur in three types of growth forms. Crustose lichens appear just as the name implies —a crust clinging tightly to a tree or rock. Foliose lichens have a loose, leafy appearance that resembles plant foliage. Fruticose lichens have a shrubby or hair-like form. See if you can identify the growth forms of the lichens on display.

## POSTLAB QUESTIONS

1. If both organisms are presumed to benefit, what type of symbiotic relationship exists in lichens between the fungus and the alga (or the cyanobacterium)?

2. How would your answer to Question 1 be affected if you consider that the algae of some lichens can live alone, whereas the fungi of those lichens cannot survive alone?

3. Why may the term "controlled parasitism" be a better description of the relationship between the fungus and alga (or cyanobacterium)?

4. Identify a–f as one of the following groups of fungi: Zygomycota, Basidiomycota, Ascomycota, or Deuteromycota.
   a. reproduce sexually by asci
   b. eukaryotic cells
   c. zygospores
   d. reproduce asexually by conidia
   e. hyphae lack cross walls
   f. four basidiospores per basidium

## FOR FURTHER READING

Ahmadjian, V., and S. Paracer. 1986. *Symbiosis: An Introduction to Biological Associations.* Hanover, N.H.: University Press of New England.

Atlas, R. 1988. *Microbiology.* New York: Macmillan.

Bessette, A., and W. J. Sundberg. 1987. *Mushrooms: A Quick Reference Guide to Mushrooms of North America.* New York: Macmillan Publishing Co.

Hale, M. E. 1983. *The Biology of Lichens.* Baltimore, Md.: University Park Press.

Hall, G. S., and D. L. Hawksworth, eds. 1990. *International Mycological Directory.* Wallingford, England: C. A. B. International.

Leong, Sally A., and Randy M. Berka, eds. 1991. *Molecular Industrial Mycology: Systems and Applications for Filamentous Fungi.* New York: Dekker.

McKane, L., and J. Kandel. 1985. *Microbiology: Essentials and Applications.* New York: McGraw-Hill.

Panchal, Chandra J. 1990. *Yeast Strain Selection.* New York: Dekker.

Phillips, Roger et al. 1991. *Mushrooms of North America.* Boston, Mass.: Little, Brown, and Co.

Ross, I. K. 1979. *Biology of the Fungi: Their Development, Regulation, and Associations.* New York: McGraw-Hill.

Tortora, G. J. et al. 1989. *Microbiology.* Menlo Park, Calif.: Benjamin-Cummings Publishing Co.

Wistreich, G., and M. Lechtman. 1984. *Microbiology.* New York: Macmillan Publishing Co.

❏

# Plant Life

1. List the distinguishing characteristics of the kingdom Plantae.

2. Summarize the features of bryophytes.

3. List the characteristics of the ferns and fern allies.

4. Diagram a generalized plant life cycle, clearly showing alternation of generations.

5. Distinguish between gymnosperms and angiosperms.

6. Contrast dicots with monocots.

---

## INTRODUCTION

The plant kingdom (Plantae) contains mainly multicellular, photosynthetic eukaryotes. All plant cells have cell walls made of cellulose and contain plastids that hold photosynthetic pigments.

Most plants have a life cycle that alternates between a multicellular haploid stage and a multicellular diploid stage. This is called an **alternation of generations.** Meiosis in plants produces haploid cells called **spores,** not sex cells **(gametes)** as in animals. Spores are produced in structures called **sporangia** (sporangium, singular). Spores grow into haploid **gametophytes,** which are plants that produce haploid gametes by mitosis. The gametes (egg and sperm) fuse. This forms a diploid zygote that grows into a multicellular diploid **sporophyte,** a plant that produces spores by meiosis, and the cycle continues (Fig. 16–1).

The sporophyte and gametophyte stages of land plants differ in their appearance. In the bryophytes (mosses and their relatives), the gametophyte is the more conspicuous, green "leafy" form. The sporophyte is greatly reduced and grows on top of the gametophyte. The opposite situation is true of higher plants in which the sporophyte is the most conspicuous and the gametophyte is reduced. The potted fern is a sporophyte, whereas the fern gametophyte is just a few millimeters in diameter. The sporophyte of gymnosperms and angiosperms is what we see in the wild. The gametophyte stage of both groups is tiny, occurring as dust-like male **pollen** grains and microscopic female gametophytes in the seed cones of gymnosperms or in the flower parts of angiosperms.

The **vascular tissue** of higher plants is comprised of a system of vessels that transports materials to all parts of the plant. Plants that have such tissue are called vascular plants. Through the course of vascular plant evolution, there has been a continued selection for the sporophyte as the dominant phase in the life cycle. This is because there are advantages within the sporophyte such as the freedom of the sporophyte from dependence on water for gamete transport. The vascular tissue of the sporophyte is an efficient means of distributing water and food throughout the plant. This promotes plant growth. In the higher plants, the **xylem** occurs as woody tissue. This strengthens the stem and allows an increase in height, which, in turn, aids in capturing sunlight. Woody tissue also annually increases in diameter, which subsequently increases water delivery to the leaves where it is used in photosynthesis.

Aquatic plants face different growing conditions than land plants. Aquatic plants have an abundance

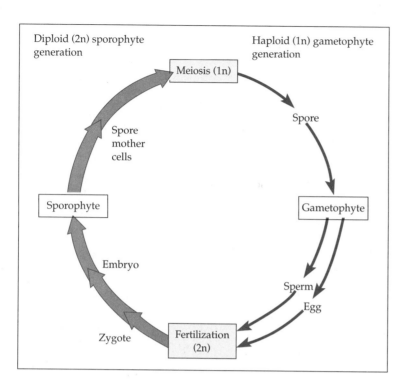

**FIGURE 16-1**

A generalized plant life cycle depicting how the plant alternates between a multicellular haploid stage and a multicellular diploid stage.

of water whereas water is not as readily available to land plants. Dissolved minerals are very accessible to aquatic plants, but land plants live in an environment where minerals are generally available only by uptake from the soil. A water environment affords buoyancy support. However, air provides no support to land plants. Aquatic plants suffer less temperature fluctuations because temperature changes occur slowly in water. This is not the case for land plants as air temperatures change quickly. Water serves as a medium for swimming gametes, but it is seldom available for swimming gametes of land plants. Water also carries offspring to new areas, but it is not likely to disperse the offspring of terrestrial plants.

The evolution of land plants involved a number of adaptations that were irrelevant to aquatic plants. Land plants, for example, were no longer surrounded by an aqueous solution of minerals. **Rhizoids,** and later roots, were adapted as a means of absorbing minerals from the soil. A further adaptation was the evolution of the vascular tissues, a system of transporting water within the plant and one which, in addition, gave support to plants no longer buoyed up in an aqueous environment. Excessive loss of water by evaporation was avoided by the evolution of the **cuticle,** a layer of wax over the plant surface that reduces evaporation. **Stomata** (air pores) developed as a means of gas exchange through the impermeable cuticle. Pollen evolved as a means of bringing gametes together in an environment where swimming gametes were of no adaptive value. Airborne spores, then later

seeds, developed as an effective way of offspring dispersal without the use of water.

As new groups of land plants appeared through evolution, each produced characteristics that made them more adapted to terrestrial life. The bryophytes (mosses and liverworts) are some of the lower land plants. They have a flagellated sperm that can reach the egg only in a watery environment. Bryophytes lack true vascular tissue. Their small size allows adequate transportation of nutrients and water via diffusion. Due to these reasons, bryophytes grow in moist areas where water is readily available. Although water is important, they do have some adaptations that enable them to survive on land. Their root-like rhizoids anchor the plants and have some water and mineral absorption capabilities. The leaf-like structures are very thin, which allows water to diffuse throughout the plant. Moss gametophytes grow close together in a configuration not unlike that of a sponge, a configuration which holds water.

The club mosses and horsetails are some of the first vascular plants with true xylem and phloem. These tissues appear in the sporophyte, which is the dominant stage in these plants and in all higher groups. The advantages of the dominant sporophyte were described above. The club mosses and horsetails are larger and more complex than the bryophytes with a **rhizome** (underground stem) that produces aerial stems.

The ferns continued the rhizome-aerial stem design, but they were the first to have large, complex

Monocots

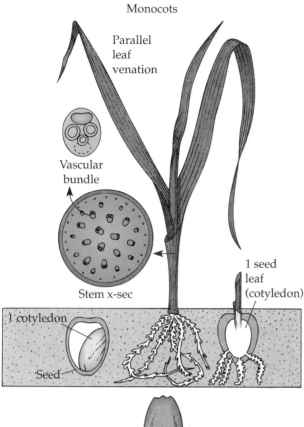

Parallel
leaf
venation

Vascular
bundle

Stem x-sec

1 cotyledon

Seed

1 seed
leaf
(cotyledon)

Dicots

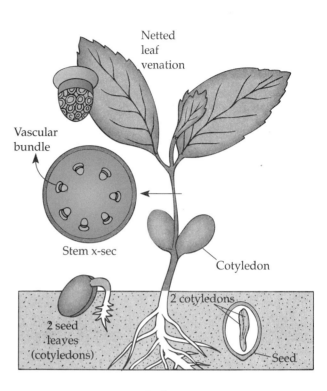

Netted
leaf
venation

Vascular
bundle

Stem x-sec

Cotyledon

2 cotyledons

2 seed
leaves
(cotyledons)

Seed

Flowers: floral parts in multiples of 3's
Seeds: contain one cotyledon
Stems: vascular bundles are scattered
Leaves: long tapering blades with parallel venation.
The base of the leaf, the sheath, encircles the stem.

Flowers: floral parts in multiples of 4's or 5's
Seeds: contains two cotyledons
Stems: organized vascular bundle arrangement
Leaves: broad to narrow leaves with petioles
and netted venation.

**FIGURE 16–2**

A generalized comparison of monocots and dicots.

leaves for efficient photosynthesis. The sporophyte is even larger in the ferns. Still, as in all groups below the ferns, the flagellated sperm requires water for swimming to the egg. This restricts these plants to somewhat moist areas.

The gymnosperms (pines, spruces, and so on) evolved several traits that made them highly adapted to dry environments. The gymnosperms were the first to evolve woody xylem tissue for a strong stem and an efficient vascular system. Air-borne pollen replaced swimming sperm, an adaptation that led gymnosperms away from watery habitats. **Pollination** occurs when wind carries pollen to a female seed cone. A pollen tube grows from the pollen grain into the female gametophyte where the sperm nucleus enters and fuses with the egg nucleus for **fertilization.** The fertilized egg **(zygote)** develops into a multicellular **embryo.** The embryo develops within a **seed,** a great advancement over spores. Seeds contain a food supply for nourishing the embryo until it can start photosynthesizing. The outer part of the seed is a seed coat for protecting its contents. Seeds are widely dispersed by wind or animals without relying on water.

The angiosperms (flowering plants) carried the advanced reproduction by seeds a step further by developing flowers. Flowers are modified sporophyte leaves that contain the highly reduced male and/or female gametophytes. Many plants have specialized flowers that attract animal pollinators. Angiosperms are the only group that evolved double fertilization, which produces the seed's food supply. The fruit is also exclusive to the angiosperms. It protects the seeds and aids in dispersing them. In addition to advanced reproduction, angiosperms also have more efficient vascular tissue and a faster growth rate than all plant groups before them.

There are two major groups of angiosperms based on differences in their morphology or form — monocots and dicots (Fig. 16–2). The term "monocot" is derived from a characteristic of seed structure in this group of plants. A monocot seed contains *one* cotyledon, which is an embryonic leaf forming part of the plant embryo. Hence the name monocot. In addition, monocots typically have flower parts occurring in groups of three or a multiple of three, veins that run parallel to one another in the leaves, and vascular tissue that is more or less equally distributed throughout the stem.

The dicots are quite the inverse of the monocots. These plants have *two* cotyledons in the seed; flower parts occurring in groups of four, five, or multiples of these numbers; veins that form a network throughout the leaves; and vascular tissue that is restricted to specific regions of the stem.

The exercises in this chapter have been selected to provide experiences distinguishing characteristics of the plants and the salient aspects of their life cycles.

## PRELAB QUESTIONS

1. What is the composition of all plant cell walls?

2. What term is used to identify the multicellular diploid stage of land plants?

3. Why are bryophytes mainly restricted to moist areas?

4. What group of land plants are characterized by flowers?

5. How many cotyledons are found in the seeds of dicots?

## LAB 16.A
## DIVISION BRYOPHYTA: MOSSES AND LIVERWORTS

### MATERIALS REQUIRED
live mosses, some with sporophytes

moss antheridia, prepared slide

moss archegonia, prepared slide

live liverworts

forceps

stereo- and compound microscopes

### PROCEDURE

Mosses grow almost anywhere there is sufficient water to meet their high moisture requirement. Even urban areas support many mosses atop old rooftops, on stone walls, and between the cracks of the sidewalk. Look for them the next time you are out for a walk.

1. Observe the mosses on display. The soft, green clump that we typically think of as a moss is the haploid gametophyte stage of the life cycle. Each moss clump consists of numerous individual gametophyte plants growing closely together. The gametophyte plant produces gametes (sperm or egg cells) by mitosis in antheridia (antheridium, singular) or in archegonia (archegonium, singular) found at the top of the gametophyte plant (Fig. 16–3).

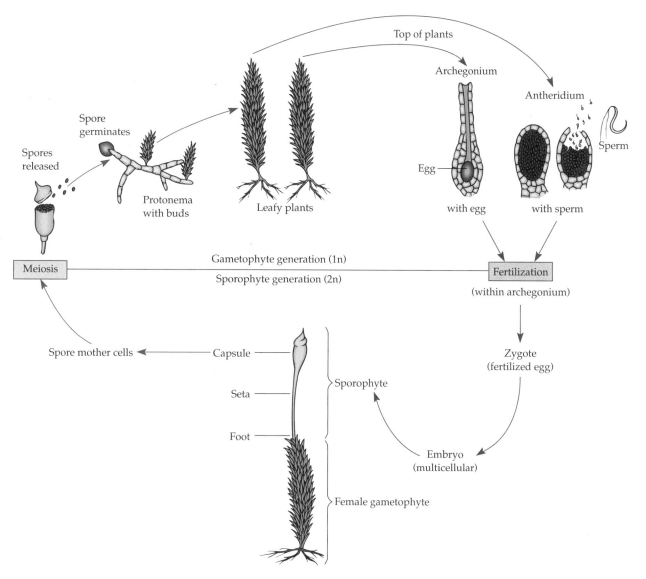

**FIGURE 16-3**

The moss life cycle.

2. Obtain a prepared slide of a moss antheridial head, which is a longitudinal section of a male moss gametophyte plant. Place the slide on the compound microscope. Then examine under both high and low power. The antheridia structures that produce sperm cells appear as stained, elongated sacs within the head of the gametophyte plant. The cells within the antheridia will give rise to moss sperm. Moss sperm cells are flagellated and actively swim when released.

3. Obtain a prepared slide of a moss archegonial head, which is a longitudinal section of a female moss gametophyte plant. Examine the slide under low power and look for elongated, bottle-shaped archegonia each of which contains a single large

egg cell within its base. Locate an egg cell and examine it under high power.

4. Examine the moss clumps for slender brown structures extending up from the green gametophyte plants. These are moss sporophytes. They are formed when a swimming sperm cell fertilizes the egg cell of an archegonium. The zygote (fertilized egg) develops into a sporophyte plant that grows from the top of the gametophyte plant. The nongreen sporophyte is dependent upon the green gametophyte plant for nourishment. Note that the sporophyte plant consists of a long, slender stalk and a capsule. The capsule contains diploid spore mother cells that divide by meiosis and give rise to numerous haploid spores. The spores are released

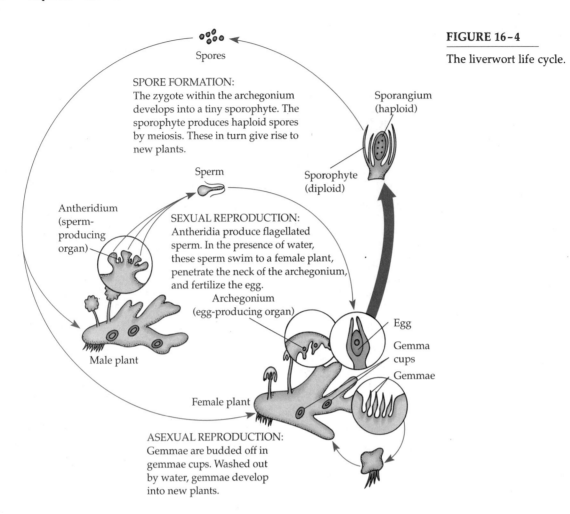

**FIGURE 16–4**

The liverwort life cycle.

Spores

SPORE FORMATION:
The zygote within the archegonium develops into a tiny sporophyte. The sporophyte produces haploid spores by meiosis. These in turn give rise to new plants.

Sporangium (haploid)

Sperm

Sporophyte (diploid)

Antheridium (sperm-producing organ)

SEXUAL REPRODUCTION:
Antheridia produce flagellated sperm. In the presence of water, these sperm swim to a female plant, penetrate the neck of the archegonium, and fertilize the egg.

Archegonium (egg-producing organ)

Egg

Gemma cups

Gemmae

Male plant

Female plant

ASEXUAL REPRODUCTION:
Gemmae are budded off in gemmae cups. Washed out by water, gemmae develop into new plants.

from the capsule and will produce young haploid gametophyte plants if they fall into a suitable environment.

5. Using forceps, gently pull away one of the **gametophyte plants** from the edge of the moss clump and examine it under your stereomicroscope. Notice what appears to be the "stem" and "leaves" of the gametophyte plant. You may see rhizoids at the bottom of the plant that resemble roots. However, these are not a true stem, leaves, and roots because they do not contain vascular tissue, a system of tubules that transport water and food throughout a plant. Vascular tissue is absent in all bryophytes. The transport of water and nutrients within mosses is accomplished by diffusion, which is adequate for a plant of small size.

Liverworts are another group of bryophytes that are also restricted to moist environments. They are not commonly seen like the mosses although they are not hard to find if you know where to look.

Typical habitats include stream banks and moist, shaded spots in woodlands. Liverworts have a life cycle very similar to that of mosses (Fig. 16–4).

6. Observe the live liverwort specimens on display. The fleshy, green plant you see here is the haploid gametophyte stage of the life cycle. Antheridia and archegonia are produced within unusual structures resembling tiny palm trees (absent in these specimens) that arise from the liverwort (Fig. 16–4). A diploid, bulb-shaped sporophyte develops from the archegonia of these structures after fertilization occurs.

7. Examine the liverwort under the stereomicroscope. Your specimen may have small cup-like structures here and there on its surface. These are called **gemmae cups** and are asexual reproductive structures. The cups contain **gemmae,** tiny clusters of cells that develop into new liverworts when they are scattered into a suitable environment by rain or some other method.

## POSTLAB QUESTIONS

1.  How does the lack of vascular tissue effect the size of bryophytes?

2.  What is the dominant stage of bryophytes? Why is that stage considered dominant?

3.  Are gemmae involved in asexual or sexual reproduction of liverworts?

4.  What type of cellular division produces spores? Gametes? Gemmae?

## LAB 16.B
## DIVISION PTEROPHYTA: FERNS

### MATERIALS REQUIRED

live and preserved fern specimens, some with sori
fern prothallus, prepared slide

### PROCEDURE

The ferns are the first group of plants encountered in our survey of the plant kingdom that possess vascular tissue. Ferns are successful plants that have survived for over 300 million years. Like the bryophytes, ferns have a high moisture requirement. Most species occur in the tropics where rainfall is abundant. Yet, ferns are also common in moist habitats of temperate forests.

1.  Observe the live and preserved specimens of ferns on demonstration. The plant commonly called a fern is the diploid sporophyte stage of the life cycle (Fig. 16–5). The striking appearance of the fern sporophyte makes it a popular houseplant. This differs from the mosses in which the gametophyte stage is conspicuous. A dominant sporophyte stage appears in the ferns and remains as the trend throughout all remaining higher plants presented in our survey.
2.  Examine the leaves (fronds) of a live potted fern. The entire structure of a frond from leaf tip to the soil is one leaf composed of many leaflets. Each frond grows from an underground stem called a rhizome that grows horizontally below the soil surface (Fig. 16–5). New leaves arise from the rhizome as a coiled shoot that uncurls as it matures. New uncurling leaves resemble the top of a violin and are called fiddleheads. Fiddleheads of some North American species can be cooked and eaten as a vegetable. Look for fiddleheads on the fern specimens.
3.  Observe the live or preserved fern fronds that possess dark spots on their underside. These structures are called sori (sorus, singular). Each sorus contains numerous sporangia that produce haploid spores by meiosis (Fig. 16–5). Sori are sometimes mistaken for insect pests or a disease on the frond by those less familiar with ferns. Sori shed numerous spores that grow into the haploid gametophyte stage of the fern life cycle. A fern gametophyte is surprisingly small as part of the life cycle of such a large plant.
4.  Obtain a prepared slide of a fern gametophyte, known as a prothallus (Fig. 16–5). Examine its structure under low power of your compound microscope. The fern prothallus, somewhat heart shaped, is mounted on the slide so that you can see the underside, which grows flat on the soil. Archegonia, which contain egg cells, are found in the notch of the "heart." Antheridia, which produce flagellated sperm cells, are located among the rhizoids that grow from the pointed end of the "heart." You may need to switch to high power to see the archegonia and antheridia. Sperm released from the antheridia swim in water to the archegonia where one of the egg cells becomes fertilized, forming a diploid zygote. A young sporophyte then develops from the zygote and eventually matures into a leafy fern.

### POSTLAB QUESTIONS

1.  What environmental requirement for reproduction is still found in ferns?

2.  What type of cellular division is required for a zygote to develop into a young sporophyte?

3.  Write a brief explanation in support of the following statement: Ferns are better adapted for terrestrial life than mosses.

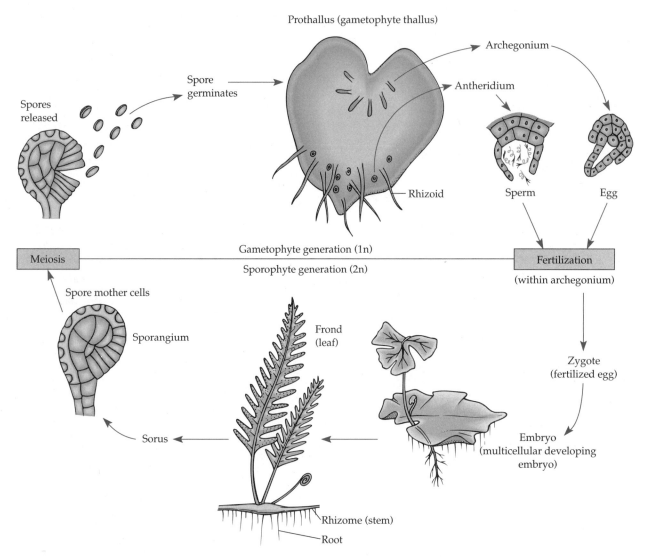

**FIGURE 16–5**

The fern life cycle.

4. What is the adaptive advantage of antheridia maturing at a different time than the archegonia within the same gametophyte?

## LAB 16.C
## FERN ALLIES: HORSETAILS AND CLUB MOSSES

### MATERIALS REQUIRED

preserved horsetail specimens, some with strobili

preserved club moss specimens, some with strobili

live club moss, *Selaginella*

### PROCEDURE

The horsetails and club mosses are vascular plants (with xylem and phloem) having a life cycle similar to that of a fern. Like the ferns, these plants also lack the characteristics of more advanced plants. For these reasons, the horsetails and club mosses have long been called the "fern allies." The fern allies living today are a relatively small group of plants that were far greater in size and species diversity millions of years ago when giant horsetails and club mosses the size of trees grew in vast forests. The accumulation and geologic transformation of these numerous plants formed the extensive coal deposits throughout the world that now fuel modern civilization.

### Division Sphenophyta: Horsetails

1. Observe the preserved horsetail specimens on display. These plants receive their name from the bushy appearance of some shoots of the plant. The shoots arise from an underground rhizome. What appear to be leaves on the bushy shoots are actually numerous slender branches that grow in whorls along the plant stem. These horsetail specimens represent the diploid sporophyte stage of the horsetail life cycle, which is similar to that of a fern.

2. Look for a horsetail shoot that appears as a bare stem. Note that it ends in a cone-like structure called a strobilus (strobili, plural). The strobilus contains numerous sporangia that produce haploid spores by meiosis. The spores germinate into tiny haploid gametophytes with antheridia and archegonia that produce gametes and again give rise to another sporophyte plant after fertilization.

### Division Lycophyta: Club Mosses

3. Observe live and preserved specimens of club mosses. *Selaginella* is a club moss that is cultivated as a houseplant. Of course, these plants are not true mosses, but their delicate appearance and smaller size suggest a similarity to the mosses. Many of the club mosses consist of an underground rhizome with short stems above ground. The leaves are small and somewhat spiny in many species, which is the origin of the common name "ground pine."

4. Look for a club moss stem that ends in a cone-like strobilus. The club shape of the strobilus contributes to the name "club moss." As in the horsetails, the strobilus contains sporangia that release numerous haploid spores that germinate into the gametophyte stage of the life cycle.

The club mosses are of special interest in one particular aspect of their reproduction. Generally speaking, the sporangia of all the plants you observed previously produce one kind of haploid spore that germinates into a gametophyte plant which produces both sperm and egg cells. These are appropriately called homosporous (one spore) plants. Some of the club mosses were the first plants to develop two kinds of spores that give rise to two kinds of gametophytes: one that produces egg cells and another that produces sperm cells. Such plants are said to be heterosporous (other spore).

Heterosporous plants have two kinds of sporangia. The microsporangia contain diploid micro- spore mother cells that divide by meiosis, which results in haploid microspores. Each microspore develops into a small microgametophyte (male) that produces sperm cells. Megasporangia contain diploid megaspore mother cells that meiotically produce megaspores. A megaspore develops into a larger megagametophyte (female) that produces egg cells. The heterosporous condition is notable because this method of reproduction occurs in the most advanced and successful groups of plants.

### POSTLAB QUESTIONS

1. _____ are to horsetails as sori are to ferns.
2. What three features of club mosses place them in a more advanced group of plants than the "true mosses"?

3. Draw a generalized life cycle of a heterosporous plant.

## LAB 16.D
## DIVISION CONIFEROPHYTA: CONIFERS

### MATERIALS REQUIRED

male pine strobilus, prepared slide (longitudinal section)

pine pollen, prepared slide

female pine strobilus, prepared slide (longitudinal section)

the following on demonstration:
a. preserved or fresh pine branches
b. mature female pine cones
c. pine seeds

### PROCEDURE

In this group of plants, you encounter an advancement not found in any of the previous plants — seeds. A seed is a self-contained package in which the plant embryo and its food source is surrounded by a protective seed coat. This nourishment and protection of the plant embryo make seeds a reproductive advantage as shown by the success of those plants that reproduce by seeds.

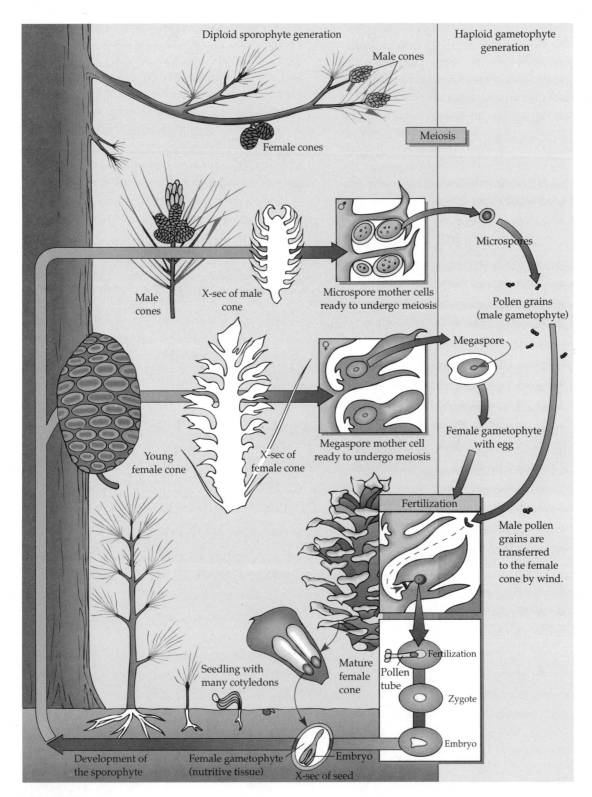

Diploid sporophyte generation

Haploid gametophyte generation

Male cones

Female cones

Meiosis

Male cones

X-sec of male cone

Microspore mother cells ready to undergo meiosis

Microspores

Pollen grains (male gametophyte)

Young female cone

X-sec of female cone

Megaspore mother cell ready to undergo meiosis

Megaspore

Female gametophyte with egg

Fertilization

Male pollen grains are transferred to the female cone by wind.

Seedling with many cotyledons

Mature female cone

Pollen tube

Fertilization

Zygote

Embryo

Development of the sporophyte

Female gametophyte (nutritive tissue)

X-sec of seed

Embryo

**FIGURE 16–6**

The pine life cycle.

There are two major groups of seed plants. You will first observe a group of seed plants known as the gymnosperms which literally means "naked seed." These plants are so named because the seeds are totally exposed on the plant or produced on the scales of cones, also called strobili (strobilus, singular). The cone-bearing plants are appropriately called conifers and they are the largest group of gymnosperm plants. The conifers include many beautiful and familiar trees most commonly thought of as evergreens—firs, pines, spruces, and so on.

1. Examine a prepared slide of a male pine strobilus (cone) under scanning or low power of your compound microscope. Note the multiple inflated compartments that compose the male cone. Each of these compartments is a microsporangium. The numerous cells within each microsporangium are microspore mother cells that will produce microspores by meiosis (Fig. 16–6). Each microspore develops into a pollen grain, which is the immature male gametophyte in seed plants such as pine. The pollen grain is considered a mature male gametophyte when it eventually produces sperm cells. The male cones of pine trees release yellow clouds of pollen grains in spring. The pollen grains are carried by the wind to other pine trees where cross-pollination occurs.

2. Examine a prepared slide of pine pollen grains. Note the two wing-like structures of the pollen grains that aid in wind dispersal. Each pollen grain consists of four cells. Two of the more important cells are the most obvious—a large tube cell and a smaller generative cell. When the wind carries a pollen grain to a female pine cone (pollination), the tube cell develops into a long pollen tube that grows out of the pollen grain toward the egg cell. The generative cell divides by mitosis forming two sperm that move down the pollen tube toward the egg. One of the sperm degenerates leaving one that fertilizes the egg cell within the female cone.

3. Examine a prepared slide of a female pine cone. Observe the oval-shaped megasporangium at the base of each scale of the pine cone. In seed plants such as the pine, the megasporangium is also called an ovule, or immature seed. Each megasporangium contains a single megaspore mother cell that produces four megaspores by meiosis (Fig. 16–6). Three of the four megaspores disintegrate leaving one that grows into the female gametophyte. This gametophyte remains within the boundaries of the megasporangium on the cone scale. Several archegonia develop within the female gametophyte tissue, each containing an egg cell. One of the egg cells will become fertilized by a sperm from a pollen grain as described above. The fertilized egg develops into an embryo. The embryo is surrounded by the female gametophyte tissue, which serves as a food source until the young plant can begin photosynthesis. The cell layers surrounding the ovule harden, transforming it into a mature seed with an embryo, a food source, and a protective seed coat.

4. Observe the preserved or fresh branches of pine and the mature, woody female cones on demonstration. These structures from the pine tree represent the sporophyte generation of the life cycle. The woody scales of the mature female cones spread apart releasing the enclosed seeds. Observe the pine seeds on display, which have a thin, wing-like structure. This structure helps disperse the seeds by wind.

## POSTLAB QUESTIONS
1. The pine tree is the ___ generation of the pine life cycle.
2. Why is gymnosperm reproduction totally adapted for life on land?

## LAB 16.E
## OTHER GYMNOSPERMS: DIVISIONS CYCADOPHYTA AND GINKGOPHYTA

### MATERIALS REQUIRED
live cycad or preserved specimens
preserved or fresh ginkgo branches

### PROCEDURE
Gymnosperm seed plants occur in some fascinating forms other than the pine-like conifer evergreens that are so familiar. In fact, cycads and the ginkgo differ so greatly in appearance from the familiar conifers that it seems unlikely they are all related. However, these plants all have "naked seeds" and similar life cycles that link them as gymnosperms.

## Cycads

Cycads are gymnosperms with a remarkable palm-like appearance. Large female strobili (cones) similar to those of conifers are produced from the growing tip of the plant. Cycads are an ancient, once widespread group of plants that flourished with the dinosaurs. Now cycads are limited to tropical regions of the world. One cycad species is native to southern Florida. Due to their attractive appearance, cycads are widely cultivated in mild climates as ornamental landscaping plants.

1. Observe preserved or live cycad specimens on display.

## Ginkgo

Like the cycads, the ginkgo tree is another gymnosperm that has a link to the distant past. Ginkgo trees were first known only as fossils until the only surviving species was discovered in China in the early 1900s. The ginkgo is known for its attractive foliage and resistance to air pollution. These qualities make ginkgo trees popular in urban areas. Ginkgo trees have separate sexes. Female trees bear naked seeds with a fleshy seed coat that gives the seeds a plum-like appearance. The seeds fall to the ground where the seed coat decays, producing a foul odor. Male trees are more commonly planted for this reason.

2. Examine preserved or fresh specimens of ginkgo on demonstration.

## POSTLAB QUESTIONS

1. Draw the distinguished leaves of a ginkgo tree.

2. Why are gymnosperms placed before angiosperms in the evolutionary time scale?

## LAB 16.F
## DIVISION MAGNOLIOPHYTA: FLOWERING PLANTS

### MATERIALS REQUIRED

various fruit and seed types

living and preserved specimens of monocot and dicot angiosperms

## PROCEDURE

The flowering plants, or angiosperms, are the other group of seed plants in addition to the gymnosperms. The angiosperms are no doubt the most familiar plants to you in all of our plant kingdom survey. The term "angiosperm" is derived from two Greek words meaning "vessel" and "seed." These plants are so named because the seeds are enclosed within a fruit (the "vessel," or container) as opposed to the naked seed of the gymnosperms.

The flower is a specialized reproductive organ found only in the angiosperms. Like the gymnosperms, the gametophyte stage of the angiosperm life cycle is reduced to microscopic size and, for the most part, remains within the flower parts. The flowering plant we recognize as a rose shrub or apple tree represents the sporophyte stage of the plant life cycle. Topic 19 presents the reproduction and development of angiosperms in greater detail. Here, we are concerned with a general introduction to the angiosperms as part of our survey of the plant kingdom.

1. Observe the various kinds of angiosperm fruit and seed types on display. Note the variation in size, shape, and other characteristics among the fruits and seeds.
2. Observe living or preserved specimens of monocot and dicot plants. Note the number and arrangement of flower parts on each plant and the kind of leaf venation (veins) present.

## POSTLAB QUESTIONS

1. Why is the flower more specialized than the cone?

2. Match the plant group or groups with the correct phrase.
   a. flowers and fruits           bryophytes
   b. homosporous method of    ferns
      reproduction           fern allies
   c. lack vascular tissue       gymnosperms
   d. sori                    angiosperms
   e. dominant gametophyte generation
   f. naked seeds

3. Match the definitions with the appropriate terms.

   a. absorbs water and minerals from the soil
   b. transports water within the plant
   c. prevents loss of water by evaporation
   d. disperses offspring without relying upon water
   e. supports plant body no longer buoyed in an aqueous environment
   f. transports food from sites of manufacture to sites of use
   g. brings gametes together without relying upon water

   xylem
   phloem
   cuticle
   pollen
   roots
   seeds

## FOR FURTHER READING

Allen, Michael F. 1991. *The Ecology of Mycorrhizae.* New York: Cambridge University Press.

Bell, P. R., and C. L. F. Woodcock. 1983. *The Diversity of Green Plants.* London, England: E. Arnold Ltd.

Bold, H. et al. 1987. *Morphology of Plants and Fungi.* New York: Harper & Row.

Briggs, D., and S. Walters. 1984. *Plant Variation and Evolution.* Menlo Park, Calif.: Benjamin-Cummings Publishing Co.

Cittadino, Eugene. 1990. *Nature as the Laboratory: Darwinian Plant Ecology in the German Empire, 1880–1900.* New York: Cambridge University Press.

Crum, Howard A. 1991. *Liverworts and Hornworts of Southern Michigan.* Ann Arbor, Mich.: University of Michigan Herbarium.

Grierson, Don, ed. 1990. *Plant Genetic Engineering.* New York: Chapman and Hall.

Hesse, M., and F. Ehrendorfer, eds. 1990. *Morphology, Development and Systematic Relevance of Pollen and Spores.* New York: Springer-Verlag.

Kramer, K. U., and P. S. Green, eds. 1990. *Pteridophytes and Gymnosperms.* New York: Springer-Verlag.

Lellinger, D. B. 1985. *A Field Manual of the Ferns and Fern-Allies of the United States and Canada.* Washington, D. C.: Smithsonian Institution Press.

Osmond, C. B. et al. 1990. *Plant Biology of the Basin and Range.* New York: Springer-Verlag.

Ray, P. M. et al. 1983. *Botany.* Philadelphia: Saunders College Publishing.

Rost, T. et al. 1984. *An Introduction to Plant Biology.* New York: John Wiley & Sons.

Scagel, R. F. et al. 1984. *Plants: An Evolutionary Survey.* Belmont, Calif.: Wadsworth Publishing Co.

Schwintzer, Christa R., and John D. Tjepkema, eds. 1990. *The Biology of Frankia and Actinorhizal Plants.* San Diego: Academic Press.

Stern, K. R. 1988. *Introductory Plant Biology.* Dubuque, Iowa: William C. Brown.

Tamm, Carol Olof. 1990. *Nitrogen in Terrestrial Ecosystems: Questions of Productivity, Vegetational Changes, and Ecosystem Stability.* New York: Springer-Verlag.

Tarchevsky, I. A., and G. N. Marchenko. 1991. *Cellulose: Biosynthesis and Structure.* New York: Springer-Verlag.

Weber, William A. 1990. *Colorado Flora: Eastern Slope.* Niwot, Colorado: University Press of Colorado.

# TOPIC 17

◻

# Animal Life

## O B J E C T I V E S

1. List the distinguishing characteristics of the kingdom Animalia.

2. Discuss the classification and proposed relationships of the animal phyla on the basis of (a) symmetry, (b) type of body cavity, (c) extent of germ layer development, and (d) organizational level (cellular, tissue, and so on).

3. Classify a given animal in the appropriate phylum and class.

4. Identify the distinguishing characteristics of each phylum.

5. List the three main characteristics of a chordate.

6. List examples of each of the seven classes of vertebrates.

## INTRODUCTION

The animal kingdom (Animalia) contains eukaryotic multicellular heterotrophs of truly phenomenal diversity. Animals live in virtually every conceivable type of habitat. The animals are classified in as many as 33 phyla, depending on the taxonomic criteria used. The **vertebrates** are animals with backbones, and the **invertebrates** are those lacking backbones.

The members of higher animal groups exhibit advanced traits not found in animals belonging to lower groups. The following is a condensed overview of major animal phyla and some classes in which the distinguishing characteristics and evolutionary advances particular to each phylum are given.

Phylum Porifera contains the simple **sessile** sponges that have no tissues, organs, mouth, or digestive cavity. Flagellated collar cells line the large internal cavity and draw water into the sponge through the porous body wall. Sponges are **filter feeders.** Tiny food particles are strained out of the water current by the collar cells which pass the food on to other sponge cells. **Plankton** (small floating organisms) are an important food source to all filter feeders. Most sponges are marine, which means they are ocean dwellers.

Phylum Cnidaria is represented by jellyfish, sea anemones, and their kin. Most cnidarians are marine. The body form appears as a free-swimming medusa or sessile polyp. Both medusa and sessile polyp forms may occur in the life cycle of some species. The mobile and sessile stages within the life cycle are an adaptation which results in the organism being able to obtain a food supply from more than a single habitat. Many sessile organisms accomplish this by a mobile **larva** that differs greatly from the adult in structure and function. This is an adaptive advantage employed throughout the animal kingdom.

The cnidarian body is composed of two tissue layers with a **gastrovascular cavity** that has no anus. The mouth is surrounded by tentacles that have stinging cells known as **cnidocytes.** Cnidarians have **radial symmetry.** This means that the body can be divided into mirror-image halves by several planes passing through its long axis. Radial symmetry is an advantage for slow or sessile animals because it enables them to detect food or danger from any side.

Phylum Platyhelminthes includes flatworms, free-living turbellarians, and parasitic flukes and tapeworms. Most flatworms are marine, but there are a few freshwater species. A digestive gut is present with no anus. The flatworms evolved **bilateral symmetry.** This means that only one plane can divide the body into two mirror-image halves. This symmetry is an advantage because it gives a more streamlined shape and concentrates muscle power into one direction of movement. Bilateral symmetry is common in all higher groups. Flatworms are also the first animals exhibiting **cephalization,** which is the concentrating of sensory and nervous tissue into the head end of the body where it can monitor the area of movement. This trait, too, is the rule for nearly all higher phyla.

Phylum Nematoda contains the omnipresent roundworms with many free-living and parasitic species. The nematodes are the first group that have a gut complete with a mouth and anus. Internal organs are suspended within a fluid-filled body cavity that is absent in the solid flatworms. The nematodes and all other higher animals have some type of body cavity that holds the internal organs.

Phylum Mollusca is represented by snails, clams, and octopods. Their body is covered by a mantle that secretes a shell in many species. The mantle cavity often contains gills for gas exchange. Others have lungs. Generally the head and body are mounted on a muscular foot. Mollusks have a circulatory system with a heart. They may be marine or fresh water, and some snails are terrestrial. Although very reduced in this group, mollusks are the first group with a **coelom,** a fluid-filled body cavity that develops as a space in the mesoderm. The coelom is important because it separates gut muscles from body wall muscles. This makes gut contractions and food movement independent of body wall contractions. Coelomic fluid serves as a simple circulatory system in some animals. The coelom provides a space for the development of a true circulatory system, which appears for the first time in the mollusks.

Phylum Annelida includes the segmented worms such as earthworms and leeches. Annelids are known for their **segmentation,** which means they are divided into repeated sections by partitions. In annelids, segmentation results in increased mobility. Segmentation was an important evolutionary step because as a consequence of it, various parts of the body became specialized with different functions. Gas exchange occurs through the skin or **gills,** extensions of the body surface that obtain oxygen from water. Annelids are found in the ocean, in fresh water, or in damp soil.

Phylum Arthropoda is the largest animal phylum. Spiders, shrimp, insects, and centipedes are all arthropods as shown by their high degree of segmentation, a major trait of this phylum. Jointed appendages and an **exoskeleton** (external skeleton) of chitin are typical for this group. The rigid exoskeleton is **molted** (shed) from time to time, which allows for growth of the arthropod. Gas exchange occurs by gills or tracheae. Arthropods are found in marine, freshwater, and terrestrial habitats.

Phylum Echinodermata is represented by sea stars, sea urchins, and sea cucumbers. They have a spiny calcareous skeleton just below the skin. Most have **pentaradial symmetry** (meaning that the body is a central disk surrounded with five parts) and tube feet for locomotion. All are marine.

Phylum Chordata contains a great range of animals from tunicates and sea lancelets to whales and humans. The chordate body at some stage has a rod-like **notochord** for internal skeletal support, a hollow dorsal nerve tube, **pharyngeal gill slits** leading into the **pharynx** (throat), and a post-anal tail. The heart is ventral. Chordates are in all marine, freshwater, and terrestrial habitats.

The chordates evolved several traits that were advantageous over other phyla. Body muscles working with the support of an internal skeleton made locomotion more efficient, particularly in the early vertebrates. This increase in locomotion resulted in chordates further adapting by developing a nervous system and sense organs in the head region with which the environment could be monitored. These advances resulted in a carnivorous diet (not, however, exclusive to vertebrates) that is more nutritious weight for weight than is an herbivorous diet. Further advances occurred in the vertebrates based on the new design of the carnivores.

Subphylum Vertebrata holds the chordates with a vertebral column (backbone). Internal organs are complex and cephalization is pronounced. This subphylum has several vertebrate classes.

Water-dwelling vertebrates acquired several significant adaptations during the evolution toward life on land. Strong backbones and appendage bones and strong muscles were important adaptive changes. Other terrestrial life adaptations included the development of respiratory structures with a moist surface (lungs) through which gas could be exchanged and a waterproof skin which minimized evaporation of water from the body.

These adaptations were initiated among the first fish-like land vertebrates. Short stubby legs developed from fin bones. The swimbladder served as a respiratory organ before the development of lungs. As evolution continued, vertebrate groups appeared that were truly independent of water, such as the reptiles with a dry waterproof skin and a shelled egg that contained its own watery environment as an embryonic development adaptation.

These, then, are the major characteristics of the animals. The following laboratory exercises will allow you to explore some of the diverse characteristics of the animals.

## PRELAB QUESTIONS

1. What phylum do sponges belong to?

2. What type of body symmetry do cnidarians have?

3. Which was the first animal phylum to exhibit cephalization?

4. Which worm phylum is known for its segmentation and was the first to have a closed circulatory system?

5. Which was the first animal phylum to have a gut complete with a mouth and anus?

6. What phylum and class do salamanders and frogs belong to?

7. Are reptiles endothermic or ectothermic?

## LAB 17.A
## PHYLUM PORIFERA: SPONGES

### MATERIALS REQUIRED
  dried specimens of commercial bath sponge

### PROCEDURE
Sponges are unusual animals found mostly in oceans, although there are some freshwater species. Regarding structure, sponges are some of the simplest animals, lacking organs or even well-organized tissues. They consist of a few cell types of various functions that are loosely organized throughout the asymmetrical (irregular) body of the sponge.

Observe the dried specimens of the commercial sponge on display. Note the numerous pores of various size throughout its surface where water enters and leaves the sponge interior. Water is moved through the sponge by currents created by active flagellated cells that line the interior channels leading into the sponge from the pores. Note the flexible, tough quality of the natural sponge. What you see here is the "skeleton" of the sponge composed of spongin, strong protein fibers (Fig. 17–1). Note how the cells of a living sponge are distributed among the fibers.

## LAB 17.B
## PHYLUM CNIDARIA: ANIMALS WITH STINGING TENTACLES

### MATERIALS REQUIRED
  living hydra
  recovery container for holding "used" hydra
  preserved *Aurelia* jellyfish
  preserved sea anemones
  coral exoskeletons
  concavity slide and coverslip
  dropper
  blunt probe
  compound microscope

### PROCEDURE
Cnidarians are mostly marine animals. The general structure of all cnidarians includes a sac-like body with one opening (the mouth) that is surrounded by tentacles. The tentacles are armed with cnidocytes (stinging cells) that paralyze prey organisms. The tentacles then draw the prey into the mouth and digestive body cavity, which is called the gastrovascular cavity. Since the gastrovascular cavity has only one opening, food wastes are expelled from the mouth. Cnidarians occur in two body forms that are variations of the basic sac-like structure: the polyp and the medusa (Fig. 17–2). Both forms occur at different stages of the life cycle in many cnidarians.

The circular arrangement of tentacles around the mouth gives cnidarians radial symmetry that is typical of this phylum. True tissues occur in cnidarians as an inner layer of cells lining the gastrovascular cavity and

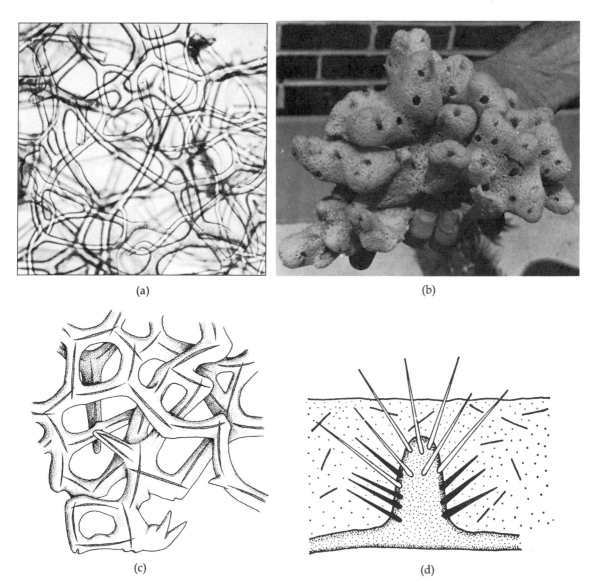

(a)          (b)

(c)          (d)

**FIGURE 17-1**

Spongin fibers as seen under a microscope.

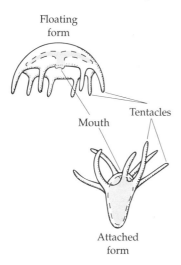

Floating
form

Mouth

Tentacles

Attached
form

**FIGURE 17-2**

Diagram of the floating medusa and attached polyp forms of the cnidarian body. The medusa form is essentially a flattened, inverted polyp. Tentacles capture food and bring it to the mouth. Digestion occurs in the gastrovascular cavity represented by the dashed line.

**FIGURE 17–3**

A hydra snares a small crustacean with its tentacles. The buds on the hydra's body will develop into new individuals and break away from the original hydra, a form of asexual reproduction.

an outer layer of cells that form the outer body wall. The two cell layers are separated by a layer of jelly-like mesoglea. In addition to tissues, cnidarians are also the first animals of our survey to have true nerve cells that are spread as a loose, simple net throughout the body.

1. Obtain a concavity slide and coverslip for observation of the freshwater hydra that represents the polyp body form of cnidarians (Fig. 17–3). The polyp form is essentially a cylindrical body with tentacles arising from the upper end. Use a dropper to pick up a live hydra from the culture jar in the following manner. Pinch the dropper bulb and hold it closed. Use the tip of the dropper to gently nudge loose a hydra from the wall or bottom of the culture jar, then draw up the hydra into the dropper by releasing the bulb. Place the hydra and a few drops of water into the concavity of the slide and cover with the coverslip. Place the hydra on your compound microscope and observe under low power. Your hydra will probably be contracted for a few minutes after handling, but it will soon relax and elongate its body and tentacles.

2. Note the presence of cnidocytes on the tentacles. These appear as tiny bumps. The mouth is on top of the slight elevation located at the base of the tentacles. Adjust the lighting of your microscope to see various details of the hydra's body and tentacles. Place your hydra in the recovery container when you finish observing it.

3. Closely observe a preserved specimen of the marine jellyfish, *Aurelia*, floating in water. *Aurelia* represents the medusa, or "jellyfish," body form of cnidarians. The medusa body form can be described as a flattened, inverted polyp with the tentacles hanging downward instead of reaching upward as in the hydra. The medusa form typically has a larger quantity of mesoglea in the body; hence the name "jellyfish."

4. Use your finger or a blunt probe to examine the body of *Aurelia*. The body is very soft and delicate, so use care when touching the specimen. Note the transparent, umbrella-shaped body. Draw the medusa form of *Aurelia*.

The jellyfish swims by contracting the umbrella. This contraction propels it forward in short pulses. The tentacles in *Aurelia* are reduced and appear as a fringe of fine threads around the edge of the umbrella. From the underside of the umbrella hang

**FIGURE 17-4**

Coral polyps that secrete the hard, white exoskeleton commonly known as "coral".

four conspicuous oral arms that capture small prey organisms from the water. The mouth is located at the base of the oral arms. Gonads (reproductive organs) are visible within the upper side of the umbrella as four crescent-shaped structures.

5. Observe the preserved specimens of sea anemones. These cnidarians are of the polyp form and resemble enlarged hydra. Notice the numerous tentacles surrounding the mouth and the thick, muscular body. Your specimens are preserved in the expanded condition. Like the hydra, anemones contract when disturbed by pulling in their tentacles and shortening the body. Anemones are common in tide pools that are formed on rocky beaches when the tide receeds.

6. Closely examine the various forms of coral on display. Note the various pores and cup-shaped indentations on the coral branches. These openings are the sites where the coral polyps lived. Note how they resemble miniature sea anemones (Fig. 17-4). The polyps secrete calcium carbonate — the white, stony substance of coral — which builds up around the animals and serves as a protective exoskeleton (external skeleton). The coral you see here is what remains after the polyps die and the exoskeleton is bleached.

## POSTLAB QUESTIONS

1. Identify the body form and the following structures: mouth, tentacles, cnidocytes, gastrovascular cavity, and mesoglea.

2. What is the advantage of radial symmetry, particularly for the sessile polyps?

3. What tissues must exist to allow sea anemones to contract when disturbed?

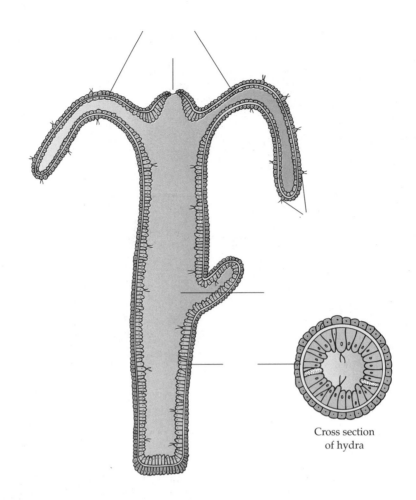

Cross section
of hydra

## LAB 17.C
## PHYLUM PLATYHELMINTHES: FLATWORMS

### MATERIALS REQUIRED

live planaria

recovery container for "used" planaria

dropper

small watch glass

planaria, prepared slide

stereomicroscope

compound microscope

preserved tapeworm

tapeworm scolex and proglottids, prepared slide

### PROCEDURE

The flatworms are so named because most members of this phylum are ribbon like in appearance. Several advances not present in the sponges and cnidarians appear in the flatworms. The flatworms are the first in our survey to have a definite head end that contains sensory organs such as eyes. The presence of a definite head and tail end results in a body form of bilateral symmetry that is the dominant trend throughout the remaining phyla in our animal survey. Animals of bilateral symmetry tend to be more active than radial or asymmetrical animals. The head end, with its concentration of sensory organs, moves first through the animal's environment detecting food or danger. The animal can then respond accordingly.

To note another advancement of flatworms, it is necessary to discuss briefly, perhaps surprisingly, an aspect of embryo development. A fertilized egg cell quickly develops by multiple cellular divisions into an early embryo with definite cell layers. These layers give rise to specific structures in the fully developed organism. These cell layers (known as germ layers) are called the endoderm, mesoderm, and ectoderm and are named according to their position within the early embryo. Endoderm is the innermost layer, which will form the lining of the digestive tract. Mesoderm, the middle cell layer, produces muscles, bones (in those organisms having them), and many other structures present in the "middle" regions of the body. The outer ectoderm produces the nervous system, if present, and the outer covering of the body. Flatworms are the first phylum to possess all three germ layers. As a

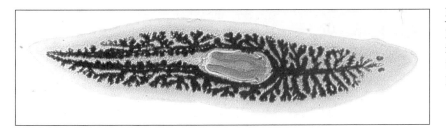

**FIGURE 17-5**

Internal structure of a planarian. The highly branched intestine is darkly stained in the preserved specimen on the left.

result, flatworms are also the first group to have true organs, such as an intestine and nervous system, composed of the various tissues that arise from the three germ layers.

1. Use a dropper to obtain a live planarian, which is a freshwater, free-living (nonparasitic) flatworm, from the culture jar. Place it in a small watch glass with enough water from the culture jar to form a small pool. Observe the planarian with a stereomicroscope. Note how the thin, flat body glides over the surface of the glass. The worm's lower body surface is covered with cilia that aid in the gliding locomotion. Observe the distinct head end that contains eyes and ear-like projections called auricles. Return the planarian to the culture jar after your observations if the organism appears to be in good condition. If necessary, check with your instructor.

2. Observe a prepared slide of a planarian under your stereomicroscope (Fig. 17–5). Note the darkly stained intestine that branches throughout the body. The large tubular structure in the center of the body is the pharynx ("throat"). The mouth is located on the underside of the body. The planarian extends its pharynx outward through the mouth when feeding to ingest food particles. The mouth is the only opening of the digestive tract; an anus is absent in this phylum. As in the cnidarians, food wastes are expelled through the mouth.

3. Observe a preserved specimen of a tapeworm, which is an intestinal parasite. The head is located at the narrow end of the worm. The body segments, called proglottids, continually grow from the neck region of the worm and increase in size as they mature. As a result, the oldest and most mature proglottids are farthest from the head.

4. Place a prepared slide showing portions of a tapeworm *(Taenia)* body on your compound microscope and examine under low power. The head region, or scolex, is the smallest structure on the slide (Fig. 17–6). Notice the suckers and hooks of the scolex that attach the worm to the intestinal

wall of the host, in this case a dog or cat. Nutrients are absorbed directly through the body wall of the proglottids from the host's predigested food, which surrounds the tapeworm. Consequently, tapeworms have no digestive organs. Observe the internal structure of the proglottids. Most of the organs you see are associated with reproduction. Tapeworms are hermaphroditic and have complex life cycles that involve more than one host. Tapeworms produce numerous eggs that fill mature proglottids. The mature segments break off and pass out of the host's body in the feces (solid food waste) where the eggs subsequently reach the first host of the life cycle.

**FIGURE 17-6**

Scanning electron micrograph of a tapeworm scolex. Note the hooks and suckers that attach the worm to its host.

## POSTLAB QUESTIONS

1. Draw a planarian and identify the following structures: eyespots, head, auricles, pharynx, mouth, and intestine.

2. What structures helped you determine which was the head end of the planaria?

3. Identify one disadvantage of the planaria's digestive system.

4. Compare the digestive system of a planaria to that of a tapeworm.

5. Where within the tapeworm would the ripest eggs be contained?

6. Argue for or against the following statement: Tapeworms are the most degenerate members of the phylum Platyhelminthes.

## LAB 17.D

## PHYLUM NEMATODA: ROUNDWORMS

### MATERIALS REQUIRED

live vinegar eel *(Turbatrix)* culture

slides and coverslips

dropper

compound microscope

preserved *Ascaris,* female and male

### PROCEDURE

The nematodes are typically very slender and circular in cross section; hence the name "roundworm" (Fig. 17–7). Nematodes are extremely numerous throughout the world in every kind of aquatic and terrestrial habitat. You usually don't see them because many are of microscopic size and hidden under the soil, within

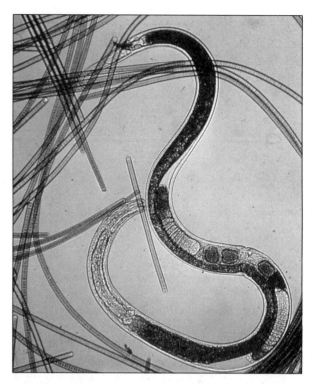

**FIGURE 17–7**

Micrograph of a freshwater, free-living nematode exhibiting typical nematode form. The green strands around the worm are cyanobacteria.

bodies of water, or within the bodies of plants and animals as parasites.

Members of this phylum have some advancements that you have not seen in the previous phyla. The nematodes are the first group of our survey to have a complete digestive tract with a one-way food passage from the mouth to the anus. Like the flatworms, nematodes have bilateral symmetry although there is no well-defined head in the roundworms. Well-developed organ systems are derived from the three germ layers, but still no circulatory system occurs in this phylum. The nematodes are the first phylum encountered to have a body cavity, which is considered as an internal organ space between the endoderm and the mesoderm. In flatworms, this region is filled with mesoglea. The mesoderm is absent in the cnidarians. Therefore, cnidarians and flatworms are acoelomate, meaning "without a body cavity (coelom)." Roundworms are considered as pseudocoelomate ("false body cavity") because tissue derived from the mesoderm does not entirely line the body cavity. Animals of higher phyla have a true coelom that is lined with tissue of mesoderm origin (Fig. 17–8).

1. Obtain a culture jar of vinegar eels, a species of free-living, microscopic nematode that lives in the

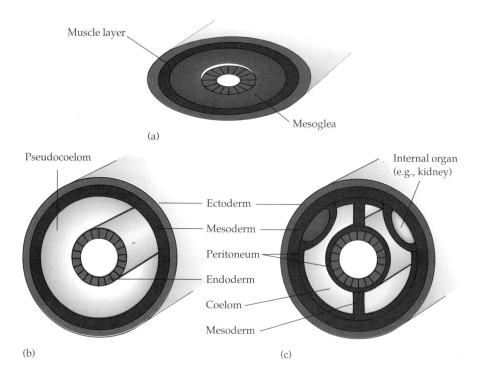

**FIGURE 17-8**

Three basic animal body plans are illustrated by these cross sections. (a) An acoelomate animal has no body cavity. (b) A pseudocoelomate animal has a body cavity that develops between the mesoderm and endoderm. The term ''body cavity'' refers to the space between the body wall and the internal organs. (c) In a coelomate animal, the body cavity, called a coelom, is completely lined with tissue derived from mesoderm.

sediment of unpasteurized vinegar. Use a dropper to draw up liquid from the surface of the culture. Place a few drops on a slide, add a coverslip, and examine under both low and high power. Note the rapid, side-to-side swimming action of the vinegar eels that is typical of many nematodes. Vinegar eels are also representative of nematodes because they have slender bodies with pointed ends.

2. Observe preserved specimens of male and female *Ascaris*, an intestinal parasite found in pigs and humans. The sexes can be determined by the larger size of the female and the hooked tail of the male. Note the worm's external cuticle, a tough, flexible covering that resists digestive enzymes of the small intestine where it feeds on partially digested food of the host. As in other intestinal parasites, female roundworms release thousands of eggs that pass

out of the body with the feces. Consequently, areas with poor sanitation have a much higher rate of roundworm infection. Draw a male and female *Ascaris.*

## POSTLAB QUESTIONS

1. Propose a muscle configuration that explains the side-to-side swimming action of the vinegar eels.

2. Describe the appearance of the cuticle of *Ascaris.*

## LAB 17.E
## PHYLUM MOLLUSCA: ANIMALS WITH SHELLS

### MATERIALS REQUIRED

preserved specimens of the following:
a. land snail
b. freshwater mussel, shell opened
c. squid
d. chitons

## PROCEDURE

The mollusks are the first animals of our survey to have all organ systems and a true coelom, which is always present in more complex animals. Most mollusks have an open circulatory system. This means that the blood does not continually stay within the blood vessels. During its circulation, blood is released into body spaces where it directly contacts the tissues for exchange of materials and gases. In cephalopod mollusks (discussed below) and more complex animals, the circulatory system is closed; the blood always flows within vessels and has no direct contact with the body tissues.

The mollusks are characterized by a shell, although in some highly modified mollusks the shell is reduced or absent. The shell, composed of calcium carbonate, is secreted by a fold of tissue called the mantle. Other mollusk traits include a muscular foot that functions in locomotion and the radula, a rasping tongue covered with teeth that tears away bits of food from a food source (Fig. 17–9).

The mollusks are an ancient, diverse phylum of numerous species and forms. Consequently, we shall look at four major classes of phylum Mollusca.

### Class *Polyplacophora:* Chitons

1. Observe the preserved chiton specimens. Chitons are strictly marine mollusks found on rocky beaches where they graze algae from rocks among the surf. Chitons are characterized by a shell of eight plates that partially cover the body. Examine the ventral (bottom) surface of a chiton. Note the large, muscular foot and the mouth. The anus is located near the posterior (rear) end of the foot. Draw a chiton.

### Class *Gastropoda:* Snails and Slugs

2. Examine the coiled shell of the preserved land snails. There are many kinds of snails that may be terrestrial, freshwater, or marine species. Gastropods are characterized by a one-piece shell that is often spirally or concentrically coiled. Note the large muscular foot on which the snail glides. The mouth, containing the radula, is found on the ventral surface of the head region. Eyes are located at the ends of the two long tentacles that protrude from the head. When disturbed, the snail contracts its body into the protective shell. Slugs are similar to snails in form, but they lack a shell.

### Class *Bivalvia:* Clam-Like Mollusks

3. Note the shell of the freshwater mussel, a representative bivalve. The shell has been opened to expose the body of the mussel. The two-piece shell characterizes the members of this class, which also includes clams and oysters. The hinge of the shell marks the dorsal (upper) surface of the mussel. In life, the mussel sits in the sand or in the mud with the hinge side up and with the edges of the shell positioned downward. Note the thick, muscular foot located near the edge of the shell. The foot extends downward out of the shell into the sand and pulls the mussel forward. Bivalve mollusks are filter feeders that continually circulate water through their bodies and filter out microscopic food particles.

### Class *Cephalopoda:* Octopods and Squids

4. Examine the preserved squid. The cephalopods are all marine species. They are the most advanced mollusks, being modified to the point that they are hardly recognizable as a mollusk. Note the squid's eight arms and two longer tentacles that have suckers that can grasp prey organisms. These appendages evolved from the foot of the mollusk. The mouth, containing a sharp, parrot-like beak, is located in the central region of the arms. Notice the well-developed eyes; their internal structure is remarkably similar to that of vertebrate animals. The mantle forms the thick wall of the tubular body. The shell of the squid is greatly reduced to a thin, plastic-like strip within the body called a pen. The fins at the tip of the body propel the squid arms first through the sea. Examine the head region for a tubular funnel that protrudes from under the mantle. For a rapid escape, the squid can forcefully expel water from the funnel, which jets the squid backward, fins first.

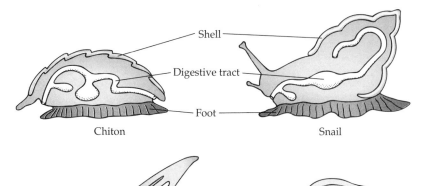

**FIGURE 17–9**

Basic features of four representative mollusks that show great variety in body form. Notice the reduced, internal shell and modified foot (arms) of the squid.

Shell

Digestive tract

Foot

Chiton                    Snail

Shell

Digestive tract

Foot

Squid                    Clam

## POSTLAB QUESTIONS

1. Identify the following mollusks and their structures.

2. Compare the foot of a mussel with that of a chiton.

3. Compare the shell of a land snail with that of a squid.

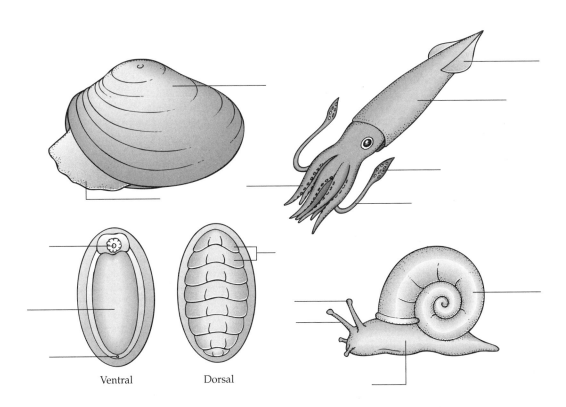

Ventral          Dorsal

## LAB 17.F
## PHYLUM ANNELIDA: SEGMENTED WORMS

### MATERIALS REQUIRED

preserved sandworm

preserved and live earthworms

preserved leech

### PROCEDURE

The segmentation of annelid worms is the beginning of a segmentation trend found in more advanced animals of higher phyla as well. The bodies of many annelids consist of nearly identical segments repeated along the length of the animal. Each segment contains internal organs that are also repeated throughout the annelid body. The annelids are the first of our survey to have a closed circulatory system. Annelid blood contains hemoglobin, the same red pigment that carries oxygen in human blood. There are three major classes of annelids, with segmentation being a major common trait (Fig. 17–10).

### Class *Polychaeta:* Sandworms and Tubeworms

1. Examine the preserved sandworm, a representative polychaete. All polychaetes are marine annelids. The name polychaete means "many hairs,"

and it is derived from the row of bristles on each side of the worm. The bristles (setae) extend from fleshy appendages called parapodia that function in locomotion. Unlike the other classes of annelids, many polychaetes have well-developed heads.

### Class *Oligochaeta:* Earthworms

2. Observe the preserved and live earthworms on display. The name oligochaete means "few hairs" and refers to the few reduced setae found on the bodies of these annelids. Note the absence of a well-developed head. The anterior (front) end of the worm can be distinguished by the clitellum, a wide band around the body located somewhat toward the anterior end. The clitellum secretes a cocoon that holds the developing eggs. Note the movement of a live earthworm. Earthworms are valuable because they are commonly sold as fish bait and their burrowing aerates the soil. Oligochaetes occur in freshwater, marine, and terrestrial habitats.

### Class *Hirudinea:* Leeches

3. Examine the preserved leech. Leeches are parasites and are found in fresh water or on land in moist tropical forests. Leeches suck the blood of fish, turtles, and other vertebrates. Like the earthworm, they do not have a well-developed head. Their

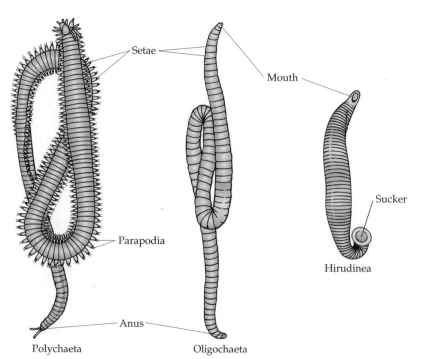

**FIGURE 17–10**

Representatives of the three annelid classes: sandworm (left), earthworm (center), and leech (right).

Setae

Mouth

Parapodia

Sucker

Anus

Hirudinea

Polychaeta

Oligochaeta

salivary glands secrete an anticoagulant that prevents blood from clotting. This facilitates blood flow when the leech feeds from its host. Leeches typically have anterior and posterior (rear) suckers that enable them to attach to their host.

## POSTLAB QUESTIONS

1. Draw the following segmented worms and related structures. Sandworm: setae, parapodia, and head. Earthworm: clitellum. Leeches: anterior and posterior sucker.

2. Propose a muscle configuration that explains the movement of earthworms.

3. Compare the movement of earthworms to vinegar eels.

4. Compare the quantity of setae in sandworms, earthworms, and leeches.

## LAB 17.G
## PHYLUM ARTHROPODA: ANIMALS WITH JOINTED APPENDAGES

### MATERIALS REQUIRED

preserved specimens of the following:
a. horseshoe crab
b. spider
c. scorpion
d. crayfish
e. crab
f. centipede
g. millipede
h. insects

### PROCEDURE

The arthropods are the most numerous animals on Earth. They have adapted to almost every kind of aquatic and terrestrial habitat. Segmentation, first noted in the annelids, is also a major trait of arthropods. Different regions of the segmented arthropod body are typically modified for various functions. Arthropods have an exoskeleton of chitin, which is a hard, chemically resistant substance. The organ systems are well developed and enable arthropods to grow to a relatively large size and to obtain high levels of activity. However, size is somewhat limited in terrestrial arthropods due to the heavy exoskeleton. Phylum Arthropoda contains such a diversity of animals that it is necessary to consider the arthropods in their various subphyla and classes.

**Subphylum Chelicerata.** The name of this subphylum is derived from chelicerae, the claw-like mouthparts that members of this subphylum possess. Arthropods of this group generally have four pairs of walking legs and bodies with two main segments—a cephalothorax (fused head and thorax) and abdomen. Antennae are absent.

### Class *Merostomata*: Horseshoe Crabs

1. Observe the preserved horseshoe crab, the only remaining representative of this very old marine group. Note the two eyes on the large, dome-shaped carapace (shell). Five pairs of walking legs are concealed below the carapace. The gills appear as a series of plates posterior to the legs. Despite its fearsome appearance, horseshoe crabs are harmless scavengers that live on the ocean floor (Fig. 17–11).

**FIGURE 17–11**

A mating pair of horseshoe crabs. Mating usually occurs on a beach where numerous individuals gather during the breeding season.

**FIGURE 17-12**

A scorpion killing a grasshopper with its venomous stinger, which is located on the end of its abdomen. The scorpion will feed on the body fluids of the grasshopper.

### Class *Arachnida:* Spiders and Scorpions

2. Examine the preserved specimens of spiders and scorpions. Both have four pairs of walking legs that are typical of arachnids. The pedipalps ("pinchers") of the scorpion function in handling food. The chelicerae of arachnids are modified for piercing prey and sucking out the body fluids as food (Fig. 17-12). Spiders and scorpions are known for their venomous bite and sting, but a great majority of species are not dangerous to humans.

**Subphylum Crustacea.**   Crayfishes and Crabs

3. Observe the crayfish and crab on display (Fig. 17-13). Crustaceans typically have two pairs of anten-

**FIGURE 17-13**

A crab (subphylum Crustacea) resting on coral.

nae and mandibles (jaws) in contrast to the chelicerate arthropods that lack these structures. Crustaceans as a group have biramous appendages that are two-branched at the ends as seen on the mouthparts and some walking legs of the crayfish. Note the variation in body form between the abdomen of the freshwater crayfish and the abdomen of the marine crab, which is curled under the body.

**Subphylum Uniramia.**   In contrast to the crustaceans, this group of arthropods has one pair of antennae and appendages that are unbranched, or uniramous. Three diverse classes are contained within this subphylum.

### Class *Chilopoda:* Centipedes

4. Observe the preserved centipede. Note that each body segment has one pair of relatively long legs that make centipedes rapid runners (Fig. 17-14).

**FIGURE 17-14**

A centipede. The head is located to the left.

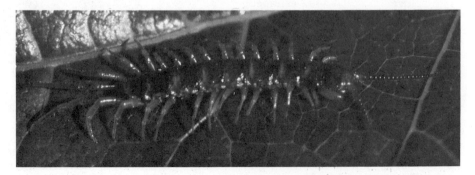

**FIGURE 17-15**

A millipede. The head is located to the right.

Centipedes are predators and feed mostly on insects that they kill with poison claws located on the first body segment.

### Class *Diplopoda:* Millipedes

5. Examine the millipede on display. Notice that millipedes have two pairs of relatively short legs per body segment, which distinguishes them from the centipedes (Fig. 17–15). Millipedes are slow moving scavengers that feed on decaying plant matter.

### Class *Insecta:* Insects

6. Observe the various insects on display. Although insects occur in many forms, they all have bodies consisting of a head, thorax, and abdomen. Six legs arise from the thorax. All but the most primitive insects also have two pairs of wings that originate from the thorax (Fig. 17–16). Class Insecta, containing over 750,000 species with more continually being discovered, is the largest group of animals.

## POSTLAB QUESTIONS

1. Why would it be incorrect to call a spider an insect?

2. Compare the size and shape of a spider's pedipalps to a spider's chelicerae.

3. Why are the gills of horseshoe crabs sometimes called "book" gills?

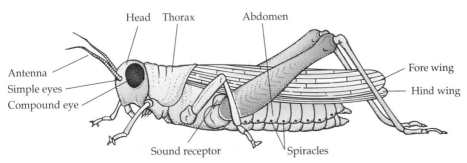

**FIGURE 17-16**

External structure of a grasshopper showing many features common to most insects.

4. Compare the abdomen of the crayfish with that of a marine crab.

5. Identify the phylum and subphylum of this organism: the organism has an exoskeleton of chitin, two pairs of antennae, mandibles, and biramous appendages.

6. Compare the following features of organisms in the subphyla Chelicerata, Crustacea and Uniramia: walking legs, antennae, main body segments, major mouth parts, and segmentation.

## LAB 17.H
## PHYLUM ECHINODERMATA: MARINE SPINY-SKINNED ANIMALS

### MATERIALS REQUIRED

preserved specimens of the following:
a. sea star
b. brittle star
c. sea urchin and sand dollar
d. sea cucumber

### PROCEDURE

The echinoderms are exclusively marine animals having some characteristics that make them unique among the animal kingdom. They are the first animals of our survey to have an endoskeleton (internal skeleton). The endoskeleton consists of calcium carbonate plates with spines that project upward from the body surface. These spines give the phylum its name, which means "spiny-skinned." Echinoderms have a water vascular system, a type of hydraulic system found in no other phylum. The water vascular system consists of internal water canals with contractile tube feet branching from them. The tube feet extend when water is forced into them from the canal. The tube feet retract when water pressure in the canal is decreased and water flows out of the tube feet. Each tube foot ends in a suction cup that adheres to a surface when the tube foot is extended. The tube feet enable echinoderms to move and obtain food.

All phyla we have seen in our animal survey, except for the sponges and cnidarians, are bilaterally symmetrical. The echinoderms interrupt this trend because the adults have pentaradial symmetry, a body plan radially arranged in five parts around a central axis where the mouth is located. Curiously, echinoderm larvae begin life as bilaterally symmetrical, then acquire pentaradial symmetry as they mature. Four classes of echinoderms that exhibit a variety of forms are presented here.

**FIGURE 17–17**

A close-up picture showing the extended tube feet of a sea star.

### Class *Asteroidea:* **Sea Stars**

1. Observe the preserved sea star on display. Sea stars in many ways are representative of echinoderms. Note the spiny skin and the five arms radiating out from the central disc giving the animal pentaradial symmetry. The lower surface of the sea star is determined by a ventral groove that runs the length of each arm. Numerous tube feet, which may be retracted, occur within these grooves (Fig. 17–17). The mouth is located ventrally on the central disc. On the dorsal surface, look for the madreporite, a small button-like structure where water enters the vascular system of the sea star. Draw a ventral and dorsal view of the sea star.

### Class *Ophiuroidea:* **Brittle Stars**

2. Examine the brittle star specimen. Note how it resembles the sea star. The brittle star differs from the sea star in that the slender arms are distinctly separate from one another on the central disc and the arms have no ventral groove (Fig. 17–18). The brittle stars are so named because the arms easily break off if seized.

### Class *Echinoidea:* **Sea Urchins and Sand Dollars**

3. Observe the sea urchin and sand dollar specimens and their lack of arms. A star-like arrangement of the skeletal plates is visible in bleached specimens with the spines removed. Note the spines of sea urchins (Fig. 17–19) compared to the spines of the sand dollar.

### Class *Holothuroidea:* **Sea Cucumbers**

4. Observe the preserved sea cucumber. These are unusual echinoderms because of their rather soft bodies and elongated shape (Fig. 17–20). Note the five rows of tube feet that run the length of the body. These give a hint of the pentaradial symmetry that is more obvious in other echinoderms. The mouth is located in the center of the branching tentacles (modified tube feet) at the anterior end of the sea cucumber. The anus is found at the opposite posterior end.

### POSTLAB QUESTIONS

1. Do brittle stars have tube feet? Were they visible?

2. Are sea urchins pentaradially symmetrical? Explain your answer.

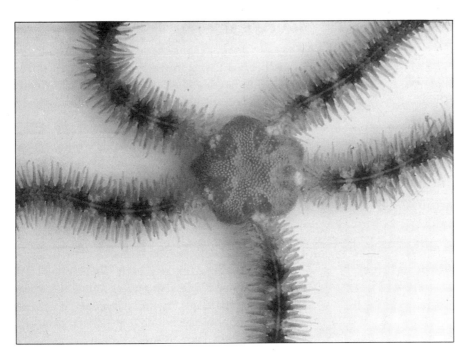

**FIGURE 17–18**

A brittle star. Note how the arms are distinctly separate from one another on the central disc.

**FIGURE 17–19**

The spines of a sea urchin vary from long needles in one species to pencil-like stubs in another.

**FIGURE 17–20**

A sea cucumber. Some species, as this one, do not have tentacles.

3. Compare the spines of a sea urchin with that of a sand dollar.

## LAB 17.I
## PHYLUM CHORDATA: ANIMALS WITH NOTOCHORDS

### MATERIALS REQUIRED

preserved or live specimens of the following:

a. tunicates
b. sea lancelet
c. lamprey
d. shark pup
e. bird
f. perch, or other bony fish
g. frog or salamander
h. lizard or snake
i. small mammal

### PROCEDURE

Phylum Chordata contains some of the most diverse animal forms found in the animal kingdom—from small, blob-like sea squirts to complex human beings. All chordates have three characteristics that appear at some time in their life history:

1. a flexible, dorsal notochord that gives body support;
2. a dorsal tubular nerve cord located just above the notochord; and
3. gill slits associated with the pharyngeal, or throat, area of the animal (Fig. 17–21).

In many chordates, such as humans, these traits appear in the embryo during early development and quickly give rise to other structures found in the fully developed offspring. Phylum Chordata is divided into three subphyla according to major characteristics possessed by the animals they contain.

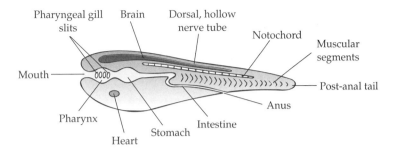

**FIGURE 17–21**

A generalized chordate body showing structures common to all members of phylum Chordata.

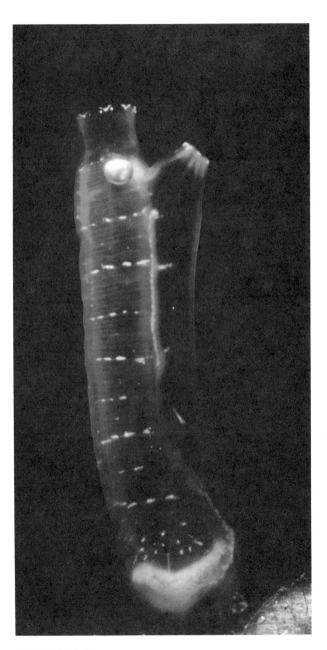

**FIGURE 17–22**

A transparent sea squirt (tunicate). Note the siphons (tubes) at the top of the animal that continually take in and expel water.

**Subphylum Urochordata.**   Tunicates

1. Observe the preserved tunicate specimens, which are sessile marine organisms. These tunicates are commonly called sea squirts because they expel a jet of water from their siphon (tube) when disturbed (Fig. 17–22). Tunicates continually pass water through their bodies for oxygen and small food particles. Before attaching to the sea bottom as an adult, the tadpole-like tunicate larva actively swims and exhibits all the major chordate characteristics, which make these unusual animals members of this phylum.

**Subphylum Cephalochordata.**   Sea Lancelets

2. Examine the specimens of *Branchiostoma,* a sea lancelet. Sea lancelets are somewhat generalized chordates having pharyngeal gill slits, a dorsal notochord that runs the length of the body, and a hollow nerve cord located just above the notochord (Fig. 17–23). The anterior end of the sea lancelet can be distinguished by its slightly larger size.

**Subphylum Vertebrata.**   Vertebrates

The vertebrates are the most advanced chordates. Vertebrates have evolved numerous body forms and occur in almost every kind of habitat. In this group of chordates, the spinal column, or backbone, develops around the notochord and eventually replaces it during embryonic development. The spinal column is composed of bone or cartilage segments called vertebrae (vertebra, singular); hence the subphylum Vertebrata. There are seven classes of vertebrates, most of which are familiar to you.

**Class *Agnatha:* Jawless Fishes**

1. Observe the lamprey specimen, a jawless fish. Notice that the mouth is simply a sucker-like disc with

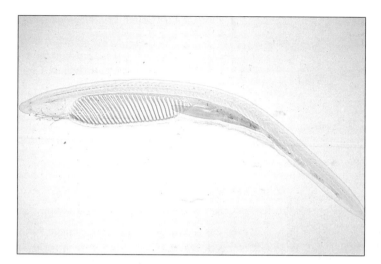

**FIGURE 17-23**

*Branchiostoma,* a preserved sea lancelet. Note the numerous gill slits and red-stained digestive tract.

numerous teeth. Some lampreys are parasitic, feeding off other fishes. They attach to a host with their sucker mouth, create a wound with the sharp teeth, then feed on the fish's body fluids (Fig. 17–24). Note the prominent gill slits and the absence of paired fins.

### Class *Chondrichthyes:* **Cartilaginous Fishes**

2. Examine the preserved specimen of a shark pup (young shark). Fishes of this class are almost exclusively marine and have skeletons composed mostly of cartilage, the flexible tissue found in the human nose and ear. Note the presence of jaws and gill slits. Well-developed paired fins on the streamlined body make sharks efficient swimmers.

### Class *Osteichthyes:* **Bony Fishes**

3. Observe the perch or other bony fish specimen. Fishes of this class have skeletons composed of bone. You are most familiar with members of this class as food and aquarium fishes. Compare the membranous fins of the perch with the fleshy fins of the shark. Notice that the gills of the perch are covered by a protective external flap called an operculum.

### Class *Amphibia:* **Amphibians**

4. Observe the frog or salamander specimen on display. Amphibians are typically covered with a smooth skin, although toads have a rough, bumpy skin. The skin is kept moist by a secretion of mucus. Although adult amphibians are adapted for life on land, they usually occur in moist habitats. Many

**FIGURE 17-24**

Lampreys attached to a host fish.

amphibians return to water to lay their eggs because the larval stage (tadpole) is aquatic and has gills. The tadpole gradually develops legs and lungs, at which time it becomes terrestrial.

## Class *Reptilia:* Reptiles

5. Observe the snake or lizard specimen. Note the scaled skin, a hallmark of reptiles. The scaled reptile skin is dry and prevents water loss from the body. Reptilian feet, exclusive of snakes, have claws—a feature absent in amphibians. Reptiles reproduce by laying a leathery shelled egg. This makes them independent of water for reproduction.

## Class *Aves:* Birds

6. Observe the bird specimen on display. The feathers covering a bird's skin enable flight and retain body heat. Birds are endothermic, which means they have a constant, high body temperature regardless of the surrounding temperature. The body temperature of all animals discussed so far is influenced by environmental temperature, which means they are ectothermic. Note the jaws of birds have been modified into a beak without teeth. Birds, like reptiles, reproduce by a shelled egg. The bird eggshell, consisting of calcium carbonate, is brittle and harder than the flexible reptilian eggshell.

## Class *Mammalia:* Mammals

7. Examine the representative mammal specimen. Mammals are endothermic; the hair covering the body helps to retain body heat. Mammals carry their developing offspring within the uterus, or womb, where the embryos are protected from the external environment. When born, the mother mammal nurses her offspring with nutritious milk secreted from mammary glands. Most mammals nurture their offspring with extended parental care, which greatly increases their chances of survival. Consequently, mammals typically produce fewer offspring than other animals that exhibit little or no parental care of their young.

## POSTLAB QUESTIONS

1. Argue for or against the following statement: Humans do not meet the criteria for chordates because they lack pharyngeal gill slits.
2. Why aren't all the vertebrate fishes grouped into one class?

3. Which of the following organisms are ectothermic?
   a. bluejay
   b. turtle
   c. salamander
   d. lion
   e. snake
4. Identify the class, subphylum, and phylum of these chordates: tunicates, sea lancelet, lamprey, shark, robin, human, perch, frog, and lizard.

5. Write the names of the appropriate phylum in the space provided: Porifera, Cnidaria, Playthelminthes, Mollusca, Arthropoda, Echinodermata, Nematoda, Annelida, and Chordata.
   a. _____ gastrovascular cavity with one opening
   b. _____ lack organs
   c. _____ notochord
   d. _____ cnidocytes
   e. _____ nerve net
   f. _____ true organs
   g. _____ pseudocoelomate
   h. _____ setae
   i. _____ open circulatory system
   j. _____ clitellum
   k. _____ digestive tract with only one opening
   l. _____ chelicerae
   m. _____ pentaradial symmetry
   n. _____ complete digestive tract with mouth and anus
   o. _____ lack tissues
   p. _____ radial symmetry
   q. _____ acoelomate
   r. _____ bilateral symmetry
   s. _____ mantle
   t. _____ tube feet
   u. _____ parapodia
   v. _____ exoskeleton of chitin
   w. _____ endoskeleton
6. Place these organisms in the appropriate phylum: sponges, hydra, sea anemones, planaria, tapeworm, vinegar eel, snail, squid, spider, sea star, earthworm, leech, crayfish, sea cucumbers, snake and bird.

# FOR FURTHER READING

Alexander, R. M. 1981. *The Chordates.* Cambridge, England: Cambridge University Press.

Bailey, W. J., and T. J. Ridsdill-Smith, eds. 1991. *Reproductive Behaviour of Insects: Individuals and Populations.* New York: Chapman and Hall.

Barnes, R. 1987. *Invertebrate Zoology.* Philadelphia: Saunders College Publishing.

Barrington, E. J. 1979. *Invertebrate Structure and Function.* Boston: Houghton Mifflin.

Betancourt, Julio L. et al., eds. 1990. *Packrat Middens: The Last 40,000 Years of Biotic Change.* Tucson: University of Arizona Press.

Borkovec, A. B., and E. P. Masler, eds. 1990. *Insect Neurochemistry and Neurophysiology.* Clifton, N.J.: Humana Press.

Corbet, G. B., and J. E. Hill. 1991. *A World List of Mammalian Species.* New York: Oxford University Press.

Daly, H. V. et al. 1978. *Introduction to Insect Biology and Diversity.* New York: McGraw-Hill.

Davis, Lloyd S. et al., eds. *Penguin Biology.* San Diego: Academic Press.

Emmons, Louise H. 1990. *Neotropical Rainforest Mammals: A Field Guide.* Chicago: University of Chicago Press.

Gibbons, J. Whitfield. 1990. *Life History and Ecology of the Slider Turtle.* Washington, D.C.: Smithsonian Institution Press.

Gilbert, Francis, ed. 1990. *Insect Life Cycles: Genetics, Evolution, and Co-ordination.* New York: Springer-Verlag.

Gilbert, Pamela, and Chris J. Hamilton. 1990. *Entomology: A Guide to Information Sources.* New York: Mansell.

Holldobler, Bert, and Edward O. Wilson. 1990. *The Ants.* Cambridge, Mass.: Belknap Press of Harvard University Press.

Johnsgard, Paul A. 1990. *Hawks, Eagles, and Falcons of North America.* Washington, D.C.: Smithsonian Institution Press.

Kent, G. 1987. *Comparative Anatomy of the Vertebrates.* St. Louis: Times Mirror/Mosby College Publishing Co.

Kitchener, Andrew. 1991. *The Natural History of the Wild Cats.* London, England: Christopher Helm.

Kozloff, Eugene N. 1990. *Invertebrates.* Philadelphia: Saunders College Publishing.

Little, C. 1983. *The Colonization of Land: Origins and Adaptations of Terrestrial Animals.* New York: Cambridge University Press.

MacCall, Alec D. 1990. *Dynamic Geography of Marine Fish Populations.* Seattle: University of Washington Press.

Maitland, Peter S. 1990. *Biology of Fresh Waters.* New York: Chapman and Hall.

Mayr, Ernst, and Peter D. Ashlock. 1991. *Principles of Systematic Zoology.* New York: McGraw-Hill.

Meglitsch, Paul A., and Frederick R. Schram. 1991. *Invertebrate Zoology.* New York: Oxford University Press.

Mitchell, L. et al. 1988. *Zoology.* Menlo Park, Calif. Benjamin-Cummings Publishing Co.

Randall, John E. et al. 1990. *Fishes of the Great Barrier Reef and Coral Sea.* Honolulu: University of Hawaii Press.

Riedman, Marianne. 1990. *The Pinnipeds: Seals, Sea Lions, and Walruses.* Berkeley: University of California Press.

Robinson, John C., ed. 1990. *An Annotated Checklist of the Birds of Tennessee.* Knoxville: University of Tennessee Press.

Schmidt-Nielsen, K. 1983. *Animal Physiology: Adaptation and Environment.* New York: Cambridge University Press.

Schreck, Carl B., and Peter B. Moyle, eds. 1990. *Methods for Fish Biology.* Bethesda, Md.: American Fisheries Society.

Seeley, R. et al. 1989. *Anatomy and Physiology.* St. Louis: Times Mirror/Mosby College Publishing Co.

Sibley, Charles G., and Burt Monroe, Jr. 1990. *Distribution and Taxonomy of Birds of the World.* New Haven, Conn.: Yale University Press.

# PART VI

❏

# Plant Structure and Function

**TOPIC 18**
The Plant

**TOPIC 19**
Reproduction and Development in Complex Plants

**TOPIC 20**
Plant Hormones and Responses

⬓

# The Plant

1. List several functions of stems and explain how the structure of stems relates to their function.

2. Compare and contrast the structure of herbaceous dicot stems and monocot stems.

3. List several functions of roots.

4. Compare the arrangement of tissues in primary dicot and monocot roots. Give at least one function for each tissue.

5. Distinguish between primary and secondary growth and between apical and lateral meristems.

6. Characterize the ground tissue system, the vascular tissue system, and the dermal tissue system of plants.

7. Identify the major tissues of the leaf and explain how its structures are related to photosynthesis.

8. Compare and contrast leaf anatomy of dicot and monocot leaves.

9. Trace the pathway of water movement in plants.

10. Distinguish between root pressure and tension-cohesion as mechanisms to explain the rise of water in xylem.

11. Outline the pressure flow theory of sugar transport in phloem.

---

## INTRODUCTION

The vascular plant is composed of three basic organs: roots, stems, and leaves. These result from the growth of **meristems,** areas of cells that retain the capability of cell division. The structure of roots, stems, and leaves reflects their basic functions in vascular plants. Roots usually have a high surface area, an adaptation that facilitates the absorption of water. Stems have evolved with cellular modifications that support the leaves in the air. Leaves have a high surface area, an adaptation that allows capture of sunlight by photosynthetic tissues that produce food. All of these structures are interconnected by **vascular tissue,** which transports materials throughout the entire plant.

Roots have four main functions. First, roots anchor the plant in soil. Second, roots absorb water and minerals from the soil. Third, roots transport water and minerals up into the **shoot system.** Absorption occurs through **root hairs,** which are extensions of root epi-

dermal cells. And fourth, roots store food for the plant in the **cortex** of many taproots. The root system occurs in two growth forms. A **taproot system** has one large main root with shorter, thinner roots that branch from its side. Taproots serve as good anchors, but exposure of root surface area to the soil is somewhat limited. A **fibrous root system** has multiple roots of about equal size. This root system offers the advantage of greater surface area.

Stems also have four main functions. First, stems support structures of the shoot system by their **turgor** (water pressure in the cells), thick cell walls, and fibers that resist breaking. The central area of most stems has varying amounts of **pith,** which is ground tissue of large **parenchyma** cells. Second, stems transport substances between roots and leaves by vascular tissue of two types—the **xylem** and **phloem.** Third, green stems produce food by photosynthesis. Fourth, stems are food and water storage sites.

Leaves are the chief food-producing organs in most plants. Diffusion of respiratory and photosynthetic gases into and out of the plant occurs through **stomata** on the surface of the leaf. Other than the stomata, a waterproof **cuticle** covers the leaf and prevents excessive water loss. **Guard cells** surrounding the stomata open and close them according to the plant's internal characteristics. Most of the leaf interior consists of **palisade** and **spongy mesophyll,** which is composed of photosynthetic cells that produce many organic compounds. **Air spaces** throughout the spongy mesophyll are continuous with the stomata to the leaf exterior and thus facilitate gas exchange.

The anatomy of a seed (and, in fact, other plant parts as well) varies between the **monocotyledons** and **dicotyledons.** A bean seed (representing a dicot) is covered by a protective **seed coat.** Most of the bean seed consists of the two **cotyledons** (seed leaves) that supply food for the embryo before photosynthesis begins. The tiny embryo plant lies between the cotyledons. It has two reduced **first foliage leaves** that will grow into the first true leaves. The short main axis of the embryo has **apical meristems** at each end that produce root and stem growth.

The protective outer covering of a corn kernel (a typical monocot) is actually a fusion of the seed coat and the structure of the mature fruit. Internally, a large **endosperm,** food for the germinating embryo, occupies much of the seed. The single cotyledon has no nutritional function. The small foliage leaves are covered by a tough **coleoptile** that protects them as they push through the soil. As in the dicots, monocot apical meristems give rise to leaves, stem, and roots.

**Primary growth** occurs as an increase in shoot and root length due to apical meristems. Roots reach deeper into the soil by primary growth. A **root cap** protects the dividing meristem tip from abrasion. Behind the meristem region is a **zone of elongation** where new cells increase in length and actually push the root tip through the soil. Behind this area is the **zone of maturation** where cells reach full size and become specialized for their specific functions. One such example is the **endodermis,** which controls the movement of substances between the root cortex and root interior.

Stems increase in length as the apical meristems produce new cells. **Secondary growth** increases stem girth produced by the **vascular cambium,** a ring of **lateral meristem** that gives rise to **secondary xylem (wood).** The stem grows in diameter as new xylem cells are produced toward the stem interior. The phloem outside the vascular cambium is crushed as stem diameter increases. Destroyed phloem is replaced by secondary phloem that is produced by cells on the opposite side of the vascular cambium. As more wood is produced, more secondary phloem is destroyed but again replaced. The process continues and the stem grows ever wider. The **cork cambium** is located on the outside of the secondary phloem. It produces a protective, waterproof layer of **cork.** Secondary phloem and the cork layer compose tree bark.

Branches are produced by different processes in roots and stems. Branch roots arise from the meristematic **pericycle** within the root interior. The branch roots grow through the cortex and emerge into the soil. Stem branches occur as **axillary (lateral) buds** become mitotically active, producing a new shoot.

In addition to seed structure (as above), other traits distinguish the monocots and dicots. Leaf venation in monocots is usually parallel in contrast to the network venation of dicots. Monocot flower parts typically occur in threes or multiples of three. Dicot flower parts typically occur in fours, fives, or their multiples. Vascular tissue is scattered throughout the stem in monocots. In dicots, it is restricted to the outer portion of the stem. Some monocots have **intercalary meristems,** which are located between leaf bases and are not found in dicots. Monocots are mostly herbaceous with few trees and no significant secondary growth. Dicots occur as herbs, shrubs, and trees with impressive secondary growth.

Transport of materials throughout the entire plant occurs via the xylem and phloem, the plant's vascular tissues. The xylem carries water and minerals from the roots to the stem and leaves. The phloem transfers food products from the leaves down to the roots. These functions are demonstrated when we girdle a tree trunk, which is when the outermost ring of bark is removed from a tree trunk. Sugary fluid appears on the bark edge above the cut, which substantiates the assertion that food moves downward via phloem located in bark. The leaves continue to grow normally since they are supplied with water by the unharmed xylem (wood). With time, however, the roots eventually die of starvation because the phloem route has been broken by the girdling. With the loss of a water and nutrient supply system, the entire tree then dies.

The movement of water up a plant stem is a result of several forces working together. Some movement occurs by **capillarity,** the rising of water up a thin tubule (such as a xylem cell) due to cohesive and polar properties of water molecules and their attraction (adhesion) to xylem cell walls. Additional force comes from **root pressure,** indirectly produced by active transport in the roots. Active transport brings necessary ions into the root and water follows from the soil by osmosis. Hydrostatic pressure builds in the root as water content increases and it rises up the xylem. If water loss from the leaves is minimal, such as on a humid day, root pressure may actually force sap from the leaf tips as **guttation.**

Another force is **tension-cohesion** (also known as **transpiration pull**) that occurs as a result of water vapor loss through the stomata. As water escapes to the atmosphere it is replaced in the leaf mesophyll by water from xylem veinlets. This continual loss and replacement of water draws water up the xylem by cohesion. Transpiration pull can vary depending on atmospheric conditions. A dry, hot, or windy day increases transpiration by increasing water loss. Little water in the soil and high humidity reduce transpiration pull.

By these three mechanisms, water can move great distances from deep roots, through a towering trunk, and up to every leaf of a tree. Root pressure and transpiration pull have been demonstrated visually by experiments. Root pressure was observed by sealing a glass tube onto a plant stump which was cut near the ground. With time, sap crept up into the tube to a maximum height where it could no longer be pushed upward by root pressure. Transpiration pull was measured by attaching a tube filled with water to the cut root of a tree. A vessel of mercury was then placed in contact with the water so that a change in mercury movement up the tube indicated the pulling of water by transpiration.

Phloem transport probably occurs by **pressure flow** (also known as **mass flow**), which works on a principle of a sucrose gradient within different areas of the plant. Sucrose sources are photosynthetic parts or food storage areas that release sucrose for use. Other areas continually use sucrose (**sucrose sinks**) either by respiration or conversion into storage forms. The phloem contents always move from a region of high sucrose concentration to a region of low sucrose concentration. As the plant's physiology lowers the sucrose concentration in any one area, the sucrose content of the phloem will always move or shift towards the lowered concentration.

The following experiments will demonstrate some of the characteristics of the plant body and the plant transport systems.

*Note:* At the beginning of the laboratory period, start Labs 18.G and 18.I, which require longer periods of time in order to obtain results.

## PRELAB QUESTIONS

1. What type of vascular tissue carries water and minerals from the roots to the stem and leaves?

2. What three mechanisms are involved in water movements in plants? Which is primary?

3. What type of gradient is involved in the pressure flow theory of phloem transport?

4. What event occurs when root pressure forces sap from the leaf tips?

5. What three basic organs compose a vascular plant?

6. Identify one major difference between a monocot and a dicot.

7. What are the two types of vascular tissue?

8. What type of growth occurs as an increase in length of shoots and roots?

9. What type of root system is characterized by having one large main root?

## LAB 18.A
## GENERAL STRUCTURE OF AN ANGIOSPERM PLANT

### MATERIALS REQUIRED
potted bean plant

### PROCEDURE
1. Obtain a potted bean plant to identify the following general structures of an angiosperm plant.
2. An angiosperm plant is composed of three basic organs: leaves, stem, and roots. The root system is below the soil level; therefore, carefully lift the plant out of the pot to view the root system. After

viewing the root system, carefully place the bean plant back into the pot.

3. The part of the plant above ground is the stem. The areas of the stem where the leaves are attached are called nodes. The areas of the stem between the nodes are referred to as internodes.

4. At the tip of the plant is the terminal bud, which contains the actively dividing apical meristem.

5. The leaf of the bean plant is attached to the stem by a leaf stalk called a petiole. The blade of the leaf is divided into three leaflets, making this a compound leaf. A single undivided leaf blade is called a simple leaf.

6. At the base of the petiole are small lateral outgrowths called stipules.

7. Between the stem and the petiole of a leaf (or in the axil of the leaf) lie the axillary bud. When triggered, the axillary bud will develop into a lateral stem (branch).

## POSTLAB QUESTIONS

1. Describe the root system you observed on this bean plant.

2. Identify the following structures of this bean plant: stem, petiole, stipule, axillary bud, terminal bud, node, and internode.

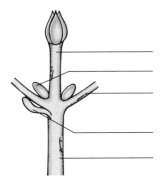

## LAB 18.B
## THE STEM

### MATERIALS REQUIRED

prepared slides of the following:
a. corn (*Zea*) stem, x. s
b. sunflower (*Helianthus*), x. s.
c. *Tilia* 3 yr. stem, x. s.

cross-section slab of small log

compound microscope

twigs of buckeye or horse chestnut (*Aesculus* sp.)

### PROCEDURE

A stem, whether it be a monocot or a dicot, is composed of three tissue systems: ground, vascular, and dermal. The ground tissue system is composed of several cell types that function in support, storage, secretions, and a variety of other functions. The vascular tissue is composed of xylem and phloem. Xylem conducts water up from the roots to the rest of the plant body, whereas phloem carries food from the leaves (or stem) to the rest of the plant. The dermal tissue system provides an outer protective covering of the plant.

1. Bring a compound microscope and a prepared slide of corn (*Zea*) stem in cross section to your work station.

2. View the slide under lowest power and note the vascular bundles scattered throughout the stem (Fig. 18–1a). Each vascular bundle is composed of xylem and phloem cells. The phloem cells are the smaller cells closer to the outside of the stem, whereas the xylem cells are generally the larger cells closer to the middle of the stem (Fig. 18–1b).

3. The single layer of cells surrounding the entire stem is the epidermis, which is part of the dermal tissue system. View the epidermal cells under high power.

4. Rotate the objective lenses to lowest power and note the remaining cells. These cells compose the ground tissue system.

5. Obtain a prepared slide of a sunflower (*Helianthus*) stem in cross section and view under lowest power. Note the distribution of vascular bundles. The peripherally arranged vascular bundles are typical of a dicot stem. Scattered vascular bundles, as you saw in the corn stem, are typical of a monocot stem (Fig. 18–2a).

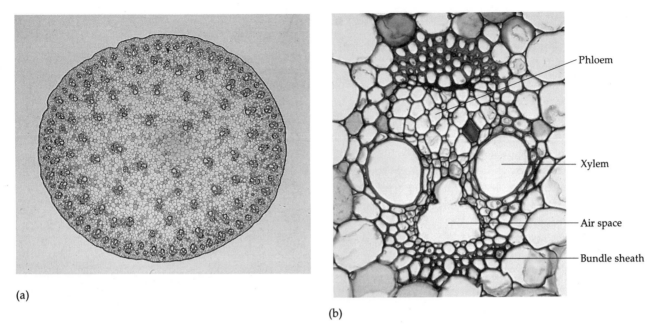

(a)

(b)

**FIGURE 18-1**

Arrangement of stem tissues in *Zea mays* (corn), a monocot. (a) Cross section of stem showing the scattered vascular bundles. (b) Close up of one of the bundles. The air space is where the first xylem cells were formed. Note that these cells are no longer functional. The entire bundle is enclosed in a sheath of sclerenchyma for additional support.

6. Again, note the smaller phloem cells toward the outside of the stem and the larger xylem cells closer to the interior of the stem (Fig. 18–2b).

7. Remember, the single layer of cells surrounding the entire stem is the epidermis, which is part of the dermal tissue system. View the epidermal cells under high power.

8. Rotate the objective lenses to lowest power and note the remaining cells. These cells compose the ground tissue system. The cells around the periphery, closer to the epidermis, are referred to as the cortex. The cortical cells usually have thicker cell walls and function in support. The rest of the cells located between the vascular bundles are the pith. These cells are usually characterized by intercellular spaces and may contain chloroplasts for photosynthesis. View cortical and pith cells under high power.

　All plants have primary growth due to the activity of apical meristems, which increases a plant in length. Some plants have secondary growth due to the activity of lateral meristems, which causes an increase in a plant's diameter. Plants that have only primary growth are called herbaceous, whereas plants with secondary growth are referred to as woody.

9. View under low power a 3-year-old stem of *Tilia* (see Fig. 18–3).

10. The vascular cambium is a lateral meristem that gives rise to secondary xylem and secondary phloem. Secondary phloem is laid down to the outside of the vascular cambium and the secondary xylem is laid down to the inside. View under low power the vascular cambium, secondary xylem, and secondary phloem.

11. Notice how the secondary xylem is divided into layers called annual rings since each ring usually represents a growing season. Also, notice within each layer two different cell sizes. The larger cells represent growth early in the growing season and are called spring wood; the smaller cells represent growth later in the growing season and are called summer wood.

12. The secondary phloem comprises the inner part of the bark and functions as an epidermis in a woody plant. The cork cambium is a lateral meristem which gives rise to the tissues that comprise the outer part of the bark. View under the microscope the secondary phloem, cork cambium, and bark.

13. View the pith cells in the middle of the stem cross section and note the rays that radiate from the

**FIGURE 18-2**

Primary growth in a dicot stem. (a) Cross section of a *Helianthus* (sunflower) stem showing the arrangement of tissues. The vascular bundles are arranged in a circle. (b) Close up of *Helianthus* vascular bundle. The xylem is to the inside and the phloem to the outside. Each vascular bundle is "capped" by a batch of fibers for additional support.

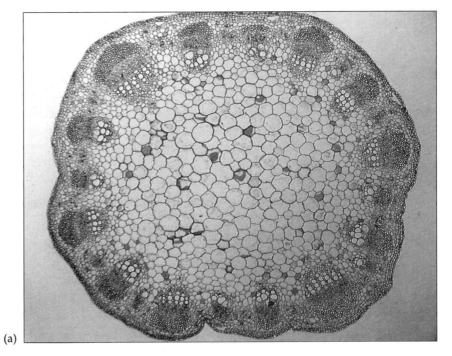

(a)

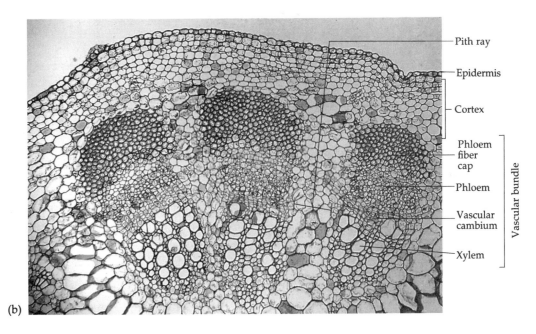

(b)

- Pith ray
- Epidermis
- Cortex
- Phloem fiber cap
- Phloem
- Vascular cambium
- Xylem

Vascular bundle

stem's center. Rays carry water and solutes laterally in the tree trunk whereas xylem and phloem cells conduct materials in a vertical direction.

14. Obtain a cross section of a log and notice that this log is composed mainly of secondary xylem, more commonly known as wood. The wood in the center is darker in color compared to the wood closer to the perimeter. The darker-colored wood is nonfunctioning xylem and is called heartwood. The cells of heartwood are clogged with resins, pigments, and other compounds that prevent sap flow through these xylem cells. The active, lighter-colored wood is referred to as sapwood because it is the xylem that transports sap up the tree trunk.

15. On this piece of log, as you did in the *Tilia* stem, locate the bark, vascular cambium, wood, annual rings, pith, and rays.

16. Obtain a woody buckeye or horse chestnut twig. Located at the tip of the stem is the terminal bud, which is covered with bud scales. The bud scales

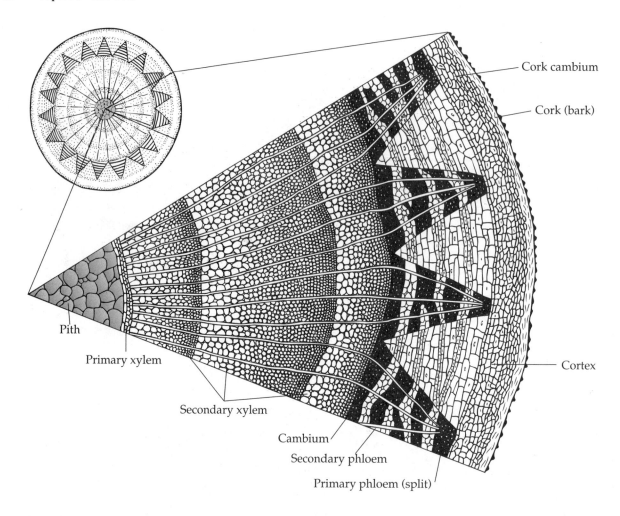

**FIGURE 18-3**

A woody twig in cross section. This twig has grown for three seasons. It has one ring of primary xylem and three rings of secondary xylem, one for each season.

protect the terminal bud during the fall and winter months. The bud scales are shed during the spring when growth begins again.

17. Locate the rings of bud scale scars along the woody twig. The distance between one ring of bud scale scars to the next represents a year of growth.

18. Locate the leaf scars (Fig. 18–4) and the vascular bundle scars (or vein scars).

19. Locate the axillary bud and the lenticels widely distributed over the woody twig. Lenticels function in gas exchange in woody twigs, much like stomates function in herbaceous plants.

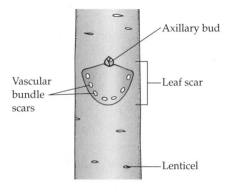

**FIGURE 18-4**

Enlarged leaf scar showing location of vascular scar (vein scar).

## POSTLAB QUESTIONS

1. If a tree is girdled (a ring of bark removed from around the tree), what happens to the tree? (Think: what tissues compose the bark?)

2. Identify an important function of secondary xylem other than the transport of water and minerals.

3. Identify whether this stem is a monocot or a dicot. Explain your answer.

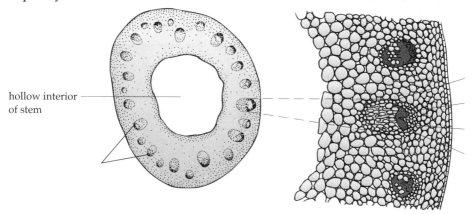

hollow interior of stem

4. Identify the structures of the stem in the previous question.

5. Identify parts of this tree trunk.

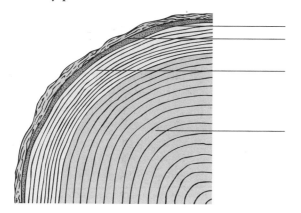

6. Identify the structures of this woody twig.

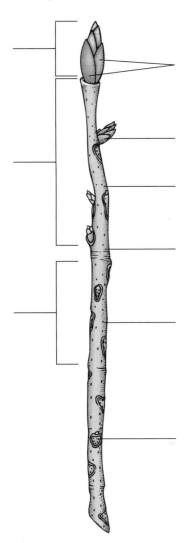

7. Would tree bark or sawdust (wood) make a better mulch for garden plants?

## LAB 18.C
## THE ROOT

### MATERIALS REQUIRED

carrot plants with greenery on demonstration

corn plants with roots on demonstration

stereomicroscope

radish seedlings

prepared slides of the following:
a. buttercup (*Ranunculus*) root, x.s.
b. corn (*Zea*) root, x.s.

### PROCEDURE

Seed plants are characterized by two types of root systems. Dicots generally have one main root or tap-root with some lateral root branching. The taproot penetrates deeply into the soil, but is not as absorptive as the fibrous root system. The fibrous root system, which is mainly found in monocots, is a finely branched network of roots.

1. Compare the root systems of the carrot with the corn plant. Note the type of root system.
2. Examine with the stereomicroscope the root hairs of radish seedlings. Root hairs are extensions of the epidermal cells and increase the absorbing surface of roots.
3. View under lowest power a prepared slide of a buttercup (*Ranunculus*) root in cross section (Fig. 18–5a). Note the epidermal layer of cells. The cortex composes the majority of this root cross section. These cortical cells contain plastids filled with starch instead of chlorophyll-containing plastids (chloroplasts).
4. The circular innermost layer of cortical cells is called the endodermis. The endodermal cells have a band of suberin (a waxy waterproof substance) within the radial and transverse walls. This band is called the Casparian strip that, in addition to the endodermal cell membranes, regulates the transport of soil solution between the cortex and the vascular tissues. View the endodermis under high power.
5. The layer of cells within the endodermal layer is called the pericycle. The pericycle still actively divides and gives rise to lateral roots that emerge through the cortex.
6. View under lowest power the vascular tissue within the endodermis. The xylem cells appear as an X and the phloem cells are between the arms of the X (Fig. 18–5b).
7. View under lowest power a prepared slide of a corn (*Zea*) root in cross section (Fig. 18–6a). Note the epidermis, cortex, endodermis, and pericycle.
8. The arrangement of the vascular tissue in the corn root differs from the buttercup root. Again, the difference between monocots and dicots are demonstrated. Dicots typically have the X configuration, whereas monocots have circularly arranged vascular tissue. Note the xylem toward the inside and the phloem toward the outside of the corn root (Fig. 18–6b).

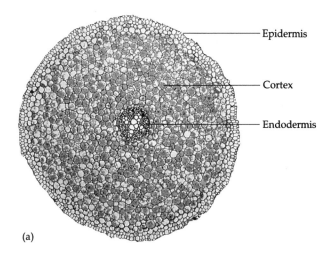

(a)

Epidermis

Cortex

Endodermis

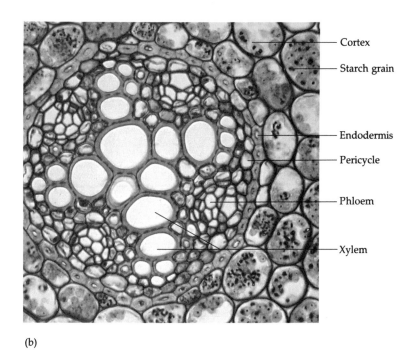

(b)

Cortex

Starch grain

Endodermis

Pericycle

Phloem

Xylem

**FIGURE 18-5**

Cross section of a buttercup *(Ranunculus)* root. Buttercups are dicots with primary growth. (a) This shows the entire root. Note that the bulk of the root is the cortex. (b) A close up of the center of the root. Note the solid core of vascular tissues, with xylem arms and phloem patches between the xylem arms.

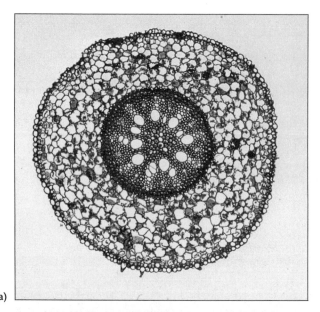

(a)

**FIGURE 18-6**

Cross section of a greenbriar *(Smilax)* root. Greenbriar exhibits the general characteristics of a monocot root. (a) This shows the entire root. (b) Close up of a portion of the center of the root showing the vascular tissues and the pith.

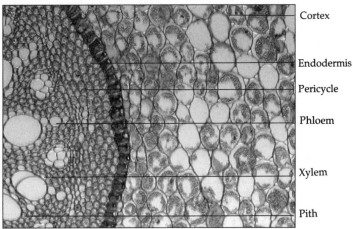

Cortex

Endodermis

Pericycle

Phloem

Xylem

Pith

(b)

## POSTLAB QUESTIONS

1. Which root system is more efficient for anchoring a plant? Absorbing water? Storage?

2. Identify whether this root is a monocot or dicot. Explain your answer.

3. Identify the structures of the root in the previous question.

4. Identify whether the following plants are monocots or dicots: carrot, corn, radish, and buttercup.

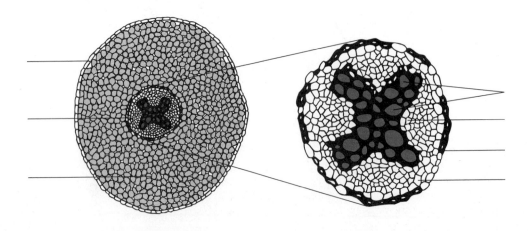

5. What type of root system did the bean plant in Lab 18.A have?

## LAB 18.D
## THE LEAF

### MATERIALS REQUIRED

prepared slide of privet *(Ligustrum)* leaf, x.s.

bean plant

corn plant

### PROCEDURE

1. View under lowest power a prepared slide of a cross section of a privet *(Ligustrum)* leaf. Locate the single layer of cells on the upper and lower surfaces of the leaf (Fig. 18–7). These layers are the upper and lower epidermis. Now, scan the lower epidermis for stomates. On each side of these stomates are specialized cells appropriately called guard cells. The swelling and shrinking of the guard cells control the opening of the stomates. The stomates allow for the exchange of gases. View the stomates and guard cells under high power.
2. The tissue between the two epidermal layers are the mesophyll layers. These layers are composed of cells that contain chloroplasts; therefore, they are the photosynthetic tissue of the leaf.

The upper mesophyll layer of compactly arranged cells is called the palisade mesophyll. The layer of mesophyll that is irregularly arranged with lots of air spaces is called spongy mesophyll. Locate the spongy and palisade mesophyll on the privet leaf.
3. Veins of a leaf are strands of vascular tissue running through the leaf. Locate the vein in the middle of the cross section of the privet leaf. Locate the xylem, which lies closer to the upper epidermis, and the phloem, which lies closer to the lower epidermis. The arrangement of nearby vascular tissue allows the transport of materials to and from mesophyll cells.
4. Compare the vein pattern in the bean plant with the vein pattern in the corn plant. Monocots generally have a parallel leaf venation; dicots commonly have a network venation pattern.

### POSTLAB QUESTIONS

1. Are the epidermal layers entirely covered with a cuticle?

2. How is the location of chloroplasts consistent with the function of leaves?

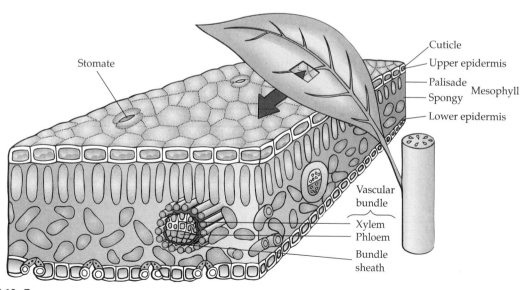

**FIGURE 18–7**

The arrangement of tissues in a typical leaf blade. The blade is covered by an upper and lower epidermis. The photosynthetic tissue, the mesophyll, is often organized into palisade and spongy layers. Veins branch throughout the mesophyll.

3. Which plant (bean, corn) is a dicot? A monocot?

4. Complete Table 18–1.

**Table 18–1** Monocot Vs. Dicot

| Plant Organ | Monocot Characteristics | Dicot Characteristics |
|---|---|---|
| Stem | | |
| Leaf | | |
| Root | | |

5. Identify two structures found in root and/or shoot epidermis that change in function from protection to some other role.

## LAB 18.E
## PLANT ORGANS AS FOOD

### MATERIALS REQUIRED

variety of vegetables

### PROCEDURE

Examine the various grocery store vegetables and identify the parts of the plant they represent. Complete Table 18–2.

**Table 18–2** Grocery Store Vegetables

| Vegetable | Plant Organ |
|---|---|
| Brussel sprout | |
| Cabbage | |
| Celery | |
| Onion | |
| Potato | |
| Radish | |
| Sweet potato | |

### POSTLAB QUESTION

1. Stems and leaves are always found growing above ground. Which grocery store vegetable(s) supports or disputes this statement?

## LAB 18.F
## ROOT HAIRS—THE SITE OF WATER UPTAKE IN ROOTS

### MATERIALS REQUIRED

petri plate of germinating radish seeds, on demonstration

stereomicroscope

### PROCEDURE

In addition to anchoring the plant in soil, the roots also absorb water found between the microscopic soil particles. Since water is located in microscopic spaces, microscopic root structures are well suited for water absorption.

1. Remove the cover of the petri plate on display that contains germinating radish seeds. Carefully focus on the young root emerging from the seed coat. Notice the delicate white fuzz in the region near the tip of the growing root. These tiny strands are root hairs and are actual extensions of the root epidermal cells (Fig. 18–8). The longer root hairs farthest from the root tip are the older root hairs.

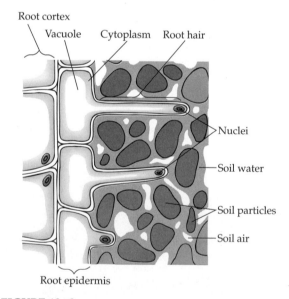

**FIGURE 18–8**

Diagram of root hairs showing their structure and relationship to the root epidermis and surrounding soil.

2. Replace the petri plate cover over the radish seeds after you have observed them.

## POSTLAB QUESTIONS

1. Why are the root hairs located farthest from the growing tip the older root hairs?

2. Besides water, what else is absorbed by roots and how?

3. Why is the cylindrical shape advantageous in carrying out root functions?

## LAB 18.G
## A DEMONSTRATION OF ROOT PRESSURE

### MATERIALS REQUIRED

severed rooted plant stems with attached 1 ml pipet, on demonstration

### PROCEDURE

Root pressure is one of the forces that moves water upward through the stem from the roots. The water collected by thousands of root hairs begins to accumulate within a plant's root system. After a while, the quantity of water becomes significant, and the building pressure pushes water (and dissolved nutrients) from the roots upward into the xylem tissue of the stem.

1. Observe the initial water volume in the pipet attached to the rooted cut stem by a piece of rubber tubing (Fig. 18–9). Record the initial measurement in Table 18–3 at Time 0.

2. Continue to read the water volume in the pipet every 20 minutes during the next hour. Record the change in water volume in Table 18–3.

**Table 18–3**  Demonstration of Root Pressure

| Time | Measurement in Mls | Change in Mls |
|------|--------------------|---------------|
| 0 |  | 0 |
| 20 |  |  |
| 40 |  |  |
| 60 |  |  |

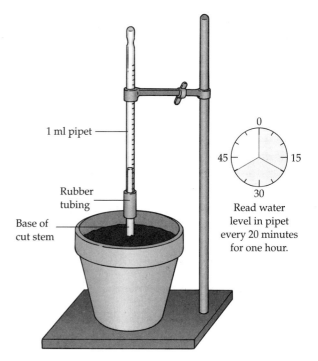

**FIGURE 18–9**

Apparatus of severed, rooted stem and attached pipet for measuring root pressure.

## POSTLAB QUESTIONS

1. What process moves water from the soil into the roots of plants?

2. Why does root pressure not explain the rise of water in the tallest plants?

## LAB 18.H
## STOMATA—SOURCE OF WATER LOSS IN PLANTS

### MATERIALS REQUIRED

fresh *Kalanchoe* leaves

forceps

slide and coverslip

distilled water in dropper bottle

10% salt solution in dropper bottle

### PROCEDURE

Stomata (stomate, singular) are microscopic pores in the surface of leaves that function in gas exchange. Carbon dioxide, essential for photosynthesis, diffuses into the leaves from the atmosphere by means of the stomata. Oxygen diffuses from the leaf tissue, out of the stomata, and into the air as a by-product of photosynthesis. Loss of water vapor from the interior of the leaf also occurs through the stomata as these gases are exchanged. However, stomata are not simply gapping holes that continually lose water from the plant—they are regulated as to how much water escapes from the leaves. This exercise introduces you to the structure of stomata and how they function.

1. Obtain a fresh leaf from a *Kalanchoe* plant. Keeping the leaf right side up, bend the leaf perpendicular to its midrib until it breaks, but do not pull apart the two pieces; they should still be attached by the lower epidermis, or "skin" of the leaf.
2. Gently pull away the two leaf pieces from one another to peel off the lower epidermis. The epidermis should appear as a thin, translucent sheet that cleanly peels from the fleshy leaf tissue.
3. Place a small sheet of epidermis on a slide. Be careful to avoid wrinkling the epidermis. Position the sheet with forceps if necessary. Add a drop of distilled water and a coverslip to the epidermis.
4. Place the slide on your compound microscope and observe under low power. The irregular epidermal cells appear as transparent pieces of a jigsaw puzzle that interlock with one another. Scattered throughout the epidermis are oval-shaped structures that resemble open mouths with two lips. The hole of the "mouth" is a stomate. The "lips" of the mouth are two guard cells that surround each side of the stomate.
5. Switch to high power and carefully observe an individual stomate and its guard cells. Notice the small green specks within the two oblong guard cells. What are the green specks? _____

_____

6. Lift the coverslip and apply a drop of 10% salt solution to the leaf epidermis. Wait a moment for the solution to penetrate the epidermal cells. Again examine an individual stomate and its guard cells under high power.

**POSTLAB QUESTIONS**

1. Why does the 10% salt solution change the size of the stomatal opening?

2. Stomatal openings on the leaf make up only about 1% of the surface area, whereas the leaf sometimes transpires half as much water as would evaporate from an equivalent area of wet filter paper. Why isn't evaporation directly proportional to area?

**LAB 18.I**

# TRANSPIRATION AS A MECHANISM OF WATER TRANSPORT IN THE XYLEM

**MATERIALS REQUIRED (per group of students)**

3 celery stalks in water, 2 with leaves, 1 without

3 small flasks of fuchsin dye, labelled A, B, C

electric fan (needed for the entire class)

gooseneck lamp with 100 watt bulb (needed for the entire class)

razor blade

metric ruler

**PROCEDURE**

Transpiration is the evaporation (loss) of water through stomata in the leaves of plants. Since water molecules cohere to one another by hydrogen bonds and adhere to xylem cell walls, water lost from the leaves by transpiration is continually replaced by water from the xylem tissue in the stem. This continual loss and replacement of water in leaf tissue results in a "pulling" of water molecules all the way from the roots and up through the stem.

1. Using a razor blade, cut about 1 cm from the bottom end of each celery stalk *as it remains submerged in water.* It is crucial that the cut be made underwater or else air will enter the xylem and interfere with the success of the exercise.
2. *Quickly* place the cut end of each celery stalk into the three flasks of fuchsin dye in the following manner: flask A, stalk with no leaves; flask B, stalk with leaves; flask C, stalk with leaves.

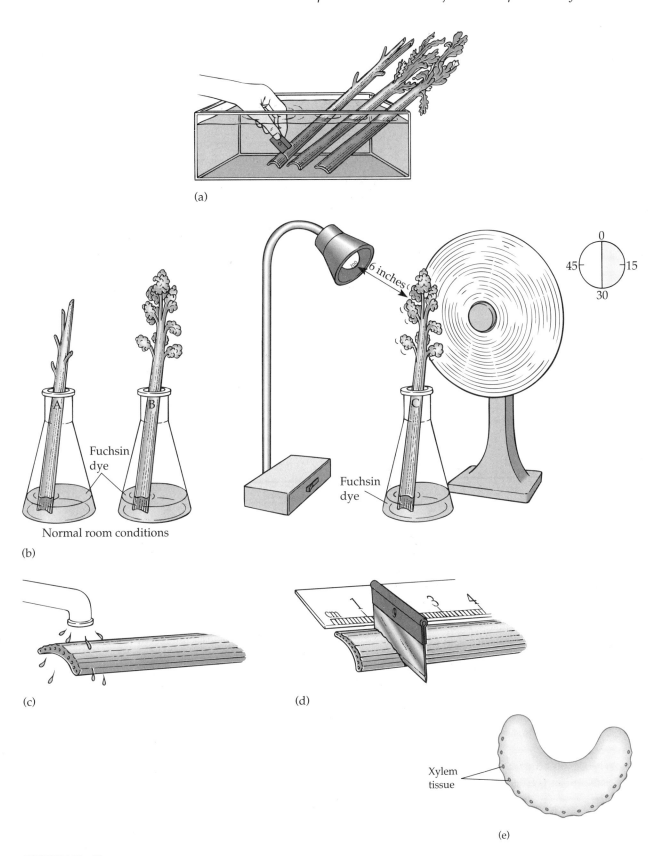

(a)

(b)

Fuchsin dye

Normal room conditions

6 inches

Fuchsin dye

(c)

(d)

Xylem tissue

(e)

**FIGURE 18–10**

Procedure demonstrating xylem transport. (e) Cross section of celery stalk showing location of xylem tissue containing fuchsin dye.

**C A U T I O N**  Fuchsin dye permanently stains clothing.

3. Place flasks A and B on a lab bench in normal room conditions.

4. Place flask C directly in front of the running fan so that air blows across the leaves of the celery stalk. In addition, turn on the gooseneck lamp and position it about 15 cm (6 in.) from the top of the celery stalk to warm the leaves.

5. Note the time and allow all celery stalks to stand for 30 minutes in their designated environmental conditions.

6. Remove the celery stalks from their flasks after timing, making sure to note which flask they came from. Rinse excess dye from the stalks under running water and place them on a paper towel.

7. Use a metric ruler and razor blade to cut 1 cm segments from the bottom end of the leafless stalk from flask A. After each cut is made, examine the cut end of the stalk for the presence of purple dye in the xylem tissue positioned along the outer edge of the stalk (Fig. 18–10). Continue cutting 1 cm segments from the stalk until the dye begins to fade in the xylem. Record in Table 18–4 the distance the dye traveled up the leafless stalk of flask A.

8. Repeat Step 7 for the remaining two celery stalks and record the results.

**Table 18–4**    Results of Transpiration Exercise

| | | |
|---|---|---|
| Flask A, leaves absent, normal environment | _____ | cm |
| Flask B, leaves present, normal environment | _____ | cm |
| Flask C, leaves present, wind and heat | _____ | cm |

## POSTLAB QUESTIONS

1. How does the environment affect water transport in plants?

2. Why doesn't the term "circulation" apply to water transport in plants?

3. Identify the two variables in flask C compared to flask B. How could this exercise more closely follow the scientific method?

## FOR FURTHER READING

Baker, D. 1978. *Transport Phenomena in Plants.* New York: Chapman & Hall.

Briggs, D., and S. Walters. 1984. *Plant Variation and Evolution.* Cambridge, England: Cambridge University Press.

Buvat, Roger. 1989. *Ontogeny, Cell Differentiation, and Structure.* New York: Springer-Verlag.

Crafts, A. S., and C. E. Crisp. 1971. *Phloem Transport in Plants.* San Francisco: W. H. Freeman and Company.

Crawford, Daniel J. 1990. *Plant Molecular Systematics: Macromolecular Approaches.* New York: Wiley.

Cronshaw, J. 1971. *Support and Protection in Plants: Topics in the Study of Life.* New York: Harper & Row.

Dennis, David T., and David H. Turpin, eds. 1990. *Plant Physiology, Biochemistry, and Molecular Biology.* New York: Wiley.

Epstein, E. 1972. *Mineral Nutrition of Plants.* New York: John Wiley and Sons.

Esau, K. 1977. *Anatomy of Seed Plants.* New York: John Wiley & Sons.

Galston, A. W. et al. 1980. *The Life of the Green Plant.* Englewood Cliffs, N.J. Prentice-Hall, Inc.

Iqbal, M., ed. 1990. *The Vascular Cambium.* New York: Wiley.

Jensen, W. A., and R. F. Salisbury. 1984. *Botany.* Belmont, Calif.: Wadsworth.

Kaufman, P. B. et al. 1989. *Plants: Their Biology and Importance.* New York: Harper & Row.

Mosbrugger, Volker. 1990. *The Tree Habitat in Land Plants: A Functional Comparison of Trunk Constructions with a Brief Introduction into the Biomechanics of Trees.* New York: Springer-Verlag.

Raven, P. H. et al. 1986. *Biology of Plants.* New York: Worth.

Ray, P. M. et al. 1983. *Botany.* Philadelphia: Saunders College Publishing.

Rollet, Bernard et al. 1990. *Stratification of Tropical Forests as Seen in Leaf Structure.* Boston: Kluwer Academic Publishers.

Sachs, Tsvi. 1991. *Pattern Formation in Plant Tissues.* New York: Cambridge University Press.

Salisbury, F., and C. Ross. 1985. *Plant Physiology.* Belmont, Calif.: Wadsworth.

Schwintzer, Christa R., and John D. Tjepkema, eds. 1990. *The Biology of Frankia and Actinorhizal Plants.* San Diego: Academic Press.

Stern, K. R. 1988. *Introductory Plant Biology.* Dubuque, Iowa: William C. Brown Publishers.

❏

# Reproduction and Development in Complex Plants

## OBJECTIVES

1. Distinguish between sexual and asexual reproduction.

2. Define each of the following structures: rhizome, tuber, stolon, corm, and bulb.

3. List the general features of an angiosperm flower.

4. Distinguish between simple, aggregate, multiple, and accessory fruits. List and define several types of simple fruits.

5. Trace the stages in embryo development in flowering plants.

## INTRODUCTION

When we think of flowering plants, what usually comes to mind are the angiosperm plants that we see outdoors or have potted inside the house. These flowering plants are, in more scientific terminology, the diploid sporophyte generation in the life cycle of the angiosperms. The sporophyte gives rise to haploid **spores** of two sizes by meiosis. The smaller spore, called a **microspore,** develops into the male gametophyte generation, called a **pollen grain.** The pollen grain produces male gametes, called sperm nuclei. Microspores and the pollen they produce are located in the **anther** portion of the **stamens,** which are the male flower parts.

The larger spore, called a **megaspore,** develops into the female gametophyte generation, called an **embryo sac.** The embryo sac produces the female gamete, an egg nucleus. The embryo sac is contained within the **ovule** (immature seed) held inside the **ovary** of the flower. The ovary is the lower, larger portion of the **carpel,** the female flower part. Above the ovary is the **style,** a neck-like elongation ending with a sticky **stigma** that traps pollen grains. The car-

pel and stamens are the actual reproductive parts of a flower. Most flowers have accessory parts surrounding the reproductive organs. **Sepals** are green leaf-like structures that protect the developing flower bud. **Petals** occur inside the sepals. They are often brightly colored so as to attract animal pollinators.

Consider for a moment how often flowers enhance our lives — as gifts, decorations, and expressions of happiness and sadness. We cultivate many plants merely for their flowers. Many of life's events involve flowers in some way. One cannot help but wonder, however, if the tables were turned, whether plants would use the reproductive organs of animals in similar fashions.

**Pollination** takes place when pollen is transferred from the anther to the stigma of a flower. Fertilization occurs when the pollen grain grows a pollen tube containing two sperm nuclei that originated from the pollen grain. The pollen tube grows down through the style tissue, through the ovary, and into the ovule where the sperm nuclei are released into the embryo sac. One sperm nucleus fuses with the egg nucleus. This produces a zygote. The other sperm nucleus fuses with two polar nuclei. This forms a triploid **endosperm** nucleus. The endosperm nucleus divides repeatedly producing starchy endosperm tissue that serves as a food source for the future plant embryo.

After the endosperm forms, the zygote develops into an embryo. The ovule matures into a **seed** that contains endosperm, an embryo, and a protective **seed coat** formed from the ovule wall.

When the ovules develop, the ovary wall begins maturing into a **fruit** that will contain seeds. Many plants have highly specialized fruits that aid in seed dispersal or that ward off predators.

There are four major types of fruit: simple, aggregate, multiple, and accessory. By far, the simple fruits are the most common. Simple fruits include blueberries, grapes, bananas, and—surprisingly, to some—the tomato. Simple fruits develop from a single ovary of a single flower. A single flower that contains many separate carpels gives rise to an aggregate fruit. Aggregate fruits include blackberries and raspberries. Fruits that form from the fusion of many flowers' ovaries are defined as multiple fruits. The pineapple is the best known multiple fruit. Finally, there are accessory fruits in which tissues, in addition to the ovary, are incorporated into the fruit. These include the edible parts of strawberries, apples, and pears.

A seed may not **germinate** (begin growing) unless the proper (environmental) conditions are met. Water, of course, is required along with oxygen. Many seeds need specific temperature and light conditions. Unusual factors, such as passage through an animal digestive tract, are involved in the germination of some seeds.

In addition to sexual reproduction, many plants reproduce asexually via **vegetative reproduction.** This type of reproduction provides an advantage to plants that are well adapted to their environment. They can vegetatively produce many copies of themselves, each with an identical set of genes that specifically adapts them to the present growing conditions.

In the asexual reproduction of flowering plants, there are a number of structures that are, essentially, modified stems that may become involved in the process of vegetative or asexual reproduction. These include rhizomes, tubers, bulbs, corms, and stolons. Rhizomes are branching horizontal underground stems that can be divided either naturally or artificially to produce new, independent plants. The iris is an example of a plant propagated by divided rhizomes.

Tubers are enlarged underground stems that function in food storage. Tubers have lateral buds (the "eyes" of the common white potato) that can grow into new plants. Bulbs contain storage leaves, a stem, and roots. A bulb can give rise to a single plant. Onions, tulips, and daffodils form bulbs.

A corm resembles a bulb, but the storage organ of a corm is the stem. The corm has scales that are modified leaves and are attached to the corm at nodes. Lateral buds can arise from these nodes. Plants that produce corms include the cyclamen, the crocus, and the gladiolus.

Stolons are aboveground runners with long internodes and adventitious buds from which new plants can arise. The strawberry plant, for example, reproduces by stolons.

Vegetative reproduction is beneficial when one wishes to reproduce the desirable traits of a specific plant. By rooting leaves or stem cuttings from the valued plant, many more individuals with the desired characteristics can be produced. **Grafting** is a vegetative technique in which a **scion** (cutting) of valued traits is grafted with a **stock** (stem or root system) of a related plant. The scion and stock are wrapped together. With time they grow together as one plant. This enables horticulturalists to combine the desirable qualities of various plant parts.

Plant tissue culture is a convenient method of producing many new individuals from small samples of meristematic tissue grown in the laboratory. This culturing process is also possible by vegetative reproduction. The new science of genetic engineering promises novel plant varieties that will arise as a result of the recombination of desired genes from different plants.

## PRELAB QUESTIONS

1. What type of cellular division gives rise to megaspores?

2. What are the reproductive parts of a flower?

3. Why is the endosperm tissue triploid?

4. What type of fruit is formed from the fusion of ovaries of many flowers?

5. How do stolons play a role in asexual reproduction?

## LAB 19.A
## SEXUAL REPRODUCTION: GROSS STRUCTURE OF THE ANGIOSPERM FLOWER

### MATERIALS REQUIRED (per pair of students)

flower of *Gladiolus*

forceps

razor blade

clean slide

stereomicroscope

### PROCEDURE

Much of a flower is composed of highly modified leaves that are produced by the sporophyte phase of the angiosperm life cycle. The flower contains the microscopic male and female gametophytes that produce egg and sperm nuclei essential for sexual reproduction.

1. Obtain an entire *Gladiolus* flower by breaking it off at the attachment point of the stalk. (Flowers that are partially open may also be used.) Discard the green bract that surrounds the base of each flower near the stalk; the bract is not part of the actual flower.

2. Note the delicate, brightly colored structures forming the bulk of the flower, usually referred to as "petals." The three outermost petals are actually sepals. They are usually smaller, green, and leaf-like in other flowers (Fig. 19–1). All sepals collectively form the calyx. The three innermost structures are the true petals.

3. Use a razor blade to carefully cut off, one at a time, the colored sepals and petals at the point where they join the green ovary of the flower. The remaining structures in the center of the flower are the true reproductive organs.

4. Note the female carpel. It is visible mostly by two of its parts: the white, slender style ending in the three-lobed stigma. The green ovary at the lower end of the carpel forms most of the flower's base.

5. Locate the three stamens surrounding the carpel. Each stamen consists of a long, slender filament supporting a narrow anther. This is where pollen is produced. Mature anthers split open and shed abundant, dust-like pollen. Touch the anthers with your finger to check for pollen in your flower.

6. Use a razor blade to cut a very thin cross section from the center of the ovary. Place the section flat on a slide and examine under a stereomicroscope. Note the small, whitish ovules (immature seeds) within the three chambers of the ovary.

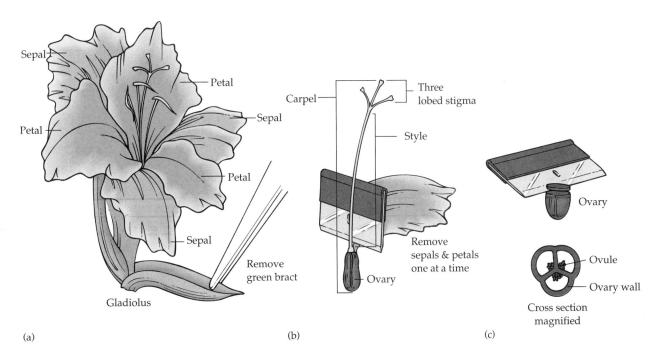

(a)  (b)  (c)

**FIGURE 19-1**

*Gladiolus* dissection.

7. Discard the flower parts and return all items to their original location.

## POSTLAB QUESTIONS

1. Based on the number of flower parts, is *Gladiolus* a monocot or a dicot plant?

2. Identify the male and female structures of the angiosperm flower.

3. How do the positions of the flower parts relate to their function?

## LAB 19.B
## SEXUAL REPRODUCTION: MICROSCOPIC STRUCTURE OF THE ANGIOSPERM FLOWER

### MATERIALS REQUIRED

lily anther c.s., prepared slide

lily ovary c.s., prepared slide

### PROCEDURE

The greatly reduced gametophyte stage of the angiosperm life cycle (Fig. 19–2) is found in the flower. Microspores are produced by meiosis within the pollen sacs (microsporangia) located inside the male anther of the flower. Each microspore develops into a pollen grain, which is the reduced male gametophyte

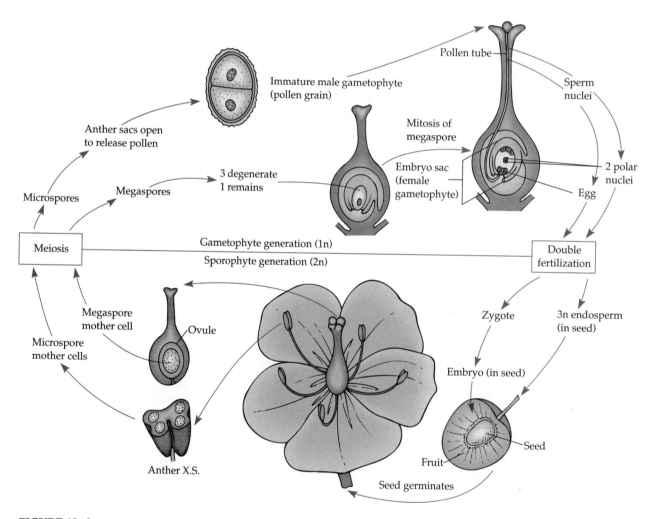

**FIGURE 19–2**

Angiosperm life cycle.

of the angiosperm plant. Megaspores are meiotically produced within the ovules (megasporangia) located inside the female ovary. Each ovule originally contains four megaspores, but all except one degenerate. The remaining megaspore develops into a microscopic megagametophyte within the ovule. In angiosperms, this is called an embryo sac. The embryo sac consists of only seven cells containing eight nuclei. One of these is a single egg cell that develops into an embryo if fertilized by a sperm cell from a pollen grain.

Pollen grains reach the stigma of the flower by wind, insects, or animals that transfer pollen from flower to flower as they search for nectar. The stigma functions as a "landing platform" for pollen grains. A pollen grain consists of two cells; a tube cell and a generative cell. The tube cell develops into a pollen tube that grows down through the carpel toward the ovary that contains the ovules. The generative cell divides by mitosis and forms two sperm nuclei. The sperm nuclei reach the embryo sac in the ovule and enter it by way of the pollen tube. One of the sperm nuclei fuses with the egg nucleus. This produces a diploid zygote that develops into a plant embryo. The second sperm nucleus fuses with two nuclei (polar nuclei) within the embryo sac. This forms a triploid cell. The resulting cell is triploid because its nucleus is a combination of three haploid nuclei—one sperm and two polar nuclei. The triploid cell divides repeatedly by mitosis. This produces a starchy tissue called endosperm that surrounds the embryo. This double fertilization, resulting in a zygote and endosperm, occurs only in the angiosperms. Endosperm functions as a food source for the embryo during germination until the young plant can begin photosynthesis. As the seed matures, the outer cell layers of the ovule harden into a seed coat that encloses the endosperm and embryo.

1. Examine a prepared slide of a lily anther cross section with your compound microscope. First look at the slide under low power. The two large chambers on either side of the anther are pollen sacs (microsporangia) where microspores arise by meiosis. Each microspore develops into a pollen grain.
2. Use high power to examine the cells in the interior of the pollen sacs. Depending on the age of the anther when the slide was made, these cells may be mature pollen grains or microspores that are in stages of pollen development (see Postlab Question 2). The walls of the anther split open when the pollen is fully developed. This exposes the pollen grains. The pollen grains may then be carried away to another flower by wind, or on the body of an insect, bat, or bird when it visits the flower for nectar.
3. Examine a prepared slide of a lily ovary cross section using low power. The three chambers within the ovary contain ovules, which are the two spherical structures that fill most of the space in the chambers.
4. Examine the interior of a few of the ovules under high power. If by good fortune one of the ovules was sectioned properly, you may be able to see the embryo sac (female gametophyte) as an oval arrangement of larger cells in the center of the ovule. Should you find an embryo sac, you most likely will see it in some stage of development other than the mature stage with eight nuclei simply due to chance. Ask your instructor for assistance in examining the ovules. Each embryo sac of an ovule contains an egg cell that may be fertilized by a sperm nucleus if a pollen tube should reach it.
5. Observe the demonstration slide of a lily mature female gametophyte on display. Using Figure 19–2 as a guide, locate the two polar nuclei and the egg nucleus (see Postlab Question 3).

## POSTLAB QUESTIONS

1. Are the following structures haploid, diploid, or triploid?
   a. egg nucleus
   b. endosperm
   c. pollen grain
   d. female gametophyte
   e. microspore
2. Sketch the pollen sacs of a lily anther in cross section. Include in your drawing sketches of microspores and mature pollen grains.

3. Sketch the lily ovary in cross section. Include in your drawing a sketch of a mature embryo sac and indicate which nuclei are involved in double fertilization.

# LAB 19.C
## SEXUAL REPRODUCTION: MONOCOT AND DICOT SEED STRUCTURE

### MATERIALS REQUIRED

soaked corn and bean seeds

razor blade

iodine solution in dropper bottle

paper towel

forceps

germinating bean seeds in petri plates

potted young bean plants

### PROCEDURE

The anatomy of a corn or bean seed clearly shows the dicot or monocot structure. Many other characteristics of the embryonic plant are also visible upon analysis.

1. Lay a soaked corn seed, which represents a monocot seed, flat on a paper towel and use a razor blade to slice it lengthwise in half (Fig. 19–3). Apply a drop of iodine to the cut surface of each half. Wait a few moments for a color reaction to occur.

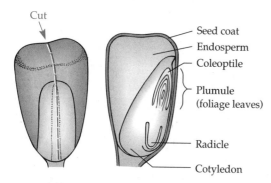

**FIGURE 19–3**

Lengthwise section of a corn seed showing endosperm and embryonic structures.

2. Note that most of the seed turns very dark once the iodine has been added. This region is the endosperm.

3. Find the lighter region of the seed interior that did not strongly react with the iodine. This is the corn embryo. It consists of several parts, some of which aren't readily apparent (refer to Fig. 19–3). Corn, being a monocot plant, has one cotyledon, an embryonic seed leaf that borders the endosperm. Closely associated with the cotyledon is the radi-

cle, an embryonic root that emerges first from a germinating (sprouting) corn seed. The radicle quickly branches and rebranches into numerous lateral roots that establish the root system of the young corn plant. The coleoptile, a protective sheath, then sprouts from the seed. It surrounds the plumule consisting of embryonic true leaves and shoot. The coleoptile pushes through the soil first, then the developing plumule grows through the coleoptile. This helps the plumule avoid any abrasive damage from soil particles.

4. Obtain a soaked kidney bean seed, which represents a dicot seed. Using fingernails or forceps, peel away the loosened seed coat from the bean. The bean may easily separate into equal halves when the seed coat is removed; if not, separate the halves. Each half is actually a large cotyledon that composes most of the bean. No endosperm is readily visible as in the corn seed. The endosperm is absorbed by the cotyledons as the bean seed matures; thus, the cotyledons function as a food reserve for the embryo.

5. Note the bean embryo that adheres to one of the cotyledons (Fig. 19–4). Locate the miniature young leaves of the plumule. The radicle occurs at the opposite end of the embryo. The cotyledons are attached to the embryo, although the attachment point is not easily visible. The portion of the embryo above the cotyledon attachment point is called the epicotyl; the portion below the cotyledon attachment point is called the hypocotyl.

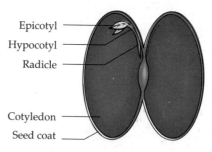

**FIGURE 19–4**

Internal structure of an opened bean seed.

6. Observe the germinating beans in petri plates. The radicle is the first structure to emerge from the seed coat and a developing root system soon becomes established. The hypocotyl begins to enlarge and elongate. It forms an arched hook that lifts the cotyledons out of the soil. The delicate plumule and apical meristem are sheltered from abrasion in

this way, similar to the protective coleoptile that is absent in dicot seeds.

7. Examine the young potted bean plants with their developing leaves. The cotyledons, now well above the soil, begin to shrivel and, eventually, to fall off. The elongated epicotyl increases stem height and the first true leaves (expanded plumule) appear. Locate the apical meristem and any axillary buds that may be present.

## POSTLAB QUESTIONS

Answer the first two questions based on the results of the iodine test on the corn seed.

1. What substance was present in the endosperm?

2. What must be the function of the endosperm?

3. What would you expect the result to be if you had applied iodine to half of a kidney bean as you did to the corn seed?

4. Why do the cotyledons wither as the young bean plant matures?

5. Indicate the presence or absence and the relative size, amount, or number of the following structures in the kidney bean and corn seed: cotyledons, embryo, endosperm, coleoptile, plumule, radicle, epicotyl, and hypocotyl.

## LAB 19.D
## ANGIOSPERM FRUIT TYPES

### MATERIALS REQUIRED

various representative fruit types, sectioned

various unidentified fruit types

### PROCEDURE

The fruit is the hallmark of angiosperm plants. A fruit is the mature, ripened ovary of the flower carpel. A ripened fruit contains mature seeds that have devel-

oped from fertilized ovules in the ovary. Fruits may be fleshy and full of juice, or hard, dry, and inedible. Several fruit types exist as a result of evolutionary variations in flower structure.

1. Examine the sectioned examples of the four basic angiosperm fruit types, giving special notice to the structural characteristics that distinguish each type (Fig. 19–5).
   a. A **simple fruit** develops from the ovary of one flower that contains only one carpel. The tomato, plum, and green bean are simple fruits.
   b. An **aggregate fruit** develops from *one* flower that contains many individual carpels. As the maturing ovaries enlarge, they merge into one structure. The blackberry, raspberry, and their similar relatives are aggregate fruits.
   c. A **multiple fruit** consists of fused ovaries from *many* flowers that grow close together. The fusion occurs as the maturing ovaries increase in size. A pineapple is a multiple fruit as seen from its many-sectioned exterior.
   d. An **accessory fruit** consists of the mature ovary plus a substantial quantity of other tissues associated with the flower. The flesh of an apple and a pear is an extensively developed floral tube surrounding the ovary. The flesh of a strawberry is the enlarged receptacle of the flower in which the various floral parts are inserted.
2. Observe the various unidentified fruits on display. Classify them according to their fruit type.

   Specimen 1 _____    Specimen 3 _____

   Specimen 2 _____    Specimen 4 _____

### POSTLAB QUESTIONS

1. A _____ is to an ovary as a seed is to an ovule.

2. What is the advantage of edible, delicious fruits to a plant?

3. What colors seem prominent among ripe fruits? Why do you think unripe fruits are often green in color?

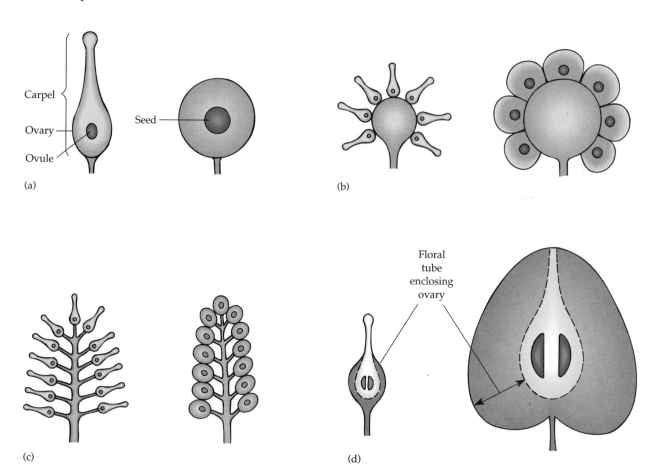

**FIGURE 19–5**

Schematic diagrams of flower and fruit structure relationships. (a) One flower with one carpel produces a simple fruit (cherry). (b) One flower with many carpels produces an aggregate fruit (raspberry). (c) Many carpels (ovaries) of individual clustered flowers fuse and produce a multiple fruit (mulberry). (d) Developed tissue of flower parts other than the ovary form a substantial part of an accessory fruit (apple).

# LAB 19.E
## ASEXUAL REPRODUCTION IN ANGIOSPERM PLANTS

### MATERIALS REQUIRED

examples of the following plants, on display:

a. tuber
b. bulb
c. corm
d. rhizome
e. stolon
f. stem cutting
g. tissue culture

### PROCEDURE

Although sexual reproduction in higher plants is important as a source of genetic diversity, many species reproduce partially or entirely by asexual means. Asexual reproduction provides an advantage by reproducing plants with genetic traits that are well suited for a particular environment. These advantageous traits could be lost with the genetic variation that occurs with sexual reproduction. Asexual reproduction occurs when vegetative, or structural, plant parts give rise to new individuals. In particular, the stems of many plants have become modified as asexual reproductive structures.

1. Examine the various types of modified stems on display that are asexual reproductive structures. Each type of specialized stem is described below.

   a. Tuber — potato. A fleshy underground stem modified for food storage. The potato on display has been allowed to sprout. Each sprout marks the location of an "eye," an axillary bud positioned where a leaf would normally appear on the swollen stem. The tiny bud produces new shoots that grow into a full potato plant. In fact, potatoes are grown by cutting "seed" potatoes into sections, each with an eye, that are buried in the soil. The eyes then develop into new potato plants that begin growth from energy stored as starch in the potato.

   b. Bulb — onion. A very short underground stem surrounded by thick fleshy leaves (the "layers" of the onion). The actual stem tissue is located in the small solid area visible just above the roots in a center lengthwise section. If planted in soil, the bulb gives rise to new green shoots and produces an entire onion plant. The fleshy leaves of the onion bulb contain food reserves that sustain early growth of new shoots.

   c. Corm — *Gladiolus*. A thick, short stem that grows vertically underground. The *Gladiolus* is a common flower popular with gardeners and florists. The corm may resemble a bulb in shape, but the sectioned corm shows that it lacks the layers of fleshy leaves present in a bulb. The bulk of the corm is composed of stem tissue. New shoots arise from small axillary buds present on the corm. The corm contains food reserves as do many underground stems.

   d. Rhizome — iris. A horizontal *underground* stem. New shoots arise from small buds along the rhizome. The iris, a common garden flower, spreads by growing rhizomes into nearby areas. In fact, irises are sold in gardening stores only as rhizomes, not seeds. Many

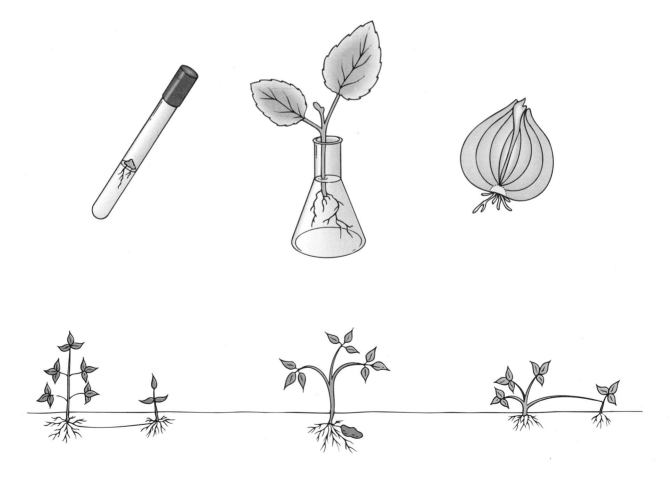

grasses and ferns also spread by extending rhizomes.

   e. Stolon—spider plant. A horizontal *aboveground* stem. Commonly called a "runner," a stolon will root and produce a new plantlet where it touches the ground. The stolons of the spider plant typically produce new plantlets at their tips. The plantlets readily become independent if placed in soil. Strawberry plants also produce stolons.

2. Observe the rooted stem cutting of *Coleus* or another plant growing in water. Many kinds of commercially important plants are propagated by stem cuttings that form roots when placed in water or in a damp potting medium. The rooted cutting continues to develop into a full-sized plant when transplanted into soil. Stem cuttings are one way of obtaining multiple individuals from one desirable plant.

3. Note the plant tissue cultures in petri plates or vials. The lump of green, undifferentiated tissue is called a callus. A callus forms after repeated mitotic divisions of just a few plant cells that were originally placed on sterile agar that contained complete nutrients and plant hormones for growth. With time, the callus will give rise to small leafy shoots that can be transferred to soil and grown normally. Many commercially important plants are now propagated by tissue culture. This enables growers to obtain new individuals that are identical to the original plant and its many desirable traits.

## POSTLAB QUESTIONS

1. What is the disadvantage of asexual reproduction?

2. Identify the following methods of asexual reproduction.

## FOR FURTHER READING

Bawa, K. S., and M. Hadley, eds. 1990. *Reproductive Ecology of Tropical Forest Plants.* Park Ridge, N.J.: Parthenon.

Bell, Adrian D. 1991. *Plant Form: An Illustrated Guide to Flowering Plant Morphology.* New York: Oxford University Press.

Bristow, A. 1978. *The Sex Life of Plants.* New York: Holt, Rinehart, and Winston.

Burgess, Jeremy. 1985. *An Introduction to Plant Cell Development.* New York: Cambridge University Press.

Chapman, G. P., ed. 1990. *Reproductive Versatility in the Grasses.* New York: Cambridge University Press.

Chopra, R. N., and S. C. Bhatla, eds. 1990. *Bryophyte Development: Physiology and Biochemistry.* Boca Raton, Florida: CRC Press.

Galston, A. W. et al. 1980. *The Life of the Green Plant.* Englewood Cliffs, N.J.: Prentice-Hall, Inc.

Haigler, Candace H., and Paul J. Weimer, 1991. *Biosynthesis and Biodegradation of Cellulose.* New York: M. Dekker.

Jensen, W. A., and R. F. Salisbury. 1984. *Botany.* Belmont, Calif.: Wadsworth.

Kaufman, P. B. et al. 1989. *Plants: Their Biology and Importance.* New York: Harper and Row.

Meeuse, B., and S. Morris. 1984. *The Sex Life of Flowers.* New York: Facts on File Publications.

Proctor, M., and P. Yeo. 1973. *The Pollination of Flowers.* Glasgow: William Collins Sons.

Raven, P. H. et al. 1986. *Biology of Plants.* New York: Worth.

Ray, P. M. et al. 1983. *Botany.* Philadelphia: Saunders.

Salisbury, F., and C. Ross. 1985. *Plant Physiology.* Belmont, Calif.: Wadsworth.

Wareing, P. F., and I. D. J. Phillips. 1981. *Growth and Differentiation in Plants.* Oxford, England: Pergamon Press, Ltd.

Weier, T. E. et al. 1982. *Botany: An Introduction to Plant Biology.* New York: John Wiley and Sons.

❑

# Plant Hormones and Responses

### O B J E C T I V E S

1. Define tropisms and give specific examples of phototropism, geotropism, and thigmotropism.

2. Identify the five classes of plant hormones and give the major effects of each.

3. Explain photoperiodism and the actions of phytochrome.

## INTRODUCTION

**Hormones** are chemical messengers produced in one part of an organism and transported to other parts. Even though hormones are produced in small concentrations, they elicit a surprisingly large response. Hormones correlate the events in a plant's life cycle with surrounding environmental conditions. Plant growth, directional movements called **tropisms,** and changes in plant parts are also determined by hormones. They occur due to a hormonal influence on enzyme activity, protein synthesis, or membrane permeability. The interaction of two or more hormones for a particular effect is common.

A plant's response to a particular hormone is influenced by many factors. Concentration of the hormone and the tissue it reaches determine the effect as does tissue age, environmental conditions, and other hormones already present.

There are five kinds of plant hormones: auxin, gibberellins, cytokinins, abscisic acid, and ethylene. A discussion of each follows.

1. **Auxin** is produced in the shoot apex and in young leaves. Auxin maintains apical dominance by stimulating meristematic growth and inhibiting growth of lateral buds. High concentrations of auxin cause production of ethylene, another plant hormone. Auxin also causes the production of new xylem as well as the development of female flower parts and fruits.

    Auxin also controls tropisms. Plants grow toward light **(positive phototropism)** by the transport of auxin to the shaded side of a plant where it stimulates cell elongation, which causes a bending of the stem toward the light (Fig. 20–1). Roots grow downward with gravity **(positive geotropism)** because of auxin, but not in the same way as do stems. Root cell elongation is inhibited by auxin except in very low concentrations. In a horizontal root, auxin is transported to the lower side by an auxin pump that is activated by **amyloplasts** falling to the gravity side of root cells. The increased auxin concentration inhibits root cell elongation. Cells on the upper side continue to grow and the root bends downward.

    Growth can also occur in response to a mechanical stimulus **(thigmotropism)** such as contact with a solid object. The twining growth of tendrils is an example of thigmotropism.

2. **Gibberellins** are produced in young leaves. A major effect of gibberellins is the elongation of stem cells that produces an increase in stem length.

227

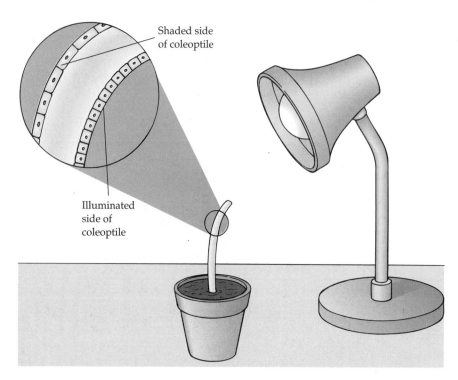

Shaded side
of coleoptile

Illuminated
side of
coleoptile

**FIGURE 20-1**

Phototropism is due to the unequal distribution of auxin. Auxin travels down the side of the stem or coleoptile away from the light, causing the cells on the shaded side to elongate. Therefore, the stem or coleoptile bends toward the light.

Gibberellins also stimulate cell division in the apical meristem.

3. **Cytokinins** originate in the roots. Cytokinins stimulate cell division, making them important to growth in the leaves and shoot meristems. Production of fruits and seeds is stimulated by cytokinins.

4. **Abscisic acid** from mature leaves is a growth inhibitor that aids dormancy in shoots and seeds. During the growing season, abscisic acid is important in plant water conservation because it causes the stomata to close when water is in short supply.

5. **Ethylene** production occurs in plant parts with high auxin concentration. Ethylene counteracts the effects of auxin and stimulates production of enzymes that ripen fruit.

**Phytochrome** is a blue-green pigment found in plasma membranes of plant cells. Phytochrome absorbs red and far-red light wavelengths. This alters the molecular structure of phytochrome to another form. This change is incorporated into the biological clock of plants, which gives them a sense of day length. Day length information perceived by the phytochrome mechanism determines when many plants flower and grow with the release of appropriate hormones.

Plant response to a particular duration of light and dark hours is called **photoperiodism,** which is regulated by the phytochrome mechanism described previously. Flowering of the so-called **short-day plants** is actually stimulated not by short days, but by a certain minimum period of uninterrupted darkness. **Long-day plants** require a certain minimum length of a light period for flowering. Some plants are **day neutral** meaning flowering is unaffected by day length of any duration.

Photoperiodic flowering is controlled by phytochrome pigment. This was determined experimentally using light flashes of different wavelengths that interrupted the dark period of short-day plants. A red light of 660 nanometers inhibited flowering most effectively. When a far-red light of 730 nanometers was used, flowering was not inhibited. Also, long-day plants on a short-day cycle were stimulated into flowering by a flash of red light during the dark period. A far-red flash reversed the effect. Eventually, it was determined that phytochrome and its two alternate forms, produced by red and far-red light, regulated the flowering response.

Phytochrome also plays a role in germinating seeds that are buried too deep or shaded by overhanging vegetation. Germination is inhibited until conditions change that allow enough light for phytochrome to convert to its other form. Following the conversion, germination begins.

The following laboratory investigations demonstrate the major principles of plant hormones and plant responses.

## PRELAB QUESTIONS

1. What is a plant hormone?

2. What type of growth occurs in response to a mechanical stimulus?

3. What hormone causes the production of ethylene?

4. What hormone is a growth inhibitor?

5. How does auxin maintain apical dominance?

## LAB 20.A
### PLANT TROPISMS

### MATERIALS REQUIRED

geotropism in radish seedlings, on demonstration

corn seeds

cafeteria trays

paper towels

water container

## PROCEDURE

A tropism is a directional movement in response to an external stimulus. When the plant grows toward the direction of the stimulus, the tropism is referred to as positive. When the plant grows away from the direction of the stimulus, the tropism is referred to as negative. In this lab exercise, you will view two tropisms one of which is demonstration: phototropism, which is directional growth in response to light, and geotropism, which is directional growth in response to gravity.

1. Observe 14-day-old radish *(Raphanus)* seedlings that were grown with a light source positioned to one side of the pot. Note the direction of the stem.
2. Place a lining of moist paper towels on a cafeteria tray (Fig. 20–2). Distribute corn seeds in different orientations on the paper towels. Cover the seeds with another layer of moist paper towels. Place another tray on top of the paper towels to hold the corn seeds in place.
3. Fill a large container with an inch of water. Place the stacked trays on edge in the container to keep the paper towels moistened. Do not submerge any seeds.

**FIGURE 20–2**

Procedure for preparing corn seeds for germination.

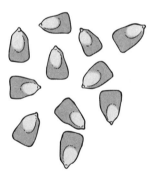

(a)  Lay seeds flat on paper towels in tray. Point seeds in various directions.

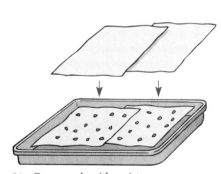

(b)  Cover seeds with moist paper towels.

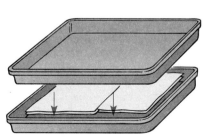

(c)  Secure seeds with second tray.

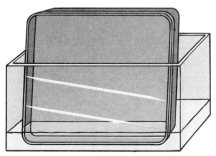

(d)  Stand stacked trays on edge in one inch water.

4. After 2–3 days, remove the trays. Note the direction of the radicles (roots) of each germinating corn seed.

### POSTLAB QUESTIONS

1. What effect does gravity have on roots? Is it a positive or negative tropic response?

2. What effect does light have on stems? Is it a positive or negative tropic response?

3. What is the adaptive significance of these tropisms?

## LAB 20.B
## THE ROLE OF AUXIN IN APICAL DOMINANCE

### MATERIALS REQUIRED (per group of students)

forty 1-week-old Alaska pea seedlings (10 seedlings per pot)

lanolin

lanolin with auxin

ruler

razor blade

### PROCEDURE

Apical dominance is the inhibiting effect of a terminal bud upon axillary bud development. Auxin, a plant hormone, is produced in cells of the apical region of the stem and inhibits the development of axillary buds (Fig. 20–3). This experiment will investigate the role of auxin in axillary bud development in a plant that normally exhibits strong apical dominance.

1. Obtain four pots of pea seedlings. Label the first pot as treatment number one. This pot will serve as a control.
2. With your second pot, use a razor blade to decapitate each pea seedling at the top of the third internode. (*Note:* Nodes are counted from the bottom of the plant, with the lowest being number one.)
3. With your third pot, decapitate each pea seedling at the top of the third internode. Apply lanolin with auxin to the cut surface of the stem.
4. With your fourth pot, decapitate each pea seedling at the top of the third internode. Apply lanolin without auxin to the cut surface of the stem.
5. Allow the plants to grow for seven days. For your group, record in Table 20–1 the number of plants in each pot in which there is axillary bud development at the second node. Measure the length of the bud at the second node of every plant in each group and record the average length in the table.

**Table 20–1** Effect of Auxin on Apical Dominance

| | Node #2 | |
|---|---|---|
| **Pot Number** | Number of Plants Released from Apical Dominance | Average Length of Axillary Bud |
| 1 | | |
| 2 | | |
| 3 | | |
| 4 | | |

### POSTLAB QUESTIONS

1. What was the effect of externally applying auxin on decapitated pea seedlings? Does this mean that auxin affects apical dominance in the intact plant?

2. What was the purpose of applying lanolin only to the decapitated pea seedlings?

3. Why must the axillary buds on the pea seedlings be dormant before beginning this experiment?

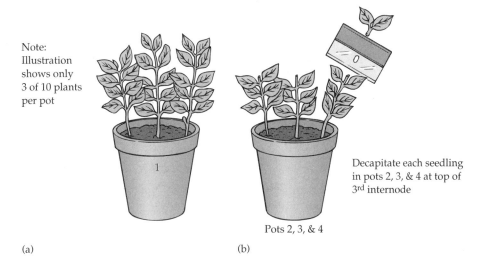

Note:
Illustration
shows only
3 of 10 plants
per pot

(a)

Decapitate each seedling
in pots 2, 3, & 4 at top of
3<sup>rd</sup> internode

Pots 2, 3, & 4

(b)

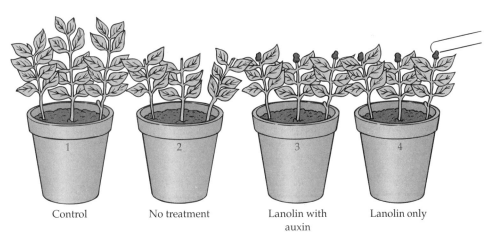

Control          No treatment       Lanolin with        Lanolin only
                                       auxin

(c)

**FIGURE 20-3**

Procedure to demonstrate apical dominance in pea seedlings. (a) Intact plant with inhibited
axillary bud development. (b) Decapitation of the pea seedling at the third internode. (c)
Experimental treatments for the four plant groups.

## LAB 20.C
## EFFECT OF LIGHT ON GERMINATION

### MATERIALS REQUIRED (per group of students)

300 lettuce (*Lactuca sativa*) seeds

red light source

far-red light source

green safe light

six 10-cm petri dishes

wax pencil or waterproof marker

eighteen 9-cm discs of filter paper, Whatman #1

forceps

aluminum foil

10 ml graduated cylinder

distilled water

darkroom

## PROCEDURE

Light plays a role in the germination of many kinds of seeds. In the lettuce (*Lactuca sativa*) seeds, red light, with a wavelength of about 600 nm, promotes seed germination above that of a dark control. Exposure to far-red light of about 730 nm will not stimulate germination. If the lettuce seeds are given alternating red and far-red light treatments, it is the last treatment that determines the germination response.

The pigment phytochrome controls the red to far-red light reaction that regulates germination in lettuce seeds. Phytochrome exists in two distinguishable forms: $P_r$, a red absorbing form, and $P_{fr}$, a far-red absorbing form. $P_{fr}$ is the active form and promotes such responses as seed germination. $P_r$ is converted to $P_{fr}$ when $P_r$ is irradiated with red light. When $P_{fr}$ is irradiated with far-red light, $P_{fr}$ is transformed to $P_r$. This experiment will investigate the effect of red light and far-red light on the germination of lettuce seeds.

1. Obtain 6 petri plates and label them with your wax pencil from 1–6 on the outside of each dish. Place 3 sheets of filter paper in the bottom of each dish.
2. Count out and place 50 dry lettuce seeds in each of the petri plates.
3. Move into the dark room with your petri plates and distilled water. Under a green safe light, pour 3–5 ml of distilled water into all the plates except petri dish 1. Petri plate 1 will remain dry.
4. After adding the distilled water, cover the petri plates with lids. Wrap each dish in aluminum foil and quickly label 1–6 again. Leave the petri plates in the dark for 16 hours.
5. After 16 hours have elapsed, give each dish the appropriate light treatment (see Table 20–2). In order to give a dish the correct light treatment, unwrap the dish and lift the lid off. Do this process in the darkroom under the green safe light. Dishes given the same light treatment can be treated together.
6. After the light treatment, rewrap the dish in foil and place the dish in the dark.
7. Determine the percentage of germination resulting from each treatment 3 to 4 days later. Count as germinated only those seeds that have definite radicle (embryonic root) protrusion. Record your results in Table 20–3.

**Table 20–3**    Light Treatment Effect on Lettuce Seeds

| Petri Plate | Number of Seeds Germinated | % of Germination |
|---|---|---|
| 1 | | |
| 2 | | |
| 3 | | |
| 4 | | |
| 5 | | |
| 6 | | |

## POSTLAB QUESTIONS

1. What do you conclude about the effect of red light on the germination of lettuce seeds?

2. What do you conclude about the effect of far-red light on the germination of lettuce seeds?

3. Were the red and far-red light treatments reversible?

4. What advantage does a light requirement offer for seed germination of a plant?

**Table 20–2**    Red and Far-red Light Treatment

| Petri Plate Number | Light Treatment |
|---|---|
| 1 | No light treatment |
| 2 | No light treatment |
| 3 | R (red light)—4 min. |
| 4 | R (4 min.) + FR (far-red)—8 min. |
| 5 | R (4 min.) + FR (8 min.) + R (4 min.) |
| 6 | FR (8 min.) |

## FOR FURTHER READING

Alscher, Ruth G., and Jonathan R. Cumming, eds. 1990. *Stress Responses in Plants: Adaptation and Acclimation Mechanisms.* New York: Wiley-Liss.

Galston, A. W. et al. 1980. *The Life of the Green Plant.* Englewood Cliffs, N.J.: Prentice-Hall, Inc.

Heath, I. B., ed. 1990. *Tip Growth in Plant and Fungal Cells.* San Diego: Academic Press.

Hill, T. A. 1980. *Endogenous Plant Growth Substances.* Baltimore, Md.: University Park Press.

Hillman, W. 1979. *Photoperiodism in Plants and Animals.* Burlington, N.C.: Oxford Biology Readers.

Jensen, W. A., and F. B. Salisbury. 1984. *Botany: An Ecological Approach.* Belmont, Calif.: Wadsworth Publishing Co.

Katterman, Frank, ed. 1990. *Environmental Injury to Plants.* San Diego: Academic Press.

Kaufman, P. B. et al. 1989. *Plants: Their Biology and Importance.* New York: Harper & Row.

Moore, D. James et al., eds. 1990. *Inositol Metabolism in Plants.* New York: Wiley-Liss.

Nickell, L. 1983. *Plant Growth Regulators—Agricultural Uses.* New York: Springer-Verlag.

Pharis, R. P., and S. B. Rood, eds. 1990. *Plant Growth Substances 1988.* New York: Springer-Verlag.

Ray, P. M. et al. 1983. *Botany.* Philadelphia: Saunders College Publishing.

Salisbury, F. B., and C. W. Ross. 1985. *Plant Physiology.* Belmont, Calif.: Wadsworth Publishing, Co.

Schonffeniels, E., and D. Margineanu. 1990. *Molecular Basis and Thermodynamics of Bioelectrogenesis.* Boston: Kluwer Academic Publishers.

Takahashi, Nobuta et al., eds. 1991. *Gibberellins.* New York: Springer-Verlag.

Wareing, P. F., and I. D. J. Phillips. 1981. *Growth and Differentiation in Plants.* Elmsford, N.Y.: Pergamon Press, Inc.

Weier, T. E. et al. 1982. *Botany: An Introduction to Plant Biology.* New York: John Wiley & Sons.

Winfree, Arthur T. 1990. *The Geometry of Biological Time.* New York: Springer-Verlag.

# PART VII

❏

# Animal Structure and Function

# Animal Tissues

## OBJECTIVES

1. Define tissue, organ, and organ systems.

2. Compare the four principal types of animal tissues — epithelial, connective, muscle, and nerve — with respect to general structure and function.

3. List the functions of epithelial tissue, describe the three main shapes of epithelial cells, and describe how these cells can be arranged into tissues.

4. Describe the main types of connective tissue and their functions.

5. Compare the three types of muscle tissue and their functions.

6. Draw and label a typical neuron.

## INTRODUCTION

Animals are multicellular. These many cells are organized into **tissues** in most animals. Tissues, in turn, are organized into **organs,** and organs are organized into **organ systems.**

A tissue is composed of a few types of cells that are closely related to one another and adapted to a specific function. Animal tissues may be classified into four basic types: **epithelial, connective, muscle,** and **nervous.** The cells of each type of tissue have characteristic sizes, shapes, and configurations.

Epithelial tissues, taken as a group, comprise the outer layer of the skin, the linings of the digestive and respiratory tracts, and the lining of the kidney tubules. Epithelial tissues are classified according to how many cells thick the epithelium is and by the shape of the epithelial cells. There are three shapes of epithelial cells. They are **squamous, cuboidal,** and **columnar.** Epithelium, which is one cell layer thick, is called **simple epithelium.** Epithelium, which is two or more cell layers thick, is referred to as **stratified epithelium.**

There are five types of connective tissue: loose, adipose, cartilage, bone, and blood. **Loose connective tissue** is located in those places of the body involved with support and elasticity. **Adipose tissue** is found as padding around various internal organs. **Cartilage** is found around the ends of bones, tip of the nose, and within the external ears. **Bones** are found throughout the body and offer skeletal support. **Blood** travels throughout the body's veins and arteries. It carries oxygen to the cells and wastes from the cells.

Muscle tissue is the most abundant tissue in most animals. There are three types: skeletal, cardiac, and smooth. **Skeletal muscle** comprises the voluntary muscles. **Cardiac muscle** is found in the heart. **Smooth muscle** makes up the walls of many organs. The three types of muscle tissue differ in a number of important characteristics. Skeletal muscle is under **voluntary control,** which means it is controlled willfully whereas smooth and cardiac muscle are under **involuntary control,** which means they are automatically controlled. Each of the three types of muscle tissue have distinctive fiber (cell) shapes. Skeletal and cardiac muscle (but not smooth muscle tissue) have cross bands called **striations.** Skeletal muscle cells are **multinucleated,** which means they have many nuclei. Smooth muscle cells have only one nucleus. Cardiac muscle cells each have one or two nuclei. The position of the nuclei in skeletal muscle cells is peripheral, while in cardiac and smooth muscle cells, the nuclei are centrally located. Skeletal muscle can **contract** the most rapidly of the three types of muscle cells. The contraction rate of smooth muscle cells is the slowest of the three types, and cardiac muscle has an intermediate contraction rate. Finally, smooth muscle has the greatest ability to remain contracted, cardiac muscle is intermediate in this criterion, and, in general, skeletal muscle has the least ability to remain contracted.

Nervous tissue receives and transmits stimuli and information in the form of nerve impulses. It consists of **neurons** and **glial cells** (supporting cells). A neuron is made up of a cell body, elongated extensions of the cytoplasm known as **dendrites,** which receive impulses, and **axons,** which transmit impulses towards another neuron or towards a muscle or a gland.

The 11 major vertebrate organ systems include the integumentary, skeletal, muscular, lymphatic, respiratory, digestive, nervous, endocrine, circulatory, urinary, and reproductive systems. The **integumentary system** is primarily concerned with protection. The **skeletal system** provides support. The **muscular system** produces movement. The **lymphatic system** offers defenses against disease. The **respiratory system** supplies oxygen and eliminates carbon dioxide. The **digestive system** processes food. The **nervous system** involves the senses and regulation. The **endocrine system** regulates metabolic activities. The **circulatory system** is involved with transportation. The **urinary system** allows for excretion. And the **reproductive systems** are involved in the production of offspring. Further discussion and study of these organ systems will follow in subsequent exercises.

The following laboratory exercises have been designed to increase your understanding of the various types of tissues.

## PRELAB QUESTIONS

1. What type of tissue functions in protection?

2. What type of tissue is bone?

3. What are the three types of muscle tissue?

4. What type of muscle tissue is multinucleated and is under voluntary control?

5. What are the three parts of a neuron?

## LAB 21.A
## EPITHELIAL TISSUES

### MATERIALS REQUIRED

prepared slides of the following:

a. squamous epithelium of mouth, smear

b. cuboidal epithelium, mammalian kidney section

c. columnar epithelium, amphibian small intestine, x.s.

### PROCEDURE

Epithelial tissues cover the surfaces of the body and line the interior of its various cavities and ducts. Epithelial tissues generally have a protective function, but they may also secrete and absorb substances depending on the type of epithelium. Among other characteristics, these tissues are classified by the shape of the cells that compose them. The three shapes of epithelial cells are demonstrated by the following prepared slides.

### Squamous Epithelium

1. Obtain a slide of squamous epithelium taken from the lining of the human mouth (cheeks). Observe the cells under low and high power. Do you observe intercellular spaces? The term "squamous" means flat, which describes these thin, transparent cells. You may see areas where cells are folded over or overlapping, which illustrates their thinness. Note the prominent nucleus. Within the mouth, these cells fit closely together as tiles on a floor, forming a protective sheet of cells (Fig. 21–1). In many areas of the body, such as the epidermis (outermost skin layer), the vagina, and the mouth, squamous epithelium is subject to friction, which removes the surface cells. The cells are continually

replaced by new cells that rise to the surface as a result of mitotic activity in the lower cell layers.

Nuclei

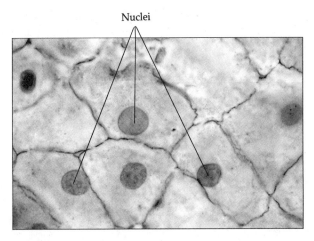

**FIGURE 21–1**

A layer of squamous epithelium cells.    *(Ed Reschke)*

## Cuboidal Epithelium

2. Observe a slide of cuboidal epithelium obtained from a section of mammalian kidney. As the name implies, the cells of this epithelium are somewhat cube shaped. The cuboidal cells line the lumen, or cavity, of the tubules that compose much of the kidney. If it were cut in cross section, the lumen of a tubule would appear as a circle (Fig. 21–2). If it

Nuclei of cuboidal epithelial cells          Lumen of tubule

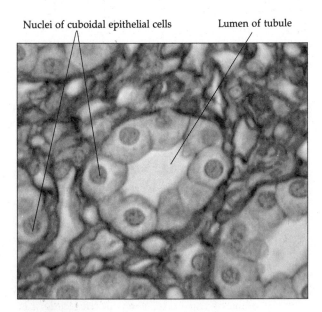

**FIGURE 21–2**

Cuboidal epithelium cells of kidney tubules.    *(Ed Reschke)*

were cut lengthwise, the lumen of a tubule would appear as a long channel. Find a tubule lumen and observe under low and high power the cuboidal cells that form its border.

## Columnar Epithelium

3. Obtain a slide of a cross section of an amphibian small intestine and view it under scanning or low power. Note the central lumen where food digestion occurs. The finger-like folds of the intestinal wall that project into the lumen are called villi (villus, singular). The columnar epithelial cells are the outermost layer of cells within the villi that borders the intestinal lumen. These cells, resembling columns, absorb nutrients from digested food in the intestinal lumen. Note the prominent nuclei. Look for enlarged vacuole-like spaces within some of the cells. These are goblet cells (Fig. 21–3), which secrete lubricating mucus onto the inner surface of the small intestine.

Goblet Cell                                     Nuclei of columnar cells

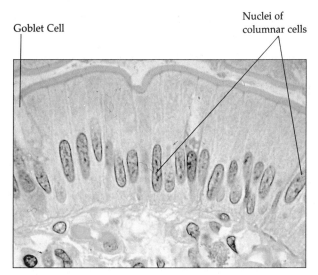

**FIGURE 21–3**

Columnar epithelium cells.    *(Ed Reschke)*

## POSTLAB QUESTIONS

1. Compare the ratio of length to width for squamous, cuboidal, and columnar epithelial cells.

2. Identify the type of epithelial tissue.

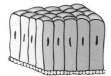

3. Why is epithelial tissue effective as a protective covering?

4. In the past, it was traditional to have students make slides from scrapings of the insides of their mouths, particularly from their cheek linings. Why is that procedure seldom used today?

## LAB 21.B
## CONNECTIVE TISSUE

### MATERIALS REQUIRED

prepared slides of the following:

a. hyaline cartilage

b. compact bone, c.s.

c. loose connective (areolar) tissue

d. human blood smear

e. adipose tissue

### PROCEDURE

As stated by its name, connective tissue holds together and supports body structures. The cells of connective tissue are typically distributed among an extra-cellular substance (matrix) with fine fibers running through it. In some highly specialized connective tissues, the extra-cellular substance may be gel-like (cartilage), inorganic mineral (bone), or liquid (blood). Since there are many kinds of connective tissue, a cross section of their diversity is presented.

### Cartilage

1. Examine the prepared slide of cartilage under low and high power. Note the small spaces distributed throughout the tissue. These are called lacunae (lacuna, singular). They contain the cartilage cells (chondrocytes) visible inside the lacunae (Fig. 21–4). The stained substance between lacunae is a firm protein matrix that gives cartilage its flexible quality. This kind of cartilage is found in the nose, external ear, and between bone joints in the human body.

**FIGURE 21–4**

Cartilage tissue. *(Ed Reschke)*

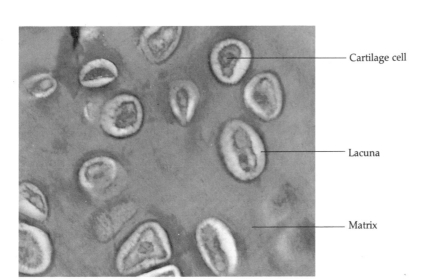

Cartilage cell

Lacuna

Matrix

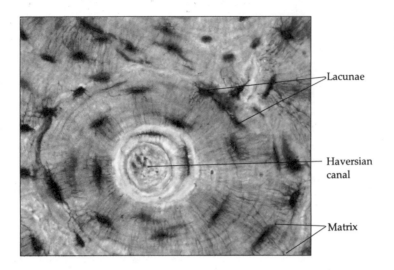

**FIGURE 21-5**

Bone tissue.    *(Ed Reschke)*

Lacunae

Haversian canal

Matrix

## Bone

2. Observe under low and high power a thin cross section of bone. Notice that the bone section is composed of a collection of circular structures that resemble the end of a sawed log (Fig. 21–5). Each one of the "logs" is a cross section of an osteon, a long cylindrical structure tapered at both ends. Osteons are the units that form compact, or dense, hard bone. Note, at the center of each osteon, a circular opening called the haversian canal. This canal holds nerves and blood vessels that run throughout the bone. The haversian canal may appear black in some slides from accumulated bone dust created when the bone tissue was ground to a thin section.

The matrix of bone tissue is a hard, white inorganic compound of calcium secreted by the bone cells (osteocytes). Small, dark lacunae that contain osteocytes are circularly arranged around the haversian canal. Use high power to focus carefully on a lacuna and note the fine canaliculi (tiny canals) that interconnect neighboring lacunae and the haversian canal. Nutrients and oxygen from blood vessels in the haversian canals reach the osteocytes by way of the canaliculi.

## Loose Connective (Areolar) Tissue

3. Examine the prepared slide of loose connective tissue noting the numerous fibers among the widely spaced cells. The darkly stained cells are fibroblasts. They produce the fibers found throughout the tissue. The thin elastic fibers, made of the protein elastin, allow the tissue to stretch. The thicker fibers are made of collagen, a tough protein that

gives the tissue strength (Fig. 21–6). The fibroblasts and fibers are suspended in a semi-fluid matrix that, in addition to the fibers, gives the tissue great flexibility. Loose connective tissue commonly occurs around blood vessels, nerves, and between the skin and underlying muscles.

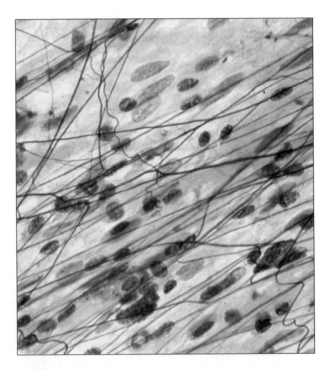

**FIGURE 21-6**

Loose connective tissue.

## Adipose (Fat) Tissue

4. Observe the many spaces throughout the slide of adipose tissue. These spaces were once filled with fat before the slide was prepared. The quantity of

fat within adipose tissue cells is so great that it pushes the nucleus and other cell structures into a thin layer along the boundary of the cell membrane. Look closely at the border between cells and you should be able to see darkly stained nuclei (Fig. 21–7). The fat of adipose tissue functions as stored energy and acts as a cushion for many internal organs. Adipose tissue below the skin functions as insulation that retains body heat, one adaptation that enables mammals to survive in extreme cold at the Earth's poles.

cell and slightly thinner in the middle than along the cell's edge.

Using low power, scan the slide for larger cells having conspicuous nuclei (Fig. 21–8). These are white blood cells (leucocytes), a broad term for various types of blood cells that perform different functions as part of the immune system.

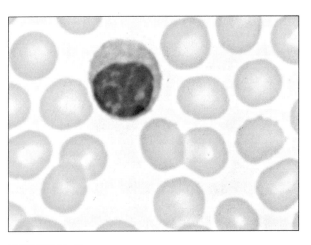

**FIGURE 21–8**

Human blood cells. One large white blood cell with a conspicuous nucleus is visible among numerous red blood cells that lack nuclei. *(Ed Reschke)*

Blood is an unusual connective tissue because the extra-cellular substance is liquid instead of solid or semi-solid as in other connective tissues. The liquid that suspends the blood cells is called plasma, a mixture of water, proteins, dissolved gases, and nutrients that flows to essentially all cells of the body. In this way, cells receive oxygen and nutrients and rid themselves of wastes and carbon dioxide, which the blood carries away.

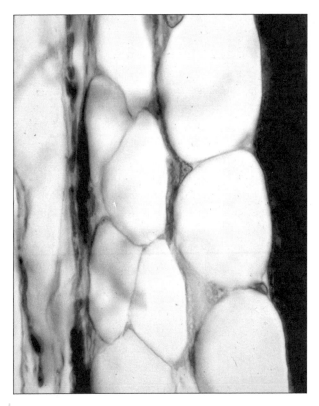

**FIGURE 21–7**

Prepared adipose tissue showing nuclei visible at cell edges and large interior spaces that once contained fat. *(Ed Reschke)*

## Blood

5. Examine a human blood smear under low power. Notice the abundance of red blood cells (erythrocytes) throughout the slide. These are the cells that carry oxygen in the bloodstream to all tissues within the body. Mammalian erythrocytes are unusual because they lose their nucleus during their development in the bone marrow; hence, you will not find a nucleus within the red blood cells. An erythrocyte is slightly concave on both sides of the

## POSTLAB QUESTIONS

1. Identify the appropriate connective tissue with its characteristic.
   a. osteon
   b. fibroblasts
   c. chondrocytes
   d. haversian canal
   e. liquid matrix
   f. storage form of energy
2. How are cartilage and bone structurally similar?

3. Compare erythrocytes with leucocytes. How do they compare in number?

4. Describe the nuclei of various leucocytes.

5. Why wasn't a fresh blood smear made and examined?

6. Identify the appropriate connective tissue.

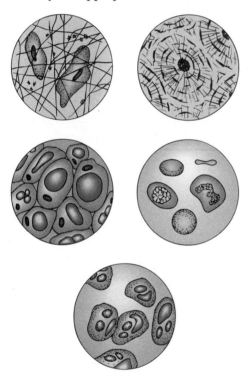

7. After identifying the tissues above, identify the appropriate structures or cells.

## LAB 21.C
## MUSCLE TISSUE

### MATERIALS REQUIRED

prepared slide of muscle types

### PROCEDURE

Muscle tissue has the special capability of contraction, which produces movement of a body part, movement of the body through its environment, or movement of contents through an organ. Muscle cells are called fibers because they are long and slender. There are three types of muscle tissue. The differences in the appearance and contraction capabilities of muscle fibers form the basis for these three types of muscle tissue. You will observe one slide that contains all three muscle types. Check with your instructor to make sure that you are viewing the proper tissue section as you proceed through the exercise. Some areas of the muscle tissue on your slide may be cross sections, with the fibers appearing as circular structures. Search for other areas of the tissue that are longitudinal (lengthwise) sections because these present a much better representation of the muscle fibers.

### Skeletal (Striated) Muscle Tissue

1. Examine the section of skeletal muscle on your prepared slide. As the name implies, this type of muscle tissue is attached to the bones of the skeleton. When the tissue contracts, the body parts move. Notice that each skeletal muscle fiber is striped (striated) with numerous fine cross bands along its entire length. These cross bands are produced by the arrangement of contractile proteins within the muscle fiber. Look for the darkly stained multiple nuclei located just under the cell membrane of each fiber (Fig. 21–9). Skeletal muscle fibers are long cells that run the entire length of

**FIGURE 21–9**

Skeletal muscle fibers showing cross bands (striations). *(Peter Arnold, Inc.)*

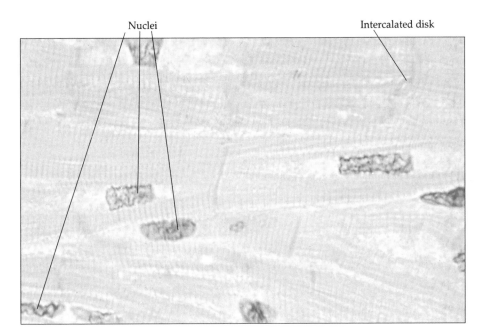

Nuclei                    Intercalated disk

**FIGURE 21-10**

Cardiac muscle tissue.

skeletal muscles in the body. Of course, skeletal muscle is under voluntary control, which means that you can consciously will these muscles to contract.

## Cardiac Muscle Tissue

2. Observe the section of cardiac muscle. Cardiac muscle occurs only in the heart of vertebrate animals. Notice that cardiac muscle is striated like skeletal muscle, but there are many more differences than similarities between these two muscle types. Each cardiac cell has only one or two nuclei compared to the multiple nuclei of skeletal muscle fibers. Cardiac fibers branch and join with neighboring fibers at junctions called intercalated discs found only in cardiac cells. Focus carefully with high power and you may be able to see intercalated discs as slightly darker cross bands in the cardiac fibers (Fig. 21-10). Cardiac muscle is involuntary tissue, meaning that you have no conscious control over its contractions.

## Smooth Muscle Tissue

3. Examine the smooth muscle tissue, which is the remaining section of muscle tissue on your slide. Smooth muscle tissue composes a portion of many internal organs, particularly in the digestive system. The thick wall of the amphibian intestine you examined for columnar epithelium is made up of smooth muscle. Contractions of smooth muscle, such as in digestive organs, typically function in

moving along contents from one organ to another, or they control the opening and closing of orifices (openings) such as the pupil of the eye. Smooth muscle fibers are spindle shaped, being long and

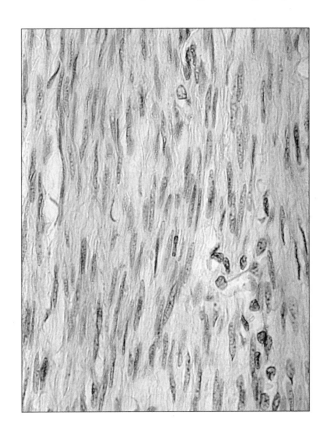

**FIGURE 21-11**

Smooth muscle tissue. *(Runk/Schoenberger/Grant Heilman Photography, Inc.)*

pointed at both ends, with a single nucleus (Fig. 21–11). Like cardiac muscle, smooth muscle tissue is involuntarily controlled.

## POSTLAB QUESTIONS

1. Identify the appropriate type(s) of muscle tissue with its characteristic.
   a. intercalated disc
   b. under voluntary control
   c. peripherally located nuclei
   d. under involuntary control
   e. single nucleus

2. Why is smooth muscle tissue sometimes referred to as visceral muscle?

3. Identify the appropriate muscle tissue.

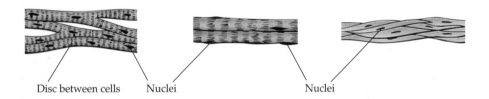

Disc between cells     Nuclei        Nuclei

## LAB 21.D
## NERVOUS TISSUE

### MATERIALS REQUIRED

prepared slide of motor neurons, spinal cord smear

### PROCEDURE

Nerve tissue has the unique capability of carrying messages throughout the body as electrical impulses. The structure of neurons, or nerve cells, reflect this function.

Examine the prepared slide of a smear of nerve tissue from the spinal cord of a cow. Observe the neurons distributed throughout the smear. They appear as large, spider-like cells with a prominent nucleus. These are motor neurons, which means that they carry nerve impulses that cause actions such as muscle contractions or glandular secretions.

Note the long extensions of the neurons that lead in several directions from the cell (Fig. 21–12). These extensions interconnect with other neurons and various organs thereby integrating the body into a large communication system that continually sends and receives nerve impulses. The neuron extensions that bring impulses toward the cell are called dendrites; those extensions that carry impulses away from the neuron are called axons. The cord-like structure commonly called a "nerve" within the body consists of many axons or a combination of axons and dendrites bound together by connective tissue. The numerous small, dark cells distributed around the neurons are glial cells. Glial cells provide support for neurons.

Cell body of neuron   Neurons   Dendrites

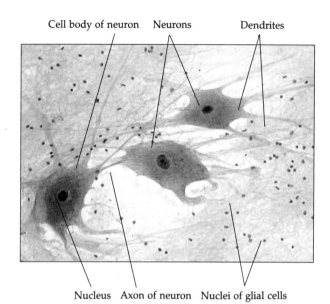

Nucleus   Axon of neuron   Nuclei of glial cells

**FIGURE 21-12**

Nerve tissue. *(Ed Reschke)*

## POSTLAB QUESTIONS

1. What system in the human body has a similar function as the nervous system?

2. Identify the following parts of the neuron: cell body, dendrite, axon, and nucleus.

Myelin sheath   Node   Synapse (gap)

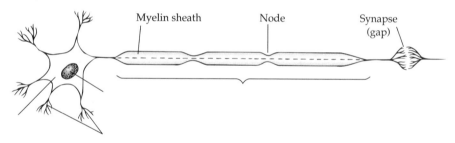

## FOR FURTHER READING

Bloom, W., and D. W. Fawcett. 1986. *A Textbook of Histology.* Philadelphia: W. B. Saunders.

DeRoberts, E. D. P., and E. M. F. DeRoberts. 1980. *Cell and Molecular Biology.* Philadelphia: W. B. Saunders.

Fawcett, D. 1981. *The Cell.* Philadelphia: W. B. Saunders.

Guyton, A. C. 1984. *Physiology of the Human Body.* Philadelphia: W. B. Saunders.

Kessel, R., and R. Kardon. 1979. *Tissues and Organs; A Text-Atlas of Scanning Electron Microscopy.* San Francisco: W. H. Freeman.

Nilsson, L. 1973. *Behold Man.* Boston: Little, Brown & Co.

Rhoades, R., and R. Pflanzer. 1989. *Human Physiology.* Philadelphia: Saunders College Publishing.

Ross, M., and E. Keith. 1985. *Histology: A Text and Atlas.* New York: Harper & Row.

Solomon, E. P., R. Schmidt, and P. Adragna. 1990. *Human Anatomy and Physiology.* Philadelphia: Saunders College Publishing.

Srivastava, Rakesh et al., eds. 1991. *Immunogenetics of the Major Histocompatibility Complex.* New York: VCH.

Wall, Robert. 1990. *This Side Up: Spatial Determination in the Early Development of Animals.* New York: Cambridge University Press.

Weiss, L. ed. 1983. *Histology: Cell and Tissue Biology.* New York: Elsevier Biomedical.

Weiss, L., and R. O. Greep. *Histology.* New York: McGraw-Hill.

❑

# Mammalian Skeletal and Muscular Systems

O B J E C T I V E S

1. Identify the major bones of the human skeletal system.

2. Describe the gross and microscopic structure of skeletal muscle.

3. Identify the major types of joints and their actions.

4. Describe the functional relationships between skeletal and muscular tissues.

## INTRODUCTION

The vertebrate skeleton supports the body, protects internal organs, and provides muscular attachment sites and joints that play an important role in movement of body parts and locomotion. Bones are attached to one another by **ligaments,** which are strong bands of connective tissue. Muscles attach to bones by **tendons** and by bands of connective tissue fibers. As part of the animal's physiology, bones produce blood cells in the **marrow** and serve as calcium reservoirs for the body.

The vertebrate skeleton is composed of **connective tissues,** specifically bone and cartilage. Cartilage cells produce tough **collagen** protein fibers, typical of connective tissue, and a firm intercelluar jelly that surrounds the cells and fibers. These traits give cartilage strength and flexibility. Cartilage occurs most commonly between the bones of joints.

The embryonic skeleton of vertebrates is, in most cases, composed of cartilage that is later replaced by bone. Bone tissue consists of collagen fibers surrounded by calcium phosphate salts, which is what gives bone its great strength. Cells that produce the collagen fibers and mineral deposits are distributed throughout the bone substance. These cells produce bone growth and repair.

The human skeleton consists of two major portions: the axial skeleton and the appendicular skeleton. The axial skeleton is comprised of the skull, the backbone, and the ribs. The appendicular skeleton involves the bones in the limbs, the pectoral girdle, and the pelvic girdle.

There are three types of muscle tissue in the vertebrate body: skeletal, smooth, and cardiac. **Skeletal (striated) muscle** is responsible for locomotion and change of body position; therefore, it is under the animal's voluntary control. These muscles generally are attached to bones in the body and produce movement via contractions. An individual skeletal muscle is composed of thousands of **muscle fibers** that run the total length of the muscle. Each fiber consists of several muscle cells that fused during embryonic development. Motor neurons from the somatic nervous system connect to skeletal muscles by neuromuscular junctions, which are synapses that stimulate contraction.

Many internal organs are composed at least partly of **smooth muscle** tissue, which occurs in sheets of

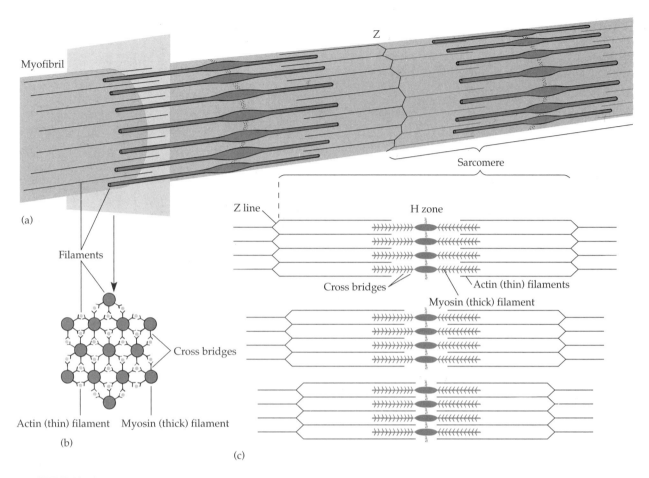

Myofibril

(a)

Filaments

Cross bridges

Actin (thin) filament   Myosin (thick) filament

(b)

Z

Sarcomere

Z line

H zone

Cross bridges

Actin (thin) filaments

Myosin (thick) filament

(c)

**FIGURE 22-1**

(a) A myofibril stripped of the accompanying membranes. The Z lines mark the ends of the sarcomeres. (b) Cross section of myofibril shown in (a). (c) Filaments slide past each other during contraction. Notice the way the filaments overlap. It is the regular pattern of overlapping filaments that gives rise to the striated appearance of skeletal and cardiac muscle. In the top drawing of (c), the myofibril is relaxed. In the middle drawing, the filaments have slid toward each other, increasing the amount of overlap and shortening the muscle cell by shortening its sarcomeres. At bottom, maximum contraction has occurred; the sarcomere has shortened considerably.

individual muscle cells. Smooth muscle is stimulated by motor neurons from the autonomic nervous system and is involuntary (not under conscious control).

**Cardiac muscle** composes only the vertebrate heart. The muscle cells are arranged in long columns of fibers. Like smooth muscle, cardiac muscle is involuntarily controlled. The rate of heart beat is regulated by a cranial nerve of the autonomic nervous system, but the heart beat originates from the **sinoatrial node** within the heart itself. The impulse spreads through the right atrium where the node is located, and proceeds to the **atrioventricular node** located on the partition between the ventricles, which stimulates their contraction.

The **sarcomere** (Fig. 22-1) is the unit of skeletal muscle contraction and occurs by a **sliding filament** mechanism. Each of the numerous **myofibrils** that compose a muscle fiber have many sarcomeres along the entire length of the myofibril. A sarcomere is the area between two dark **Z lines** crossing the width of the myofibril. The Z lines serve as an anchoring point for **thin filaments** made of the contractile protein **actin.** The thin filaments extend partially between the Z lines, leaving a central space. A fine strand of the protein **tropomyosin** winds around each actin filament. Short segments of **troponin,** another protein, occur at regular intervals along the tropomyosin strand. **Thick filaments** made of the contractile protein **myosin** occur in the central space between the thin filaments. The free ends of the thick and thin filaments overlap slightly. The free end of each myosin molecule has a club-like head capable of attaching to a neighboring thin filament.

Contraction is stimulated by a nerve impulse arriving from a motor neuron at a neuromuscular junction. Neurotransmitter molecules (acetylcholine) cross the neuromuscular junction and depolarize the **sarcolemma** (muscle fiber membrane). The T tubules carry the impulse from the outer sarcolemma down into the **sarcoplasmic reticulum** that stores calcium ions. The impulse makes the sarcoplasmic reticulum permeable to calcium ions that bond to troponin. This causes a movement of the tropomyosin strand on the thin filament, which exposes actin binding sites for the myosin heads. The myosin heads attach to the actin and pivot toward the center of the sarcomere by energy released from ATP. As a result, the thin filaments are drawn toward the center of the sarcomere and the unit contracts, bringing the Z lines closer together. Contraction continues as myosin heads release and bind another site further up the actin filament and the process repeats. Contraction of the entire muscle occurs as thousands of sarcomeres contract simultaneously.

The sarcomere relaxes when the motor neuron stops firing. Active transport, requiring ATP, in the sarcoplasmic reticulum quickly reclaims the calcium ions. Troponin is no longer bound by calcium, causing a shift in tropomyosin position that blocks the actin binding sites. Use of additional ATP detaches the myosin heads from the actin filaments and relaxation occurs.

**Tetanus** in skeletal muscle is a smooth, steady muscle contraction that is produced when constant nerve impulses cause the continual contraction of muscle fibers. Constant stimulation over a long period causes **fatigue,** which occurs when a muscle can no longer respond to stimulation and the fibers relax.

Skeletal muscle is arranged in **motor units,** each made of a motor neuron and the muscle fibers it stimulates. The number of motor units stimulated at any one time determines the strength of a muscle contraction. Sustained contractions are possible by the contraction of only some muscle fibers as others remain relaxed. Later, the relaxed fibers contract as the working fibers reach fatigue. In this way, an entire muscle can remain partially contracted for long periods.

Body movement is produced by **antagonistic** muscles—muscles with opposite effects. When one member of an antagonistic muscle pair contracts, its "partner" relaxes, and vice versa. A muscle that causes a joint to stretch out is an **extensor.** The opposite muscle causing the joint to close is a **flexor.** Simultaneous contraction of an extensor and its opposing flexor is avoided by **reciprocal inhibition,** a reflex arc that inhibits the antagonist of any contracted muscle. Therefore, only one muscle of the pair contracts at any one time.

The exercises of this chapter provide laboratory investigations of the major aspects of the mammalian skeletal and muscular systems.

## PRELAB QUESTIONS

1. What is the appendicular skeleton?

2. What type of muscle tissue is responsible for locomotion?

3. Where is cardiac muscle located?

4. What is the unit of contraction in skeletal muscle?

5. What is an extensor muscle?

# LAB 22.A
## THE ARCHITECTURE OF BONE

### MATERIALS REQUIRED

section of ground bone, prepared slide
sections of sawed bones

### PROCEDURE

You observed the intricate structure of bone tissue in Topic 21. Now is an appropriate time to review the cellular basis of bone tissue before you investigate the macroscopic aspects of the skeleton.

1. Examine under low power the prepared slide of ground bone tissue on demonstration. Using Figure 21–5 (see page 240) as a guide, locate a haversian canal that held small blood vessels in the live bone. The small dark lacunae (lacuna, singular) are

tiny chambers within the hard bone matrix that contained osteocytes, or bone cells. The hard white bone matrix is composed of inorganic mineral salts, calcium carbonate, and calcium phosphate. The skeleton of an embryo consists of cartilage that is gradually replaced by bone as bone cells deposit the mineral matrix throughout childhood and the teen years.

2. Observe the sawed bone sections that reveal internal bone structure. The hard, solid outer walls of the bone are composed of compact bone that has the microscopic structure you examined previously. The interior of the bone ends consists of porous spongy bone that is spongy only in appearance; it, too, is composed of calcium salts. In life, spongy bone in the ends of long bones is filled with red bone marrow, which produces red blood cells, some types of white blood cells, and platelets that function in blood clotting. The central cavity in many bones is filled with fatty yellow bone marrow.

## POSTLAB QUESTIONS

1. Identify the internal structures of a bone.

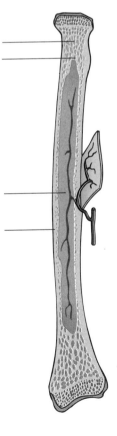

2. What are the architectural characteristics of bone that afford it its strength and flexibility, but yet minimize its weight?

## LAB 22.B
## THE HUMAN SKELETON

### MATERIALS REQUIRED

mounted human skeleton, or human skeleton model

cat skeleton

### PROCEDURE

All mammalian skeletons have essentially the same bones. Although these basic bones in many species have become modified through natural selection, the bones retain their general shape and position among many species of mammals. While becoming acquainted with the major bones of the human skeleton you will also become acquainted with the general skeletal structure of most other mammals.

The human skeleton consists of two major portions: the axial skeleton, so named because it forms the main axis of the body, and the appendicular skeleton, which is composed of the appendages and their supporting structures that attach to the axial skeleton. We shall consider each skeletal portion separately. As you negotiate this exercise, refer to Figure 22–2 and the skeleton(s) in your laboratory to locate the various bones.

### The Axial Skeleton

1. Identify the skull, the rib cage, and the vertebral column (spine) that compose the axial skeleton. Locate the mandible, or lower jaw, of the skull. A majority of the human skull consists of a large cranium, or brain case, that holds and protects the delicate brain. The size of the cranium is a direct indication of brain size. Generally speaking, a proportionally larger brain size compared to body size in mammals corresponds with a higher degree of intelligence. Notice how the vertebral column connects with the skull on the ventral surface of

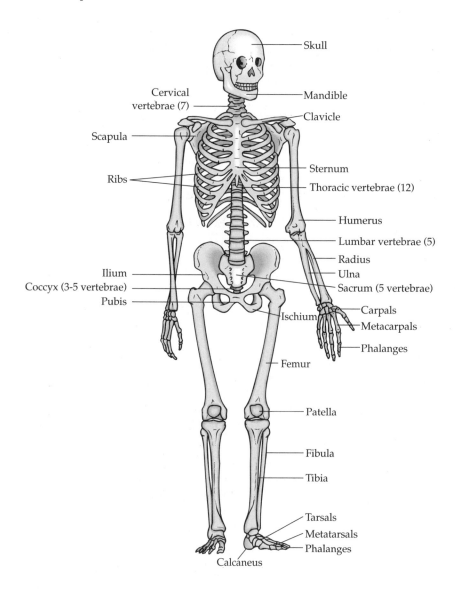

**FIGURE 22-2**

The human skeleton.

the cranium. This arrangement occurs because humans are bipedal—we walk upright on two legs. In quadruped mammals, such as the cat, the vertebral column connects with the skull more toward the back of the head instead of beneath it. This arrangement properly positions the head fully forward when walking on four legs. Compare the attachment of the skulls and vertebral columns of the cat and human skeletons.

2. Observe the numerous bone segments that compose the vertebral column. Each segment is called a vertebra (vertebrae, plural). Each vertebra has a central hole where the spinal cord of the central nervous system passes through the length of the vertebral column. The vertebrae of the vertebral column vary in size and shape in different regions of the body. Seven cervical vertebrae form the neck. Twelve thoracic vertebrae, joined by the ribs, are found in the chest region. Five large lumbar vertebrae form the lower portion of the vertebral column (the "small" of the back). The various projections of the vertebrae are sites of muscle attachment. Each vertebra is separated from its neighbor by tough cartilage discs that give the vertebral column flexibility and that absorb shock and compression forces. The sacrum, formed of 5 fused vertebrae, is located below the lumbar region of the vertebral column. Nerves originating from the spinal cord emerge from the holes of the sacrum. The coccyx, consisting of 3 to 5 reduced verte-

brae, forms the terminal portion of the vertebral column. The coccyx is all that remains of a tail in humans.

3. Examine the rib cage that protects the lungs, heart, and other vital organs. All 12 pairs of ribs attach to the thoracic vertebrae of the spinal column. All but the last 2 pairs of ribs are attached to the sternum, or breast bone, by a strip of cartilage that extends from the rib bone. The eleventh and twelfth pairs of ribs are known as floating ribs because they do not attach to the sternum.

### The Appendicular Skeleton: Upper Appendages

4. Locate the clavicle (collar bone) and scapula which supports the bones of the arm. The humerus of the upper arm attaches directly to the shoulder joint that is formed by the union of the clavicle and scapula.

5. Note the 2 bones of the lower arm below the elbow joint. These are the radius and ulna. The radius is located on the "thumb side" of the arm, and the ulna on the opposite side.

6. Observe the small carpal bones that form the wrist. The metacarpals are longer bones that compose the palm of the hand. The phalanges (phalanx, singular) are the short bone segments found in each finger. When you make a fist, the knuckles mark the end of the metacarpals and the beginning of the phalanges.

### The Appendicular Skeleton: Lower Appendages

7. Notice the large pelvic girdle that supports the bones of the legs. The pelvic girdle consists of 3 separate bones fused into 1 large bone, the pelvis. The ilium is the largest of the 3 bones; the ischium is the strongest and is located below the ilium; the pubis is the most anterior of the 3 bones. The pelvis differs between the sexes making it possible to determine the sex of a skeleton. The female pelvis is rather broad and somewhat shallow as an adaptation to childbearing. The male pelvis is narrower and deeper than the female pelvis.

8. Locate the femur, the long singular bone of the upper leg. The knee joint is marked by the patella, or knee cap. The lower leg consists of two bones — the larger tibia and the smaller fibula.

9. Observe the tarsals, the small bones that form the complex ankle. The metatarsals compose the main structure of the foot. Again you see phalanges of the toes. As you may have noticed, the bones of the hand and foot are similar and have been similarly named.

## POSTLAB QUESTIONS

1. Identify the largest and strongest bone of the face. (*Hint:* It joins with the skull at the only movable joint of the skull.)

2. Why is it mechanically important that each rib curve outward, then forward, and then downward?

3. What bone is the longest and heaviest of the body?

4. When a woman is wearing high heel shoes, what foot bones bear most of the body's weight?

5. Do the phalanges of the hand have the same function as the phalanges of the feet? Why or why not?

6. When a person has fallen arches, what bones have weak tendons and ligaments?

7. Identify the bones of the human skeleton.

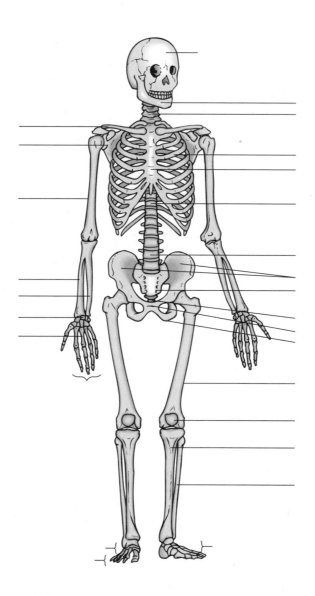

8. Identify the bones of the cat skeleton.

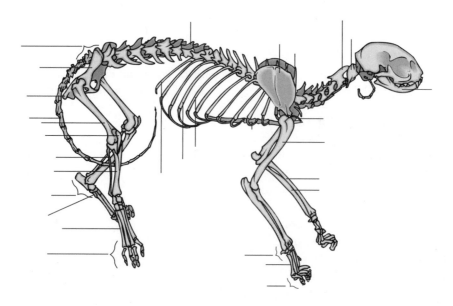

9. After identifying the bones of the human skeleton, why were the bones of the cat skeleton easy to identify?

## LAB 22.C
## JOINTS OF THE SKELETON

### MATERIALS REQUIRED

mounted human skeleton or human skeleton model

your limbs and joints

### PROCEDURE

Movement of the skeleton is made possible by joints (Fig. 22–3), or articulations, that occur where bones join one another. However, not all kinds of joints are movable. Movable joints are surrounded by a capsule membrane that contains a lubricating fluid. The ends of the joining bones are covered with cartilage that prevents wear. The bones of most joints are held together by strong bands of connective tissue called ligaments. We shall consider some major types of joints in this exercise. As the various joints are described

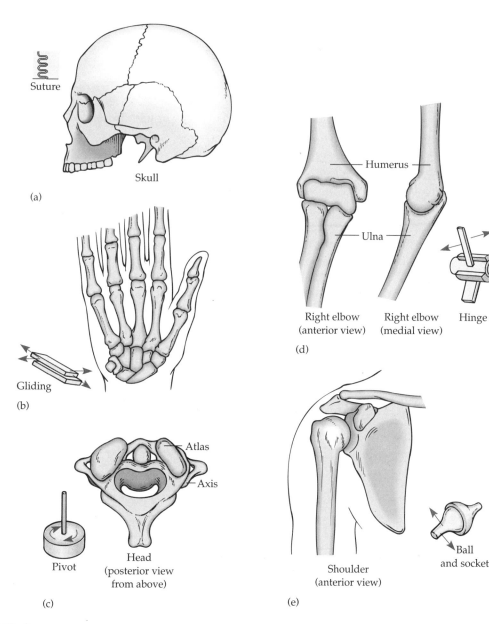

**FIGURE 22–3**

Joint types: (a) suture, (b) gliding, (c) pivot, (d) hinge, and (e) ball and socket.

below, study them on the skeleton and notice how they function in your own body.

1. Search the cranium of the skull for fine irregular lines formed where plates of bone join together. These immovable joints are called **sutures,** and are found only in the skull. The irregular sutures interlock like pieces of a jigsaw puzzle giving strength to the skull bones.

2. Move your wrist (carpals) and ankles (tarsals). These are examples of **gliding joints.** Gliding joints contain bones that slide over one another, permitting smooth side-to-side and back-and-forth movements.

3. Rotate your head from side to side. This is an example of a **pivot joint** formed by a concave surface of one bone that pivots on a projection of another bone. The first cervical vertebra, appropriately called the atlas, supports the skull. The second cervical vertebra, called the axis, has a "pin" of bone that projects upward into the atlas forming a pivot point on which the atlas turns. The atlas and axis are best observed when viewed through the open mouth of the skull if you are observing a mounted skeleton or full-sized model.

4. Bend your elbow or knee to bring the appendage closer to your body. These are examples of a **hinge joint,** so named because their movement is similar to that of a door hinge. Hinge joints also occur in the fingers and toes.

5. Extend your arm from your side and move it in a circular motion. This type of movement is possible due to the **ball-and-socket joint** present in the shoulder. As the name implies, this joint consists of a ball-like end of one bone fitting into a cup-like depression of another. The rolling freedom of the ball within the socket allows movement in many directions. The hip is another example of a ball-and-socket joint.

### POSTLAB QUESTIONS

1. What type of joint permits no movement between the articulating bones?

2. What type of joint permits the widest range of movements?

3. Complete Table 22–1 by providing an example for each joint type.

**Table 22–1**   Major Types of Joints

| Joint Types | Examples |
| --- | --- |
| Sutures | |
| Gliding joints | |
| Pivot joints | |
| Hinge joints | |
| Ball-and-socket joints | |

4. What might be the consequences if cartilage were absent in the joints between the bones of the vertebrate skeleton?

5. Explain anatomically the mobility of the shoulder joint.

6. What type of joint exists between adjacent vertebrae?

7. Why is the knee joint injured more often than the hip joint?

## LAB 22.D
### THE MUSCULAR SYSTEM

### MATERIALS REQUIRED

striated muscle tissue, prepared slide on demonstration

motor end plates, prepared slide on demonstration

preserved frog with hind leg musculature exposed, on demonstration

your leg and arm

### PROCEDURE

Before you explore how the skeleton and muscles work together, you must first investigate a few structural and functional aspects of the muscular system.

1. Review under high power the prepared slide of striated (skeletal) muscle tissue you first examined as part of Topic 21. Each long, slender muscle cell is called a muscle fiber. Numerous darkly stained nuclei may be visible throughout the slide; each

Motor end plates                    Nerve fiber

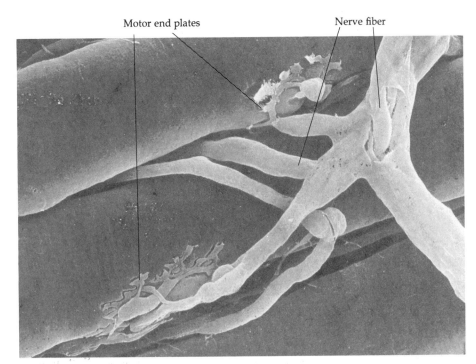

**FIGURE 22-4**

Scanning electron micrograph of nerve fibers and motor end plates connecting with skeletal muscle fibers.

striated muscle fiber contains several nuclei. The minute light and dark striations, or cross bands, characteristic of skeletal muscle are due to the specific arrangement of unique contractile protein filaments contained within the muscle fiber. What is typically called a "muscle," such as the biceps, is actually a collection of numerous long muscle fibers. The biceps, then, is the organ composed of muscle tissue and, of course, connective and nervous tissue.

2. Examine under high power the prepared slide on demonstration of motor end plates. Motor end plates (Fig. 22-4) are the junction between the nervous system and the muscular system. Nerve impulses travel to the muscle by a nerve fiber that branches to individual muscle fibers. Locate the tiny branches of the nerve fiber as they reach the muscle fibers. Each nerve branch ends as a motor end plate. The word "motor" implies that the end plates are associated with some kind of action. The nerve impulse is delivered to the muscle fiber by the motor end plate, which stimulates contraction of the contractile proteins within the fiber. When the contractile protein molecules of many muscle fibers contract in unison, the muscle as a whole contracts.

3. Observe the skinned leg of a preserved frog specimen on demonstration. Locate the large gastrocne-

mius muscle that forms the calf of the lower leg. The gastrocnemius is a good example of gross muscle structure. Locate the attachment of this muscle behind the knee of the frog. Follow the muscle down the lower leg. Notice that it narrows into a white band of tissue that attaches to the heel of the frog's foot. The white band is a tendon, a strap of tough connective tissue that attaches muscle to bone. A short tendon is also present at the other end of the gastrocnemius where it attaches to the leg bone behind the knee. The bulk of the muscle located between the attachment points is called the belly. The attachment point behind the knee is called the origin of the muscle—it *does not move* when the muscle contracts. The muscle attachment at the heel is called the insertion—it *moves* with muscle contraction.

4. Demonstrate the difference between the origin and insertion of the gastrocnemius in your own leg in the following manner. When seated, cross one leg over the other and feel for the large tendon of the gastrocnemius just above the heel in the crossed leg. This is also known as the Achilles tendon. It feels like a thick cord at the back of the ankle. While holding the tendon between your thumb and fingers, contract the gastrocnemius by extending your foot as if standing tiptoe. Try the motion several times. Can you feel the tendon move when the

muscle contracts? Keep your leg crossed and try the following.

Locate with your fingers the top of the gastrocnemius where it attaches behind the knee. Keep your fingers on this area, then extend your foot as you did previously. Try the motion several times. Can you feel any movement equivalent to what you did above the heel?

5. Extend your arm out in front of you, then cup your free hand over the biceps, the muscle of the upper arm between the inside of the elbow joint and shoulder. Contract the biceps by bending the elbow to bring your hand toward your head. Notice the shortening and tightening of the biceps when it contracts. With your biceps still contracted, cup your free hand over the triceps, the muscle on the back of the upper arm directly opposite the biceps. Contract the triceps by extending your arm in front of you again, then forcefully maintain the contraction to make the triceps more noticeable to your hand that covers it. Notice the tightening of the triceps in its contracted state.

With the triceps contracted, feel the biceps with your free hand and compare it to when your arm was flexed. Contract the biceps again and feel the triceps with your free hand. Notice that the triceps muscle has lost the tightness it had when contracted. This exercise reveals that muscles work in opposing pairs at joints, which is the elbow in this case. A muscle (biceps) that causes a desired movement (bending of the elbow) when it contracts is called the agonist. The muscle (triceps) that causes the opposite movement (straightening of the elbow) is called the antagonist. When an agonist contracts, the antagonist relaxes. Note that no muscle is permanently designated as an agonist or an antagonist. In extending the arm, the triceps becomes the agonist and the biceps becomes the antagonist. The role of a muscle changes regarding a particular body movement.

## POSTLAB QUESTIONS

1. What specialized function is highly developed in skeletal muscle tissue?

2. What is needed to activate a skeletal muscle?

3. Identify the gastrocnemius and the origin and insertion of the Achilles tendon.

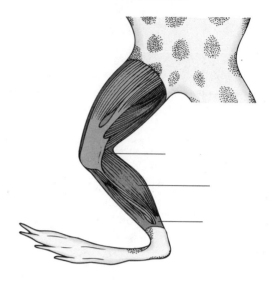

4. Identify the joints, bones, and muscles of the human arm.

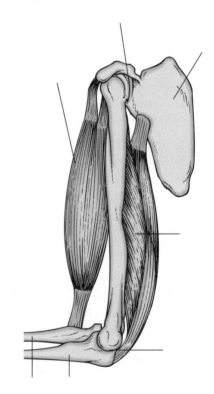

5. Identify the origins and insertions of the muscles in the previous question.

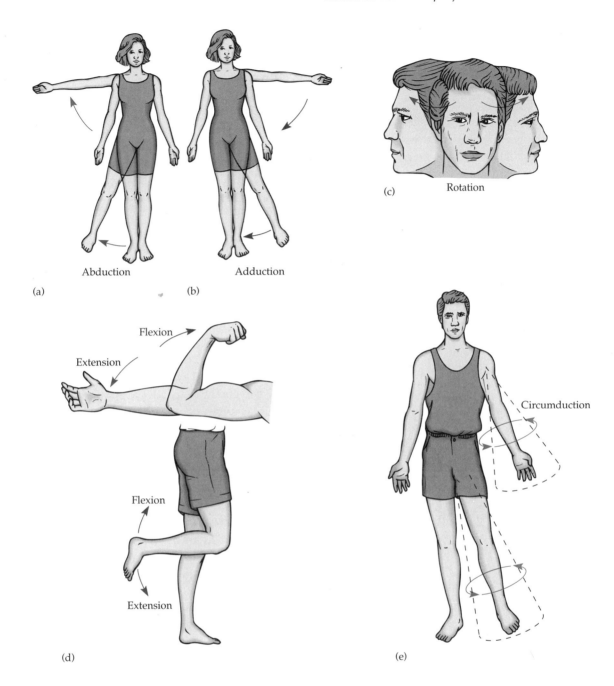

## FIGURE 22-5

Basic joint movements: (a) abduction, (b) adduction, (c) rotation, (d) extension and flexion, and (e) circumduction.

## LAB 22.E
## FUNCTIONAL RELATIONSHIPS OF THE SKELETON AND MUSCLES

### MATERIALS REQUIRED
your body's joints and muscles

### PROCEDURE

The skeletal and muscular systems work as a team to produce body movement. The skeleton cannot move without muscles; muscles cannot move the body unless they have the skeleton as an anchoring framework. Together the two systems produce a great variety of bodily movements that also reflect body flexibility due to the various types of joints. Although body movements are diverse, they can be classified into several kinds of basic movements—abduction, adduction, flexion, extension, rotation, and circumduction (Fig. 22–5). You will demonstrate these movements in the following exercises. As you perform these movements, try to be aware of the opposing muscle groups involved that cause a particular movement and the opposite effect, such as the one you observed in the earlier biceps and triceps exercise.

**Abduction.**　Movement of a limb away from the body midline.

1. Demonstrate abduction of a limb by standing and then moving one leg outward to the side so that you stand on one foot. As another example, abduct one of your arms to the side so that it extends at a right angle from your body. Spreading the fingers is abduction relative to the middle finger.

**Adduction.**　Movement of a limb toward the body midline.

2. Repeat the movement of your arm or leg as you did above, then adduct the limb back toward your body. Holding the fingers tightly parallel to one another is adduction relative to the middle finger.

**Flexion.**　Movement resulting in a decrease of the angle of a joint.

3. Extend your arm in front of your body, then bend the elbow to bring the hand toward your head. This is flexion, which decreases the angle of the elbow joint. Flex your knee joint by standing and then lifting up one foot behind you. The angle of the knee joint decreases as movement proceeds.

**Extension.**　Movement resulting in an increase of the angle of a joint.

4. Demonstrate extension by repeating the movements of Step 3. Then return the limb to its original position. After the elbow or knee joint is flexed, extension increases the angle of the joint when the limb is returned to resting position.

**Rotation.**　Movement of a bone around its own longitudinal (lengthwise) axis.

5. Turn your head from side to side. This is rotation of the head, made possible by the pivot joint of the atlas and axis cervical vertebrae described previously. As another example, straighten your arm at your side, then rotate the entire arm from the shoulder so that the palm of your hand faces forward, then backward. This movement is rotation of the humerus.

**Circumduction.**　Movement in which the free end of a limb moves in a circle as the base of the limb remains stable; involves a 360° rotation.

6. Demonstrate circumduction by standing and then rotating your outstretched arm at the shoulder as if winding up to throw a ball underhanded. Circumduction is a good example of how a seemingly simple movement is actually a complex combination of other types of movements. Circumduction, to some degree, typically involves flexion, abduction, adduction, extension, and rotation.

## POSTLAB QUESTIONS

1. Identify the following types of movement.
   a. bending the head forward as in prayer
   b. returning the head from the forward position to the upright position
   c. moving the arms straight out to the sides
   d. moving the toes toward the second toe
   e. holding the head in an upright position and turning it from one side to the other
   f. dropping the head to one shoulder, then to the chest, to the other shoulder, and then backward
2. What type(s) of movements is (are) the human thumb capable of?

3. When the flexor muscle of the upper arm contracts, the extensor muscle relaxes. Which muscle is the agonist? The antagonist?

4. If the abductor muscle of the upper arm is an agonist, what muscle of the upper arm would be its antagonist?

## FOR FURTHER READING

Alberts, B. et al. 1989. *The Molecular Biology of the Cell.* New York: Garland.

Gamble, James G. 1988. *The Musculoskeletal System: Physiological Basics.* New York: Raven Press.

Guyton, A. C. 1984. *Physiology of the Human Body.* Philadelphia: Saunders College Publishing.

Hole, J. W. 1981. *Human Anatomy and Physiology.* Dubuque, Iowa: William C. Brown.

Hoyle, G. 1983. *Muscles and Their Neural Control.* New York: Wiley.

Huxley, A. 1990. *Reflections on Muscle.* Princeton: Princeton University Press.

Kessel, R., and R. Kardon. 1979. *Tissues and Organs: A Text-Atlas of Scanning Electron Microscopy.* San Francisco: W. H. Freeman.

Luttgens, K., and K. F. Wells. 1982. *Kinesiology: Scientific Basis of Human Motion.* Philadelphia: Saunders College Publishing.

Murray, P. D. F. 1985. *Bones: A Study of the Development and Structure of the Vertebrate Skeleton.* New York: Cambridge University Press.

Novelline, Robert A., and Lucy F. Squire. 1987. *Living Anatomy: A Working Atlas Using Computed Tomography, Magnetic Resonance, and Angiography Images.* Philadelphia: Hanley and Belfus.

Romer, A. S., and T. S. Parsons. 1986. *The Vertebrate Body.* Philadelphia: Saunders College Publishing.

Vander, A. et al. 1989. *Human Physiology: The Mechanisms of Body Function.* New York: McGraw-Hill.

Wilson, Frank C., ed. 1983. *The Musculoskeletal System: Basic Processes and Disorders.* Philadelphia: Lippincott.

# Mammalian Nervous System

## OBJECTIVES

1. Draw a neuron. Label its parts and give their functions.

2. Describe the mechanism by which an impulse is transmitted along a neuron and from one neuron to another.

3. Describe the process of reception and the anatomy of the mammalian eye.

4. List the functions of the spinal cord and describe its structure.

5. Locate the following parts of the brain and give the functions for each: medulla, pons, midbrain, thalamus, hypothalamus, cerebellum, and cerebrum.

---

## INTRODUCTION

**Neurons** are the cells that transmit messages in the nervous system. A neuron consists of a **soma,** which is a cell body, and **dendrites** and **axons,** which are long, thin extensions that carry electrical impulses. Dendrites receive information from other cells or the external environment. Axons carry information to the next cell.

Information travels through the nervous system in the form of an electrical **nerve impulse,** which is created by an unequal distribution of ions on both sides of the neuron plasma membrane. The ion imbalance is produced by a **sodium-potassium pump,** which pushes sodium ions out of the cell and pumps potassium ions into the cell by active transport. Both ions diffuse back through the membrane due to their concentration gradients, but potassium diffuses out easier than sodium diffuses in because of membrane properties. As a result, a net positive charge forms on the outside of the membrane compared to a net negative charge inside the membrane. The membrane is now **polarized.** This ion arrangement is termed a **resting potential,** which "sets the stage" for a nerve impulse.

A stimulus that contacts a particular area of a polarized membrane temporarily changes the membrane's permeability to sodium ions. Sodium rushes into the neuron. This produces a **local potential,** which is a defined area of **depolarization.** If a local potential is reinforced by additional stimulation, or if a stimulus is of sufficient intensity, the neuron will fire a nerve impulse as a wave of depolarization through its entire membrane. This is an **action potential,** or nerve impulse.

Transmission speed of a nerve impulse along the axon of some neurons is increased by a **myelin sheath.** The sheath covers the axon in the form of segments of lipid-rich membrane of **glial cells** that wrap around the neuron. Small gaps of bare axon are exposed at the **nodes of Ranvier,** which are spaces between the glial cells. The fatty myelin prevents continuous depolarization along the axon. Depolarization can occur only at the nodes. This makes the impulse jump from node to node, which sends the message at a faster rate.

A nerve impulse is conveyed to other cells by a synapse. A **synapse** is an area where the membrane of an axon ending lies very close to the membrane of a neighboring cell. A nerve impulse crosses the small space between the membranes by **neurotransmitters,** which are chemicals contained in vesicles within the axon terminal. An action potential causes release of neurotransmitter molecules into the narrow space between the membranes, or **synaptic cleft.** Transmitter molecules cross the cleft and bind with receptors on the **postsynaptic membrane.** This causes depolarization of the neighboring neuron and the impulse continues. In some synapses, the electrical impulse jumps directly from one cell to another without chemical transmitters.

The **receptors** in the sense organs of the body detect stimuli, which are classified, generally, as changes in the internal and/or external environment of the body. The detected information is converted into electrical information, which is processed by the nervous system. The nervous system sends out signals to the **effectors** of the body, the structures that carry out a response. The five major types of receptors, classified according to the stimuli to which they respond, are mechano-, chemo-, thermo-, electro-, and photoreceptors.

The vertebrate nervous system consists of the **central nervous system** (brain and spinal cord) and the nerves known as the **peripheral nervous system,** which connect the various body areas to the central nervous system. The string-like nerves throughout the body consist of axon bundles and dendrite bundles. They are covered with protective connective tissue. Peripheral nerves originating from the brain are termed **cranial nerves,** and those nerves branching from the spinal cord are called **spinal nerves.** The peripheral nervous system is divided into two parts based on function. The **somatic nervous system** controls body parts under conscious control. The **autonomic nervous system** controls unconscious body functions such as digestion.

The vertebrate brain consists of three basic regions —the forebrain, the midbrain, and the hindbrain. Each region has given rise to specific brain structures with functions that vary somewhat among the different vertebrate classes. The forebrain includes the **cerebral hemispheres,** the **thalamus,** and the **hypothalamus.** The cerebral hemispheres function for sensory and motor association plus visual and auditory processing. Mammals have the greatest cerebral development in both size and functions. This endows mammals with intelligence, which is most evident in humans. The cerebral hemispheres are least developed in fishes. The thalamus is an area of sensory integration in higher vertebrates. The hypothalamus controls homeostasis of many vital body functions in all vertebrates.

The **cerebrum** is the most prominent portion of the human brain containing in excess of 70% of the cells of the brain. It is divided into a right and left hemisphere by a longitudinal fissure. The cerebrum is thought to be the center of intellect, memory, consciousness, and language. In addition, the cerebrum is thought to control sensation and motor functions.

In mammals, the midbrain forms the **colliculi** that control reflexes of the iris and eyelids and receive sensory information from the ear.

The hindbrain has produced the **cerebellum** and **medulla.** The cerebellum in all vertebrates coordinates equilibrium and body movement. The medulla functions similarly in all vertebrates. It relays messages between the brain and spinal cord plus it controls reflexes in many organs. In mammals, the **pons** forms a bridge connecting the spinal cord and the medulla with the upper portions of the brain.

A **reflex arc** is a simple nervous mechanism that automatically and consistently performs an unconscious function. The sudden withdrawal of the hand away from an uncomfortably hot or burning object is caused by a reflex arc. Elements of walking and breathing are also based on simple reflex arcs. A reflex arc consists of a sensory neuron for detecting a stimulus, a motor neuron for producing an appropriate reaction, and usually one or more association neurons that make connections between the two within the central nervous system (many times the spinal cord). Reflexes are advantageous because they produce a fast response by involving only a few synapses. A response produced by a particular reflex is always unvarying because higher nerve centers are not involved. One can easily see the adaptive value of the reflex arc.

We have designed the following laboratory exercises to assist you in learning the salient aspects of the mammalian nervous system.

## PRELAB QUESTIONS

1. What are the three parts of a neuron?

2. What are neurotransmitters?

3. What is the most prominent portion of the human brain?

4. What part of a vertebrate brain functions in coordinating equilibrium and body movement?

5. What is a reflex arc?

## LAB 23.A
## THE NEURON, BASIS OF THE NERVOUS SYSTEM

### MATERIALS REQUIRED
smear of spinal cord neurons, prepared slide
compound microscope

### PROCEDURE
Neurons are highly modified cells that conduct nerve impulses within the nervous system. Their specialized structure and function are the working basis of the nervous system. Neurons are so specialized that they have lost the capability of mitosis and cannot replace themselves if damaged or destroyed as can most other cells of your body. You are born with one set of neurons that live throughout your life span. You observed neurons previously in Topic 21. During this laboratory, as you study the nervous system, is an appropriate time to review the form and function of neurons.

1. Examine under low power the prepared slide of a neuron smear from the spinal cord of a cow. The neurons are the large, spider-like cells with an obvious nucleus. The numerous and smaller darkly stained cells around the neurons are glial cells, which support and nourish the neurons.

2. Focus on an individual neuron (Fig. 23–1). The cell body is the main part of the neuron that contains the nucleus. Note the branch-like appendages that spread out from the cell body. These branches are called axons and dendrites, which carry nerve impulses. On this slide it is somewhat difficult to distinguish axons from dendrites, but there is a functional difference based on the direction of nerve impulse flow. Dendrites carry impulses *toward* the cell body of the neuron. Axons carry impulses *away* from the cell body and are generally longer than dendrites. Impulses are transmitted throughout the body by the axon of a neuron that connects with the dendrites of one or more neighboring neurons that subsequently connect to other neurons further along in the same manner. The normal arrangement of these neurons in the spinal cord was disrupted when the slide was prepared. Return your slide and microscope to their proper place when finished.

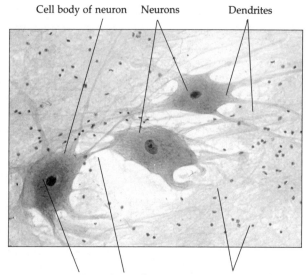

**FIGURE 23–1**

Neurons on a slide of a spinal cord smear.    *(Ed Reschke)*

### POSTLAB QUESTIONS
1. Identify the parts of a neuron.

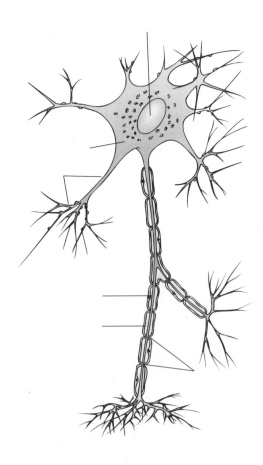

2. How does the cell body of a neuron compare to other body cells?

3. Compare the diameters of various axons. How might the diameter of an axon affect the function of neurons?

4. Many medicines used to effect brain function block neurotransmitter uptake. What part(s) of the neuron is (are) involved?

## LAB 23.B
### ANATOMY OF THE SHEEP BRAIN

### MATERIALS REQUIRED (per pair of students)

preserved sheep brain, whole

preserved sheep brain, longitudinal midsection

preserved sheep brain, cross section, on demonstration

dissecting tray

blunt probes

## PROCEDURE

The sheep brain is a convenient specimen for examining the structure of the mammalian brain. Sheep brains are easily obtained as a by-product of slaughterhouses, and are large enough for easy observation, yet small enough for easy handling. Except for minor differences, the sheep brain is similar in structure to the human brain. Refer to Figures 23–2 and 23–3 for guidance as you examine the sheep brain. Your specimens are strictly for observation and are not intended for dissection. Handle the specimens carefully and use a light touch with the blunt probe when locating structures.

1. Obtain a whole sheep brain and place it dorsal side up (see Fig. 23–2a) in your dissecting tray. Three major structures of the brain are visible from this view. Locate the large cerebrum that forms the bulk of the brain. The cerebrum is divided into right and left cerebral hemispheres that are separated by the cerebral fissure. The cerebrum is the center of intelligence, consciousness, and memory. The cerebrum also controls voluntary muscle movement and receives information from the sense organs. Notice that the cerebrum is convoluted, or "coiled."

2. Examine the cross section of a sheep cerebrum on demonstration. Notice the darker outer layer of the cerebrum where it has been cut. This darker layer is called gray matter (also known as the cerebral cortex). It contains numerous nerve cell

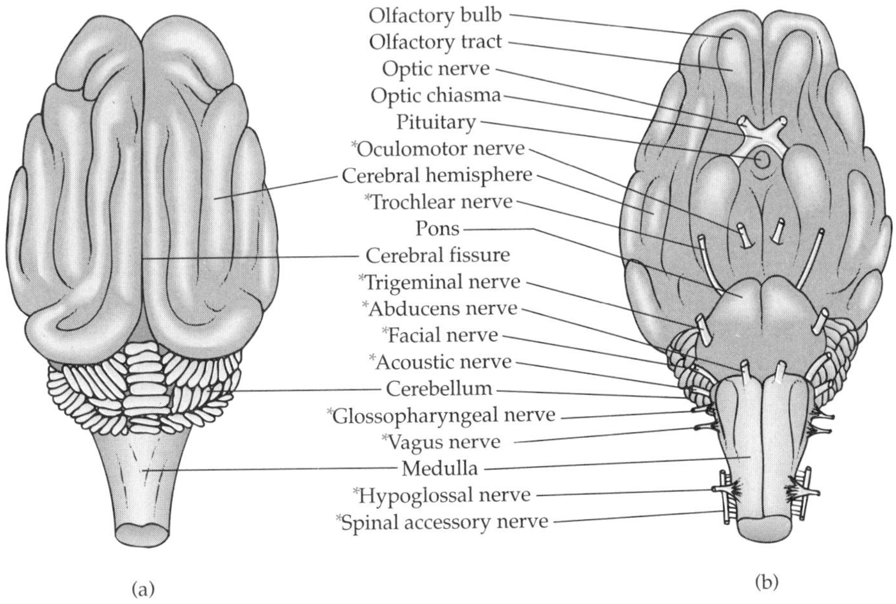

Olfactory bulb
Olfactory tract
Optic nerve
Optic chiasma
Pituitary
*Oculomotor nerve
Cerebral hemisphere
*Trochlear nerve
Pons
Cerebral fissure
*Trigeminal nerve
*Abducens nerve
*Facial nerve
*Acoustic nerve
Cerebellum
*Glossopharyngeal nerve
*Vagus nerve
Medulla
*Hypoglossal nerve
*Spinal accessory nerve

(a)

(b)

**FIGURE 23–2**

(a) Dorsal view of a sheep brain. (b) Ventral view of a sheep brain. Structures marked with an asterisk (*) are specific cranial nerves.

bodies and their axons that lack a myelin sheath. The thicker, lighter layer below the gray matter is white matter. It consists of myelinated axons that interconnect the regions of the brain.

3. Return to the dorsal side of the whole sheep brain. Locate the second largest portion of the brain, the bulbous cerebellum positioned directly behind the cerebrum. The convolutions of the cerebellum are smaller than those of the cerebrum. The cerebellum maintains equilibrium and coordinates muscular movements of the body.

4. Identify the medulla, the third and most posterior region of the brain. The medulla is continuous with the spinal cord. It contains important control centers that regulate the unconscious body functions of respiration, blood pressure, and heart rate.

5. Turn over the sheep brain to expose its ventral surface (Fig. 23–2b). Locate the olfactory bulbs and olfactory tracts near the anterior end of the brain. The olfactory bulbs, functioning in the sense of smell, receive information from nerves leading from the nasal areas of the head.

6. Locate the optic chiasma just posterior to the olfactory tracts. The optic chiasma is an X-shaped structure formed by the intersection of the two optic nerves leading from the eyes to the vision centers of the brain.

7. Find the position of the pituitary gland immediately posterior to the optic chiasma. The pea-sized pituitary gland that hangs down from the brain by a short stalk of tissue is usually lost when the brain is removed from the skull. The attachment point of the absent pituitary gland is marked by a hollow tube-like depression. The pituitary is an endocrine gland that secretes several important

hormones that regulate growth, physiology, and reproduction.

8. Search along the medulla for the small paired cranial nerves that arise from each side of the medulla and other areas of the brain as shown by the asterisks in Figure 23–2b. Twelve pairs of cranial nerves originate directly from the brain and supply areas of the head, face, and pharynx (throat). Cranial nerves bring information from the many sense organs of the head region to the brain. The cranial nerves also carry impulses from the brain to the voluntary muscles that move many structures of the head, face, and pharynx. The olfactory bulbs and optic nerves are the largest cranial nerves.

9. Locate the pons, an oval bulge just anterior to the medulla. The pons contains nerve tracts that interconnect various parts of the brain.

10. Obtain the longitudinally sectioned brain and place it in your dissecting tray as shown in Figure 23–3. The brain section enables you to see internal structures and different views of structures located previously. The cavities inside the brain are called ventricles and are continuous with the central canal (discussed later) of the spinal cord. These spaces are filled with cerebrospinal fluid that also surrounds the brain and spinal cord just below the protective membranes (meninges) that cover them. The fluid functions as a shock absorber and helps protect the brain and spinal cord from injury.

11. Find the corpus callosum, a wide band of white matter that connects the two cerebral hemispheres.

12. Note the massa intermedia, a somewhat circular structure located near the center of the brain sec-

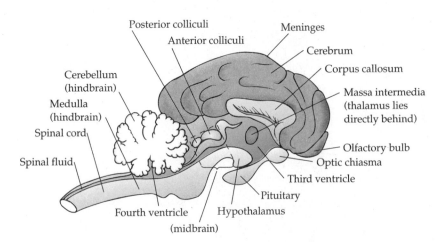

**FIGURE 23–3**

Major internal structures of the sheep brain.

tion. The massa intermedia serves as a landmark to locate the thalamus, which is the wall of the third ventricle directly posterior to it. The thalamus sorts incoming information from sensory organs and then relays the information to sensory areas of the cerebrum.

13. Locate the hypothalamus, which is found below the thalamus and just above the position of the pituitary gland. The hypothalamus controls many important body functions such as sleep, water balance, body temperature, sex drive, and appetite. As another important function, the hypothalamus secretes releasing factors that cause the pituitary gland to release several important hormones into the bloodstream.

14. Find the midbrain, which is located anterior to the cerebellum. The midbrain consists of the anterior colliculi, which controls visual reflexes such as change of pupil size, and the posterior colliculi, which control auditory reflexes such as adjustment of the ear to loud noises.

## POSTLAB QUESTIONS

1. Identify the parts of the human brain based on your knowledge of a sheep brain.

2. Complete Table 23–1.

**Table 23–1**  Mammalian Brain

| Structure | Location | Function |
| --- | --- | --- |
| Cerebrum | | |
| Cerebellum | | |
| Medulla | | |
| Pons | | |
| Midbrain | | |
| Thalamus | | |
| Hypothalamus | | |

3. Alzheimer's disease causes the cerebral cortex to become shrunken as the neurons degrade and lose their function. Based on the function of the cerebrum, what symptoms do you think this would cause?

4. Why would an injury to the medulla kill a person?

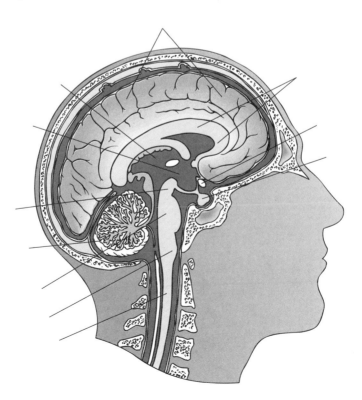

## LAB 23.C
## STRUCTURE OF THE SPINAL CORD

### MATERIALS REQUIRED
    spinal cord, cross section, prepared slide
    compound microscope
    spinal cord model

### PROCEDURE
The brain and spinal cord are the two components of the central nervous system. Higher functions of the nervous system such as learning and thought are limited to the brain. The spinal cord functions mainly as a "switching center" where nerve impulses are directed up and down the spinal cord to and from the brain. The spinal cord also controls many unconscious body reflexes.

1. Examine the spinal cord cross section under low power of your microscope (Fig. 23–4). The specimen was obtained from a rat or other small mammal. Notice the darker inner region composed of gray matter. As in the brain, gray matter consists of nerve cell bodies and their dendrites, which lack myelin. The lighter outer region of the spinal cord is composed of white matter, which consists of numerous myelinated axons that run up and down the length of the spinal cord.
2. Look for the central canal visible as a small opening in the center of the spinal cord section. The central canal runs the length of the spinal cord and is continuous with the ventricles of the brain. The central canal is filled with cerebrospinal fluid as are the ventricles within the brain.

3. Observe the spinal cord model showing an enlarged cross section of the cord and structure of spinal nerves that branch from it. Note that each spinal nerve is branched near the spinal cord. The branches are known as the dorsal and ventral roots of the spinal nerve, named according to their relative position in the body—dorsal toward the back, ventral toward the abdomen. The dorsal and ventral roots merge forming the spinal nerve. Each spinal nerve contains both sensory and motor neurons that carry impulses toward and away from the spinal cord, respectively.
4. Note the enlargement in the dorsal root of each spinal nerve. This is the dorsal root ganglion (ganglia, plural), which is a cluster of nerve cell bodies of sensory neurons that bring impulses from receptors in the body toward the spinal cord. The dorsal root contains only sensory neurons. The ventral root contains only motor neurons. Note the absence of a ganglion in the ventral root of the spinal nerve; the cell bodies of motor neurons are located in the gray matter of the spinal cord.

    Return your slide and microscope to their proper places when finished.

### POSTLAB QUESTIONS
1. Compare the locations of myelinated and unmyelinated nerve fibers in the brain and spinal cord.


2. If disease in your spinal bones causes compression of your dorsal root ganglion, what symptoms would you have?

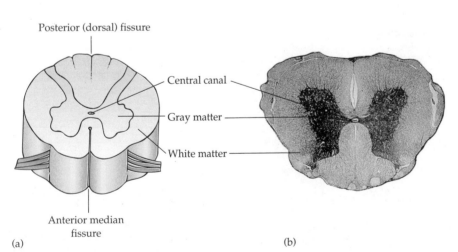

(a)    (b)

Posterior (dorsal) fissure

Central canal

Gray matter

White matter

Anterior median
fissure

### FIGURE 23–4

Light micrograph of a spinal cord cross section. *(Manfred Kage/Peter Arnold, Inc.)*

## LAB 23.D
## THE REFLEX PATHWAY

### MATERIALS REQUIRED (per pair of students)

rubber percussion hammer (or edge of hand)

### PROCEDURE

A reflex is an involuntary response to a particular stimulus. Many reflexes occur within the body as part of its normal function. For example, the presence of an object placed into the mouth, whether it be a piece of hard candy or a stone, results in the flow of saliva. In simplest form, a reflex pathway requires only three neurons: a sensory neuron, an association neuron, and a motor neuron. The sensory neuron fires a nerve impulse when the stimulus occurs. The association neuron is located in the gray matter of the spinal cord. It receives the impulse from the sensory neuron and transfers the impulse to a motor neuron. The motor neuron produces an appropriate reaction to the stimulus by causing muscle contraction, glandular secretion, and so on.

### Patellar Tendon Reflex

1. Have your lab partner sit on the table so his or her feet hang freely off the floor. Locate the patellar tendon found in the depression just below the patella (kneecap) (Fig. 23–5).

2. Use the rubber percussion hammer or the edge of your hand to tap the patellar tendon firmly; watch for a response of the leg. If no response occurs, you may have to tap the knee area a few times to locate the exact area of the patellar tendon.

3. Switch places with your partner after making your observations so that you both have the experience of being the subject and the investigator.

### Static Equilibrium Response

4. Have your lab partner stand straight and still with eyes closed and feet together. The subject must maintain the position for two minutes. Stay near your standing lab partner to prevent a fall if necessary! Carefully observe the body movements of your partner as the standing position is blindly maintained for the two-minute period.

5. Switch places with your partner after she or he has finished the two minutes.

### POSTLAB QUESTIONS

1. Describe the patellar tendon reflex.

2. Describe the static equilibrium response.

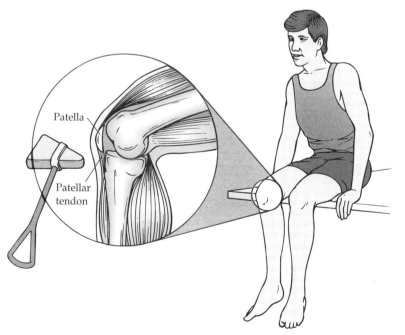

**FIGURE 23–5**

Demonstration of the knee jerk (patellar tendon) reflex.

Patella

Patellar tendon

3. What advantage does a reflex action offer an individual?

4. No reflexes will be seen in the patellar tendon if the subject is specifically thinking about keeping his or her leg still. Where in the reflex arc is this interference taking place?

## LAB 23.E
## DISSECTION OF THE SHEEP EYE

### MATERIALS REQUIRED (per pair of students)

preserved sheep eye

dissecting tray

razor blade

scissors

blunt probes

forceps

### PROCEDURE

The sheep eye is a good representative of mammalian eye structure. It is proportional to the human eye in size and has a similar anatomy. Much of the organ called an "eye" mostly functions in the capture and focusing of light. The actual light receptor cells are limited to the retina, a thin layer of tissue that lines the interior of the eyeball. Refer to Figure 23–6 for guidance as you dissect the sheep eye.

1. Obtain a preserved sheep eye. Notice the whitish fat deposits around the eye that cushion it within the skull. Search among the fat tissue for short segments of firmer, darker striated (skeletal) muscle tissue that are remnants of the six muscles that move the eyeball. These long, strap-like muscles were cut when the eye was removed from the skull.

2. Locate the cornea, the large convex area of the eyeball not covered by fat. The cornea is the transparent "window" of the eye that is fitted for contact lenses in humans. Light enters the eye through the cornea. The cornea of the sheep eye has become clouded due to the preservative.

3. Find the optic nerve on the side of the eyeball opposite the cornea. It appears as a short segment of cord about 3 mm in diameter within the fat tissue. The optic nerve carries nerve impulses from the eye to the brain where the visual image is formed.

4. Hold the sheep eye between forefinger and thumb with your forefinger on the cornea and your thumb on the rear of the eye (Fig. 23–7). Begin to open the eye with the corner of a razor blade by starting a short incision that divides the eye into equal anterior (front) and posterior (rear) halves. Insert your scissors into the cut and continue the incision around the circumference of the eyeball. Separate the halves in the dissecting tray to expose the internal structure of the eye.

5. Notice the clear gel called the vitreous body within the interior that helps to maintain the spherical shape of the eye. Locate the conspicuous lens in the anterior half of the eye. (The lens sometimes becomes displaced and may be free within the vitreous body.) Remove the lens to expose the dark, circular ciliary muscles that focus the lens. The lens is attached to the ciliary muscles by fine ligaments that change the shape of the lens as the ciliary muscles contract or relax. The lens has lost its elasticity due to the effect of the preservative. The

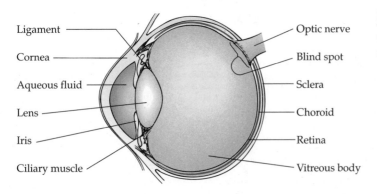

Ligament
Cornea
Aqueous fluid
Lens
Iris
Ciliary muscle

Optic nerve
Blind spot
Sclera
Choroid
Retina
Vitreous body

**FIGURE 23–6**

General structure of the mammalian eye.

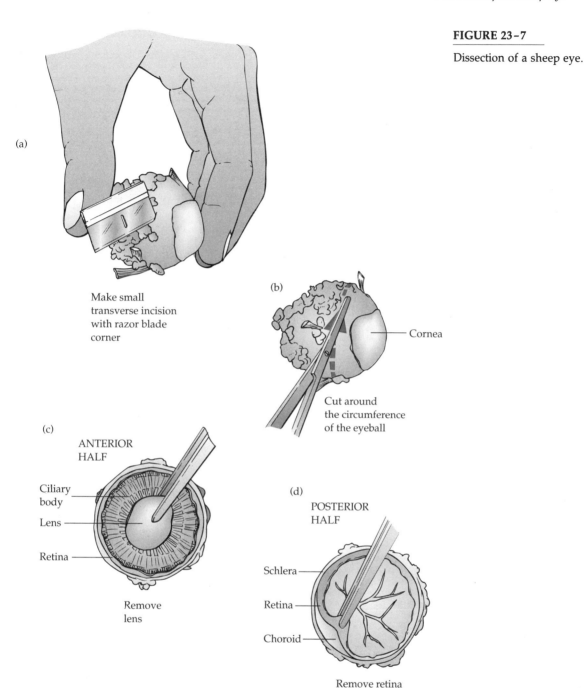

**FIGURE 23-7**

Dissection of a sheep eye.

(a)

Make small
transverse incision
with razor blade
corner

(b)

Cornea

Cut around
the circumference
of the eyeball

(c)

ANTERIOR
HALF

Ciliary
body

Lens

Retina

Remove
lens

(d)

POSTERIOR
HALF

Schlera

Retina

Choroid

Remove retina

dark, circular iris, located just in front of the ciliary muscles, regulates the amount of light entering the eye. The hole within the iris is the pupil, which changes in size as the iris dilates and contracts. The chamber between the iris and cornea is filled with a watery liquid called aqueous fluid.

6. Remove the vitreous body and locate the retina, a thin membrane that lines the interior of the eye. The retina contains rods and cones, the actual light receptor cells in the eye. Carefully lift up the retina from the posterior half of the eye. Its attachment point is the blind spot, a small area lacking rods and cones where the optic nerve passes through the retina.

7. Remove the retina from the posterior half of the eye to expose the choroid layer that appears dark in cross section. Note the reflective, blue-green layer of the choroid called tapetum, which improves

night vision. Tapetum, lacking in human eyes, produces the green shine of animals' eyes at night when spotted with bright light. The red shine of humans' eyes, commonly seen in flash photographs, is produced by reflected light from the retina, which contains many blood vessels.

8. Examine the outer layer in the cross section of the eye wall. This is the tough, exterior sclera or "white" of the eye that is mostly covered by fat.

When you have identified the structures of the sheep eye, dispose of it as directed by your instructor. Clean and dry your dissecting tray and instruments, then return them to their proper location.

## POSTLAB QUESTIONS

1. Identify the parts of a human eye based on your knowledge of a sheep eye.

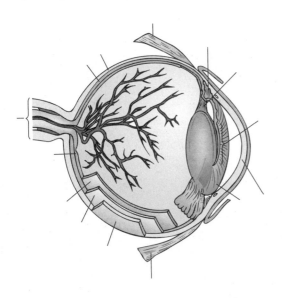

2. Cataracts in humans result in cloudiness of the lens. How would this change affect vision?

3. Given what you have learned in this lab, by what route does the reflex arc travel when pupil size changes with variations in light intensity?

## FOR FURTHER READING

Brown, Alan G. 1991. *Nerve Cells and Nervous Systems: An Introduction to Neuroscience.* New York: Springer-Verlag.

Brown, Michael C. et al. 1991. *Essentials of Neural Development.* New York: Cambridge University Press.

Bullock, T. et al. 1977. *Introduction to Nervous Systems.* San Francisco: W. H. Freeman.

Coleman, James R., ed. 1990. *Development of Sensory Systems in Mammals.* New York: Wiley.

Eccles, J. C. 1977. *The Understanding of the Brain.* New York: McGraw-Hill.

Eckert, R., and D. Randall. 1988. *Animal Physiology: Mechanisms and Adaptations.* New York: Freeman.

Guyton, A. C. 1984. *Physiology of the Human Body.* Philadelphia: Saunders College Publishing.

Hole, J. W. 1981. *Human Anatomy and Physiology.* Dubuque, Iowa: William C. Brown.

Joh, Tong H., ed. 1990. *Catecholamine Genes.* New York: Wiley-Liss.

Kandel, E. et al, eds. 1991. *Principles of Neural Science.* New York: Elsevier.

Kosower, Edward M. 1990. *Molecular Mechanisms for Sensory Signals: Recognition and Transformation.* Princeton, N.J.: Princeton University Press.

Kuffler, S. W., and J. G. Nicholls. 1984. *From Neurons to Brain: A Cellular Approach to the Function of the Nervous System.* Sunderland, Mass.: Sinauer Associates, Inc.

Romer, A. S., and T. S. Parsons. 1986. *The Vertebrate Body.* Philadelphia: Saunders College Publishing.

Shepherd, G. 1988. *Neurobiology.* New York: Oxford University Press.

Smith, Vana L., and John Dedman, eds. 1990. *Stimulus Response Coupling.* Boca Raton, FL.: CRC Press.

Vander, A. et al. 1989. *Human Physiology: The Mechanisms of Body Function.* New York: McGraw-Hill.

❏

# Mammalian Respiratory System

## OBJECTIVES

1. Identify the external and internal structures of the mammalian respiratory system.

2. Trace the route traveled by a breath of air through the mammalian respiratory system from nose to air sacs and, finally, to recipient cells.

3. Describe the mechanics of breathing.

4. Explain the function of alveoli.

## INTRODUCTION

All animals have a mechanism by which there is an exchange of gases, usually oxygen and carbon dioxide, between the cells of the animal and the environment. This process occurs by diffusion across a **respiratory surface.** Air and water serve as **respiratory media** (oxygen sources) for terrestrial and aquatic animals, respectively. Air and water differ greatly in their qualities as a respiratory medium. Air contains more oxygen than water, and oxygen diffuses much faster through air than through water. Air also weighs less than water, a property that facilitates **ventilation.** Ventilation is the movement of the respiratory medium past the respiratory surface, a movement that ultimately brings a fresh supply of oxygen to the animal.

Ventilation allows the continuance of cellular respiration, the oxidation of food molecules into carbon dioxide with a release of energy, energy which can then be used in cellular work. Terrestrial animals expend relatively little effort for ventilation as compared to aquatic animals that work harder to move heavy water past their respiratory surface. Animals that use air as a respiratory medium have advantages, but they face a danger not faced by aquatic animals—dehydration of moist respiratory surfaces.

Small **homeothermic** (warm-blooded) animals have high oxygen requirements because of their high **metabolic rate,** which is the rate of cellular respiration. A high metabolic rate is adaptive for small homeothermic animals because they have a high surface-to-volume ratio that increases loss of body heat to the environment. A high metabolic rate helps keep the body at a constant temperature regardless of environmental temperature.

The type of respiratory organ found in an animal is determined by its respiratory medium. Generally, terrestrial animals have lungs, and aquatic animals have gills. Gills and lungs differ in their structure and function. Gills are feathery tissue outgrowths that exchange gases between the blood and water by diffusion across thin gill membranes. Efficient gill structure is usually one in which a large surface area is involved in gas exchange. Aquatic animals have adapted to the problem of ventilating their gills in heavy water by evolving a one-way passage of water over or through the gills by separate entrance and exit openings.

Lungs are efficient for gas exchange between blood and air. They usually are a thin-walled, sac-like structure that is divided into multiple air sacs, thereby increasing surface area. This respiratory surface is kept moist, which facilitates gas diffusion. However, the moist lungs are a source of water loss for terrestrial animals. Ventilation of the lungs with air requires less energy than would be required if water were used as the respiratory medium. This makes possible an in-and-out air flow that enters and exits the body through one opening.

Air breathing vertebrates have evolved two ways of ventilating their lungs. Many amphibians breathe by a positive pressure mechanism in which air is pumped forcibly into the lungs by movements of the mouth floor. Contraction of abdominal muscles removes air from the lungs.

Mammals breathe by a negative pressure mechanism involving pressure changes in the chest cavity created by the **diaphragm** and rib muscles. When an animal inhales, the diaphragm and rib muscles contract, which enlarges the chest cavity and lowers its air pressure slightly below atmospheric pressure. Air rushes into the lungs toward this lower pressure. When the animal exhales, the rib muscles and diaphragm relax, which increases the chest cavity air pressure slightly above atmospheric pressure and air is expelled out of the lungs.

The following laboratory exercises are designed to demonstrate some of the important aspects of the mammalian respiratory system.

## PRELAB QUESTIONS

1. What is one disadvantage of air as a respiratory medium?

2. What type of respiratory organ is found in mammals?

3. What are gills?

4. What is negative pressure breathing?

## LAB 24.A
## INTRODUCTION TO FETAL PIG DISSECTION

### MATERIALS REQUIRED (per pair of students)
preserved fetal pig

dissecting tray

razor blade

scissors

forceps

blunt probes

dissecting pins

heavy string

labelling tag

### PROCEDURE

Before you actually observe the respiratory system of the fetal pig, a brief introduction to dissecting methods and the fetal pig itself is in order. Fetal pigs are obtained from slaughterhouses as a by-product of the meat processing industry. They provide an opportunity to observe firsthand mammalian anatomy and to see the many similarities to human organ systems. The blood vessels of your fetal pig were injected with colored latex to distinguish the arteries (red) and veins (blue). This is particularly helpful when studying the circulatory system.

Your fetal pig specimen was probably not preserved in formalin (formaldehyde), which is irritating to the skin, nose, and eyes. Biological supply companies now typically use a combination of preservative solutions that are safer and much less irritating than formalin. However, you may wish to treat your hands with an application of hand cream or petroleum jelly before handling your fetal pig to prevent the preservative from drying your skin. Some individuals may notice a slight irritation of the eyes caused by vapors from the preservative solution. Therefore, the wearing of contact lenses during dissection exercises is not recommended. Rinsing the pig with cold tap water after removing it from the preservative liquid and gently flushing the internal body cavity under running water after the pig is opened will reduce potentially irritating vapors.

The dissecting tools you will need are a single edge razor blade, scissors, forceps, and a blunt probe. Razor blades should be used sparingly because they can cause excessive damage to organs which may prevent you from seeing their true structure. The objective of any dissection is to open and reveal, not to destroy. Use scissors as much as possible for cutting. Forceps and blunt probes enable you to seek and handle structures too small for your fingers. Fingers are the most efficient tools for exploring fetal pig anatomy.

Several convenient anatomical terms are typically used in dissection procedure. These terms and their meanings include the following:

Anterior—toward the head

Posterior—toward the tail

Dorsal—toward the back

Ventral—toward the belly

Lateral—toward the side

Proximal—nearest a point of reference

Distal—Farthest from a point of reference

1. Remove the fetal pig from its plastic bag, if presented to you in this way; the pigs may be distributed from a large bulk container. Discard most of the preservative liquid in the bag, but save a little liquid and the bag for continued storage of your specimen for future use. Your instructor will supply you with a bag if your pig was packed in bulk. Thoroughly rinse your pig under cold running water to remove excess preservative.
2. Note the external features of the pig's head, such as eyes (not yet open), mouth, ears, and nostrils. Locate the umbilical cord of the abdomen, which connects the fetal pig to the placenta in the uterus of its mother. Blood vessels within the umbilical cord furnish the pig with nutrients, oxygen, and waste removal via the mother's blood circulation in the placenta. Also note the rows of nipples on the abdomen present in both sexes. Depending on its age, your pig may be covered with hair.
3. Determine the sex of your fetal pig by first locating the anus found just below the base of the tail. A small, fleshy protrusion called the urogenital papilla located immediately below the anus indicates a female. The urogenital opening within the urogenital papilla is the exit for urine and the reproductive tract.

   A male pig is easily determined by the absence of the conspicuous urogenital papilla. The urogen-

ital opening of the male fetal pig is located slightly posterior from the base of the umbilical cord. The scrotum, an external sac containing the testes, may be easily visible below the anus in older male fetal pigs. Determine the sex of your pig, then examine a pig of the opposite sex belonging to one of your classmates.

4. Secure your fetal pig specimen in the dissecting tray with string as shown in Figure 24–1. Spread the front and rear legs as far apart as possible and tie them tightly with two pieces of string that pass under the dissecting tray. This position stabilizes the pig for convenient dissection. When storing your specimen, simply slip the string from under the dissecting tray. The string allows easy repositioning of the pig the next lab period.

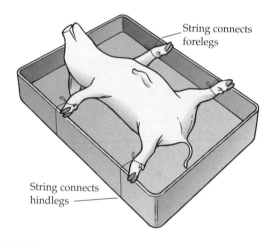

String connects forelegs

String connects hindlegs

**FIGURE 24–1**

Method for securing the fetal pig in a dissecting tray.

## POSTLAB QUESTIONS

1. In the female fetal pig, the urogenital opening is the opening to what two systems?

2. Identify the external structures and sex of these two fetal pigs.

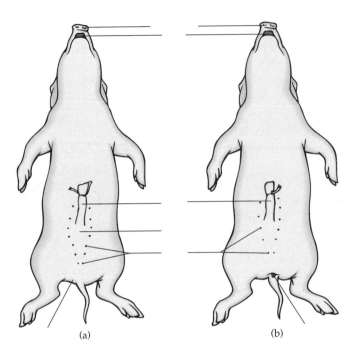

(a)                              (b)

## LAB 24.B
## RESPIRATORY SYSTEM OF THE FETAL PIG

### MATERIALS REQUIRED (per pair of students)

fetal pig specimen and

dissecting tools from above

### PROCEDURE

At this point, you have been introduced to your fetal pig specimen and have secured it with string in the dissecting tray. You are now ready to explore the respiratory system, a good place to begin dissection because the structures of this organ system are relatively large and easily located.

1. Begin your observations of the respiratory system by first using scissors to extend the opening of the mouth as shown in Figure 24–2. Start at the corners of the mouth and cut through the angle of the jaw on both sides of the head toward the ear. These cuts should allow the mouth to open widely exposing the structures of Figure 24–3. Refer to this figure as you explore the mouth region.

2. Note the conspicuous tongue in the floor of the mouth. The roof of the mouth consists of the ridged hard palate and the soft palate located

toward the rear of the mouth. Notice the teeth that are starting to push through the gums. All these structures will be discussed more thoroughly in Topic 25 when you examine the digestive system.

Locate the external nares (naris, singular)—the nostrils in the pig's snout. Air inhaled through the external nares eventually reaches the nasopharynx, an air chamber located above the soft palate. Split the soft palate with a razor blade to expose the nasopharynx. Air enters the nasopharynx by the internal nares, two openings that connect with the external nares. The nasopharynx opens into the pharynx (throat) at the rear of the mouth, a region shared by both the respiratory and digestive systems.

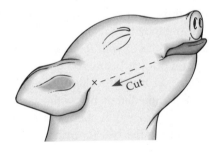

**FIGURE 24–2**

Incision line for exposing internal structures of the mouth.

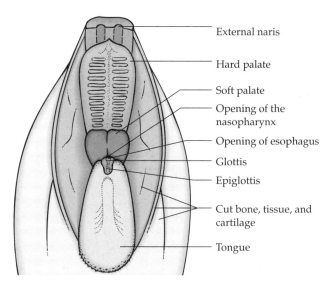

**FIGURE 24-3**

Structures in the mouth of the fetal pig.

- External naris
- Hard palate
- Soft palate
- Opening of the nasopharynx
- Opening of esophagus
- Glottis
- Epiglottis
- Cut bone, tissue, and cartilage
- Tongue

3. Locate the epiglottis, a small flap of tissue, at the far rear of the tongue in the pharynx. Extend the cuts of the mouth if it does not open wide enough to reveal the pharynx region. When food is swallowed, the epiglottis covers the glottis, the opening of the larynx (voice box), which prevents choking. Air from the nasopharynx enters the glottis by an opening located at the rear of the soft palate. The relationships of the respiratory and digestive systems in the pharynx are illustrated in Figure 24–4.

24–5. Use scissors to cut through the rib bones and cartilage of the center line. Use dissecting pins to hold back the tissue flaps of the chest area. Carefully cut the skin and muscle layers of the neck region to expose the trachea (windpipe) that delivers air to the lungs. The trachea resembles a vacuum cleaner hose due to numerous rings of cartilage that support the walls of the trachea. The larynx is a hard, conspicuous enlargement at the anterior end of the trachea.

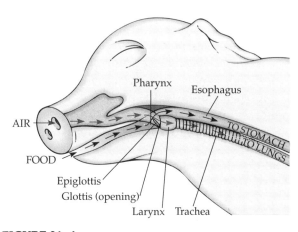

**FIGURE 24-4**

The path of food and air through the pharynx. The epiglottis closes with swallowing, preventing food from entering the respiratory tract.

4. Expose the lower respiratory organs of the fetal pig by following the incision lines presented in Figure

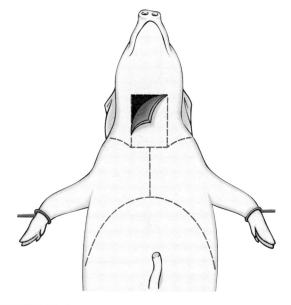

**FIGURE 24-5**

Incision lines for exposing the lower respiratory organs of the fetal pig.

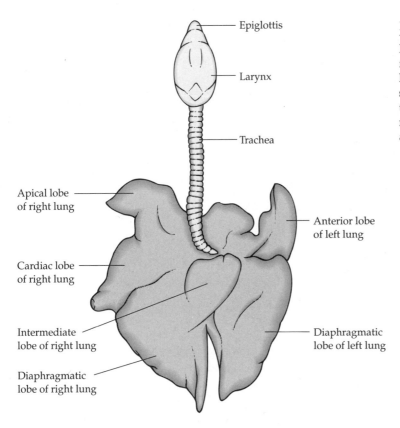

Epiglottis

Larynx

Trachea

Apical lobe
of right lung

Anterior lobe
of left lung

Cardiac lobe
of right lung

Intermediate
lobe of right lung

Diaphragmatic
lobe of left lung

Diaphragmatic
lobe of right lung

**FIGURE 24-6**

Major respiratory organs of the fetal pig. Terms for the various lobes of the lungs are presented only as a matter of interest and are not intended as major structures of fetal pig anatomy.

Notice the soft, glandular tissue on each side of the larynx and near the heart, which is centrally located between the lungs. These tissues are lobes of the thymus gland; they produce antibodies, which are special blood proteins that provide immunity against diseases.

5. Observe the compact lungs that fill most of the chest cavity. The lungs are divided into various lobes. The right lung has four lobes and the left has two lobes, as pictured in Figure 24-6, which also illustrates the major respiratory organs. Within the lungs, the trachea branches into right and left bronchi (bronchus, singular), which are not practical to reveal by dissection. Each bronchus continually branches into smaller and smaller bronchioles that finally end at microscopic air sacs called alveoli (alveolus, singular). Each alveolus is surrounded by capillaries (microscopic blood vessels) where oxygen diffuses across the thin walls of the alveolus and capillaries into the blood that reaches the cells in all body tissues. Carbon dioxide, picked up by the blood from the body's tissues, diffuses out of the capillaries into the alveolus, where it is exhaled out of the body.

6. Locate the posterior end of the lungs and notice that they are seated on a thin sheet of muscle called

the diaphragm. The diaphragm separates the thoracic (chest) and abdominal cavities. The diaphragm is essential to the breathing mechanism of mammals. Unlike other essential body functions such as a heartbeat, breathing can be voluntarily controlled because the diaphragm consists of striated (skeletal) muscle tissue.

During inhalation, the muscular diaphragm contracts and moves downward toward the abdomen. Simultaneously, the muscles of the rib cage contract, moving the ribs upward and outward. The size of the thoracic cavity expands as a result, and the elastic lungs also increase in size as they follow the movements of the diaphragm and rib cage. The air pressure within the lungs now becomes less than atmospheric pressure, and air "falls" into the lungs due to the pressure difference. During exhalation, the diaphragm and rib muscles relax; the diaphragm becomes dome shaped as it moves upward toward the chest and the ribs lower, which decreases the size of the thoracic cavity and the volume of air within the lungs is forced outward.

7. When you have located the above respiratory structures of the fetal pig, wrap your specimen in wet paper towels and place it in its original plastic

bag or in one supplied by your instructor for storage. Identify your fetal pig by writing your names in PENCIL on the tag (preservative leaks will leach ink from the paper tag). Place the tag on a piece of string and tie the bag closed. Store the bag according to your instructor's directions for use during the next lab period.

8. Rinse your dissecting tools and tray, carefully wipe them dry, and return them to their original location.

## POSTLAB QUESTIONS

1. Using the following terms, describe the path of air from the nose to the lungs (alveoli, bronchi, bronchioles, larynx, nose, pharynx, and trachea).

2. Identify the type of breathing mechanism.

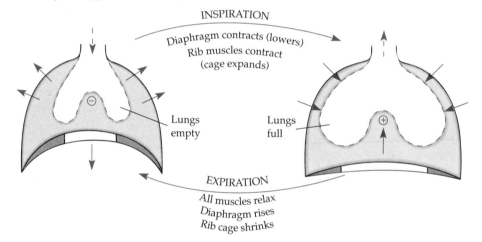

3. Identify the following respiratory structures of a fetal pig.

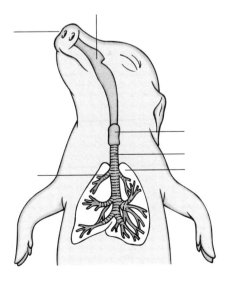

4. Why don't the rings of cartilage extend all the way around the trachea?

5. What is the common name of the largest cartilage of the larynx?

6. Aspirated foreign objects usually lodge themselves in the right bronchus instead of the left bronchus. Give an anatomical reason for this phenomenon.

# LAB 24.C
## ALVEOLI, SITE OF GAS EXCHANGE BETWEEN THE LUNGS AND BLOOD

### MATERIALS REQUIRED

prepared slide of lung tissue section

compound microscope

### PROCEDURE

The organs of the respiratory system you observed are simply a mechanical means of moving gases into and out of the mammalian body. Very little, if any, gas exchange occurs between the blood and organs of the respiratory system because oxygen and carbon dioxide cannot diffuse across their many layers of tissue. The actual exchange of respiratory gases between the bloodstream and lungs occurs at the alveoli and their capillaries, which both have extremely thin walls that allow the diffusion of gases.

1. Examine the prepared slide of a section of lung tissue under low and high power. Note its spongy appearance. Each of the numerous small air spaces throughout the tissue is an alveolus.
2. Switch to scanning or low power and search the lung tissue for sections of blood vessels and bronchioles that appear oval or round in cross section, or as channels in longitudinal section. Ask your instructor for assistance in identifying these structures if they are present on your slide.

### POSTLAB QUESTIONS

1. Identify the parts of the respiratory system that function as air distributors. As gas exchangers.

2. Identify the two thin membranes that separate the air in the alveolus from the blood.

### FOR FURTHER READING

Eckert, R., and D. Randall. 1988. *Animal Physiology.* New York: Freeman.

Gordon, M. et al. 1982. *Animal Function: Principles and Adaptations.* New York: Macmillan.

Guyton, A. C. 1984. *Physiology of the Human Body.* Philadelphia: Saunders College Publishing.

Haddad, Garriel G., and Jay P. Farber, eds. 1991. *Developmental Neurobiology of Breathing.* New York: Decker.

Hole, J. W. 1981. *Human Anatomy and Physiology.* Dubuque, Iowa: William C. Brown.

Levitzky, Michael G. 1991. *Pulmonary Physiology.* New York: McGraw-Hill.

Murray, John F. 1986. *The Normal Lung: The Basis for Diagnosis and Treatment.* Philadelphia: Saunders College Publishing.

Randall, D. J. et al. 1981. *The Evolution of Air Breathing Vertebrates.* Cambridge, England: Cambridge University Press.

Romer, A. S., and T. S. Parsons. 1986. *The Vertebrate Body.* Philadelphia: Saunders College Publishing.

Smith, H. M. 1960. *Evolution of Chordate Structure.* New York: Winston.

Spence, A. P., and E. B. Mason. 1987. *Human Anatomy and Physiology.* Menlo Park, Calif.: Benjamin-Cummings Publishing Co.

Vander, A. et al. 1989. *Human Physiology: The Mechanisms of Body Function.* New York: McGraw-Hill.

West, J. 1990. *Respiratory Physiology: The Essentials.* Baltimore, Md.: Williams & Wilkins.

Whipp, Brian J., and Karlman Wasserman, eds. 1991. *Exercise and Pathophysiology.* New York: M. Dekker.

❏

# Mammalian Digestive and Excretory Systems

O B J E C T I V E S

1. Identify the structures of the mammalian digestive system and the functions of those structures.

2. Trace the pathway traveled by an ingested meal, describing each change that takes place en route. Use the following terms in your discussion: ingestion, digestion, absorption, and elimination.

3. Draw and label a diagram of an intestinal villus and describe its function.

4. Identify the structures of the mammalian excretory system and the functions of those structures.

5. Draw and label the principal parts of the kidney and give the relative locations of the following structures within the kidney: Bowman's capsule, glomerulus, collecting duct, proximal and distal convoluted tubules, and loop of Henle.

## INTRODUCTION

The human (mammalian) digestive tract can illustrate the changes that occur to food as it is processed by the digestive organs. Food enters the mouth where it is broken into smaller particles by the jaws and teeth. Here food mixes with **saliva,** which is secreted from salivary glands in the mouth. The saliva moistens the food and begins the chemical digestion of the starches found in the food. Swallowing begins **peristalsis,** a muscular contraction that moves food through the digestive tract. Food passes through the pharynx, over the epiglottis, down the **esophagus,** and into the **stomach.** The muscular stomach stores and mixes the food with gastric juice, a fluid composed of hydrochloric acid, mucus, and **pepsin** enzymes from the stomach wall that begin protein digestion. Absorption in the stomach is minimal and limited to lipid soluble substances such as drugs and alcohol.

Food is released gradually into the **duodenum** by a

**sphincter** muscle. The duodenum is the first portion of the small intestine where the food is mixed with more enzymes secreted from the intestine itself and the pancreas. The pancreas produces many enzymes, such as **trypsin,** a protein-digesting enzyme. Bile from the liver also enters the small intestine. The various digestive juices chemically break down carbohydrates, fats, and proteins into glucose, amino acids, peptides, and lipid soluble molecules, which are all absorbed through the small intestine wall where most nutrient absorption occurs. Water is also absorbed from the intestinal contents.

The structure of the small intestine, with its extensive internal folding and finger-like projections (the **villi**), make it exceptionally adaptive for absorption. In fact, the surface of each villus is covered with tiny **microvilli,** which are extensions of the plasma membranes of the cells lining the central cavity, called the **lumen.** These folds and folded folds tremendously increase the absorptive surface of the small intestine.

Any undigested food continues into the **large intestine (colon),** which absorbs more water, minerals, and vitamins, particularly vitamin K, from the remaining material. Vitamin K is produced by the bacteria living in the large intestine. The food has become fecal matter that is held in the **rectum** until voided.

The mammalian liver is vital for life because of its many important functions. Its main role is to control organic components in the blood, which it does by detoxifying poisonous substances, changing the concentration of a substance in the blood, or altering the chemical structure of a compound. The liver also stores food (glycogen), produces bile that aids digestion, synthesizes blood proteins, and converts toxic nitrogenous wastes into urea that can be safely excreted by the kidneys. We turn now to a discussion of the excretory system.

Animal excretory organs have three common functions:

1. they collect fluids from a location within the body, usually the blood or spaces between organs;
2. they modify the fluid by resorbing essential materials, or by actively transporting wastes into the excretory product; and
3. they provide a means of removing the excretory product from the body.

Humans excrete four principal wastes that are also important in many other animals. Nitrogenous wastes are excreted mainly by the kidneys with a small amount leaving through the skin in sweat. Water leaves by the kidneys, by the skin as sweat, and by the lungs as vapor in the breath. Salts are excreted by the kidneys, and through the skin as part of sweat. Carbon dioxide leaves by the lungs.

A distinction must be made between **excretion** and **egestion.** Excretion applies only to substances that cross plasma membranes and leave the body. Egestion applies to material that is expelled from the body without ever passing through a plasma membrane, such as feces.

Every vertebrate has a pair of kidneys that are composed of numerous **nephrons** (Fig. 25–1) that form urine and control the salt, water, and nitrogenous waste content of the blood. These processes begin as blood containing wastes arrives at the kidney by the **renal artery.** Blood eventually reaches the **glomerulus,** a cluster of capillaries surrounded by a cup-shaped **Bowman's capsule,** which is the first part of the nephron. Blood pressure forces fluid out of the capillary walls into the Bowman's capsule. Blood cells and large molecules remain in the capillary. The fluid, now called **filtrate** (filtered fluid), passes to the **proximal convoluted tubule** where some substances are **resorbed** by active transport through the tubule wall and into nearby capillaries. Inversely, wastes are **secreted** by active transport from the same capillaries and into the proximal convoluted tubule.

The filtrate continues to the next nephron portion, the **loop of Henle,** where water is removed and salt

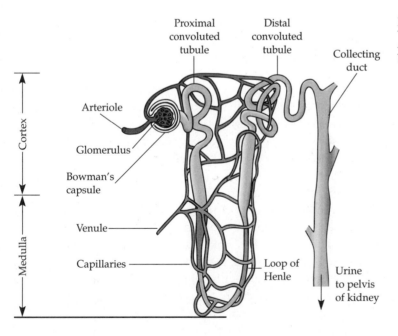

Proximal convoluted tubule

Distal convoluted tubule

Collecting duct

Arteriole

Glomerulus

Bowman's capsule

Venule

Capillaries

Loop of Henle

Cortex

Medulla

Urine to pelvis of kidney

**FIGURE 25–1**

A nephron and its associated blood supply.

ions are actively transported out of the filtrate into the extracellular fluid of the kidney medulla. This creates a high salt concentration in this region. From the loop of Henle, the filtrate continues to the **distal convoluted tubule,** where more resorption and secretion occurs between the tubule and nearby capillaries. The purified blood eventually flows to the **renal vein** that carries blood from the kidney.

By this time, the filtrate (urine) has been greatly modified by the reclamation of essential substances and the secretion of excess salts and wastes into it. The urine is concentrated as it flows from the distal convoluted tubule down the **collecting duct** that passes through the high salt concentration produced by the loop of Henle. Water moves out of the urine through the collecting duct by osmosis into the medulla where it is reclaimed. The urine becomes more concentrated as the collecting duct continues to pass through the medulla. Urine leaves the collecting duct and is eventually stored in the bladder. The urine is expelled from the body by passing through the urethra.

The following exercises have been selected to demonstrate some of the more important aspects of the mammalian digestive and excretory systems.

### PRELAB QUESTIONS

1. What is the function of the stomach?

2. What makes the small intestine exceptionally adaptive for absorption?

3. What is the function of the nephrons?

4. What action moves food down the esophagus toward the stomach?

## LAB 25.A
## THE DIGESTIVE SYSTEM OF THE FETAL PIG

### MATERIALS REQUIRED (per pair of students)

fetal pig from previous exercise

dissecting tray

blunt probes

razor blade

scissors

dissecting pins

forceps

### PROCEDURE

1. Open the mouth of your fetal pig and study the structures involved in the holding, chewing, and mixing of food; these are the lips, cheeks, tongue, and teeth, previously presented in Topic 24 during discussion of the oral cavity as part of the respiratory system (see Fig. 24–3, page 275). The milk teeth of the fetal pig are probably beginning to emerge in your specimen. Later in life, the milk teeth will be lost and replaced by permanent adult teeth as in humans.

2. Note the muscular tongue and the presence of numerous nipple-like papillae (papilla, singular) along its sides and back surface. Microscopic taste buds, which function as taste sensory receptors, are located among these papillae and over the tongue's surface.

3. Observe the anterior ridged portion of the roof of the mouth called the hard palate, which is a partition of bone that separates the oral cavity from the nasal cavity. The posterior roof of the mouth that lacks ridges and contains no bones is the soft palate.

4. Find the salivary glands by using your blunt probe to tease apart the tissue lateral to the larynx. There are three pairs of salivary glands in the head region. They secrete saliva, a mixture of water, mucus, and the enzyme amylase that begins starch digestion. Saliva also lubricates food in the mouth, which aids in swallowing.

5. Open the abdomen of your fetal pig by following the sequence of incision lines illustrated in Figure 25–2. To avoid damage to the internal organs, cut only deep enough to penetrate the body wall. Extend incision lines 3 and 4 well down the sides of the pig. Open the flaps of the abdomen and gently flush the internal cavity under running water.

6. Refer to Figure 25–3 as you locate the following organs. The large liver is positioned next to the

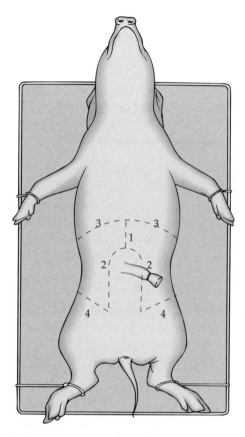

**FIGURE 25–2**

Follow this sequence of numbered incision lines to open the abdomen of your fetal pig.

diaphragm. The liver consists of four lobes and has several vital functions such as detoxification of poisons, formation of urea, production of proteins essential to blood clotting, breakdown of worn red blood cells, vitamin storage, and storage of glycogen as an energy source. The liver also secretes bile that aids in the absorption of fats.

7. Lift the liver to find the stomach on the left side of the pig. The esophagus leads from the anterior stomach to the pharynx. The esophagus lies along the dorsal body wall behind the trachea and passes through the diaphragm. Swallowing of food starts a wave of muscular contraction, called peristalsis, along the length of the esophagus that "pushes" food into the stomach. Peristalsis also moves food through all other digestive organs. Peristalsis is a very efficient process, making it possible to drink liquids while standing on your head!

Use scissors to open the stomach and expose the inner surface. Note the numerous ridges. The stomach temporarily stores food before entering the small intestine. Contractions of the stomach mix food with hydrochloric acid and enzymes, secreted by the stomach wall, that begin protein digestion. The posterior stomach and small intes-

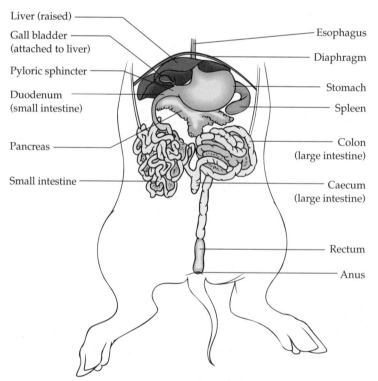

Liver (raised)
Gall bladder (attached to liver)
Pyloric sphincter
Duodenum (small intestine)
Pancreas
Small intestine

Esophagus
Diaphragm
Stomach
Spleen
Colon (large intestine)
Caecum (large intestine)
Rectum
Anus

**FIGURE 25–3**

The digestive organs of the fetal pig.

tine are separated by the pyloric sphincter, a ring of muscle that regulates the passage of food into the small intestine.

Lateral to the outer curvature of the stomach is the dark red spleen, a part of the circulatory system that filters debris and bacteria from the blood.

8. Locate the highly coiled small intestine that leads from the posterior stomach. Approximately the first inch is called the duodenum. It contains many enzymes that digest the various food groups. Some of these enzymes are secreted from the wall of the duodenum. Bile from the liver enters the duodenum from the bile duct that leads from the gall bladder where bile is stored. Lift the liver and locate the gall bladder embedded in the posterior side of the liver. The gall bladder has a greenish color because of the green bile within it.

Digestive enzymes, in addition to those produced by the duodenum, enter the duodenum from the pancreas by the pancreatic duct (difficult to see in the fetal pig). The pancreas is a soft, granular-like organ located between the stomach and duodenum. In addition to digestive enzymes, the pancreas also produces the hormone insulin, essential for the uptake of glucose by cells. Before human insulin could be produced by genetically engineered bacteria, the pancreas of slaughter-house animals was the only source of commercial insulin for diabetic humans. In addition to insulin, the hormone glucagon is produced by the pancreas. Glucagon tends to increase blood glucose concentration.

9. Note how the small intestine and other internal organs are held in position by a thin, transparent membrane called mesentery. The mesenteries also contain the blood vessels and nerves that support the internal organs. The same membrane that forms the mesenteries also lines the body cavity. Here, the membrane is called peritoneum. It is derived from the mesoderm of the early embryo. Recall from Topic 17 that a body cavity lined with tissue of mesodermal origin is called a coelom, or true body cavity.

Beyond the duodenum, the small intestine consists of two rather indistinguishable regions called the jejunum and the ileum. Further digestion and absorption of nutrients occurs in these regions as food material passes through them. The internal wall of the small intestine is cov-ered with millions of microscopic, finger-like projections called villi. Each villus is covered with still smaller microvilli, which greatly increase the contact of intestinal surface with digested food for maximum absorption of nutrients. During absorption, nutrients cross the cell membranes of columnar epithelial cells that form the outer cell layer of each villus. Nutrients enter the blood stream by capillaries within the villi.

10. Pause for a moment to examine the prepared demonstration slide of a cross section of the duodenum from a rabbit or other small mammal. Notice the prominent villi that protrude into the lumen (cavity) of the small intestine. The outer cell layer of each villus consists of columnar epithelium first described in Topic 21. The lighter, oval spaces between the columnar cells indicate goblet cells that secrete mucus onto the inner intestinal surface as lubrication.

The outer surface of each columnar epithelial cell is covered with still smaller microvilli that are beyond the magnification of your microscope. The microvilli further increase the contact of intestinal surface with digested food for maximum absorption of nutrients. During absorption, most nutrients cross the cell membranes of columnar epithelial cells and enter the blood stream by microscopic capillaries within the villi. Each villus also contains a tiny lymph vessel that absorbs lipids. The lymphatic system (see your text for a description of this part of the circulatory system) eventually empties into the bloodstream, releasing lipids into general circulation. Return to your fetal pig and continue following the digestive tract.

11. Follow the small intestine to where it joins the large intestine, or colon. At the junction of the two organs, notice the caecum, a short, blind pouch. Animals that feed strictly on plant material typically have a very large caecum, which contains numerous bacteria that aid in the digestion of cellulose, the tough material composing plant cell walls. Obviously, the caecum is of little significance in the pig. The human caecum is also very small, ending in the appendix, which may become inflamed and require surgical removal.

A majority of the food's nutrients are absorbed by the small intestine. Absorption does occur in the large intestine, but it is limited to minerals, vitamins, and water. The remaining undigestable material becomes feces (solid food waste) as water is continually absorbed from food residues by the wall of the colon. In humans, and other mammals, the colon is home for numerous bacteria that contribute as much as 50% of the dry weight of feces. These intestinal bacteria are beneficial because they synthesize vitamin K that is absorbed through the colon wall. Vitamin K is essential to blood clotting and prevents excessive bleeding from wounds.

Feces is temporarily stored in the short, straight region of the lower colon known as the rectum. The rectum opens to the outside of the body by the anus, a sphincter of striated (skeletal) muscle that permits voluntary control of defecation (elimination of feces).

## POSTLAB QUESTIONS

1. Identify the structures of the digestive system.

2. Complete Table 25–1.

**Table 25–1**   Structures and Functions of the Digestive System

| Structure | Function |
|---|---|
| Gall bladder | |
| Large intestine | |
| Liver | |
| Pancreas | |
| Salivary glands | |
| Small intestine | |
| Stomach | |
| Teeth | |
| Tongue | |

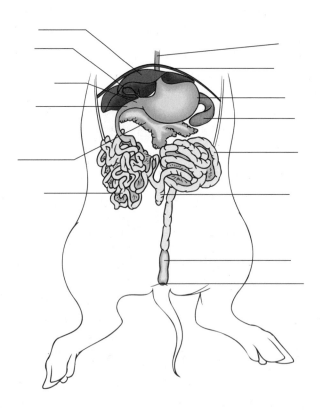

3. Why are the milk teeth sometimes referred to as deciduous teeth?

4. In a mirror, observe the papillae of your own tongue. How many different types of papillae do you notice? Describe the types. Are they located in specific regions of your tongue?

5. If a person had a cleft palate, what two cavities would not be separated?

6. Identify the childhood disease that is an acute infection of one pair of salivary glands.

7. If bile does not exit a person's body, but is absorbed by the blood, what color does their skin turn?

8. Why is the esophagus a collapsible tube and the trachea is not?

9. Explain in detail why digestion is both a mechanical and a chemical process.

10. Identify the route that food travels during the digestive process.

## LAB 25.B
## THE EXCRETORY SYSTEM OF THE FETAL PIG

### MATERIALS REQUIRED

| | |
|---|---|
| fetal pig from previous exercise | razor blade |
| | scissors |
| dissecting tray | dissecting pins |
| blunt probes | |
| forceps | |

### PROCEDURE

The mammalian excretory system has two main and vital functions: removal of nitrogenous (nitrogen-containing) wastes, and maintenance of proper water balance of the blood and, consequently, all of the body's tissues. The principal organs of the excretory system are the kidneys, which function as living filters. Simultaneously, the kidneys remove nitrogenous and other metabolic wastes from the blood, and also maintain its water content. The organs of the excretory system are closely associated with those of the reproductive system in vertebrates. Therefore, they have been illustrated together in Topic 27 (see Figs. 27–1 and 27–2, pages 299 and 300, which depict the excretory and reproductive systems of the female and male fetal pig. Refer to these figures as you locate the following structures of the excretory system. Of course the reproductive organs differ between the sexes, but the excretory system is essentially the same in the male and female pig.

1. Push aside the intestinal mass of the fetal pig to expose the dorsal body wall. Locate the kidneys, two conspicuously large, bean-shaped organs positioned on each side of the spinal column. Note the peritoneum that covers the kidneys and separates them from the abdominal cavity. Try to locate the inconspicuous adrenal gland that lies on the anterior portion of each kidney. The adrenal gland is part of the endocrine system, which is composed of various glands throughout the body that produce hormones. The adrenal gland secretes epinephrine (adrenalin), which greatly increases body metabolism and prepares the body for threatening ''fight-or-flight'' situations. The adrenal gland also produces other hormones that regulate the body's salt balance and adapt the body to long-term stress.

2. Carefully peel away the peritoneum to expose one of the kidneys. Note the two large red and blue blood vessels that lead to each kidney. These are the renal artery (red) and the renal vein (blue) that supply blood to and from the kidney. Blood circulation of the kidney will be discussed again in Topic 26. The concave region of the kidney where the blood vessels enter is called the hilum. Locate a thin, lightly colored tubule that leads out of the hilum. This is the ureter, which carries urine formed in the kidney down to the bladder (discussed below).

3. Remove a kidney and carefully section it length-

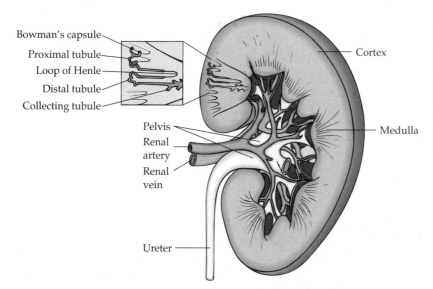

Bowman's capsule
Proximal tubule
Loop of Henle
Distal tubule
Collecting tubule

Cortex

Pelvis
Renal artery
Renal vein

Medulla

Ureter

**FIGURE 25–4**

Lengthwise section of a kidney revealing major internal features. A single nephron is represented on the left to demonstrate the positioning of nephrons within the cortex (stippled) and medulla (striated) regions of the kidney.

wise down the center plane as shown in Figure 25–4. The outer layer of the kidney is the cortex, the inner layer is the medulla. The numerous microscopic nephrons, the functional units of the kidney, are embedded within the cortex and medulla. The glomeruli and proximal and distal convoluted tubules of the nephrons are positioned in the cortex (Fig. 25–5). The loops of Henle and collecting ducts are positioned in the medulla. These relationships are illustrated in Figure 25–4. Urine produced by the nephrons empties into the central pelvis that funnels into the ureter.

4. Follow a ureter down to the urinary bladder that is joined to the umbilical cord in the fetal pig; this association is soon lost after the pig is born. Urine is temporarily stored in the bladder. Urine leaves the bladder by the urethra which leads out of the body. In the female pig (see Fig. 27–1), the urethra is easily seen leading from the base of the bladder down into the pelvic area where it briefly joins the vagina at the urogenital canal just before reaching the exterior of the body. Do not separate the bones of the pelvis at this time; you will examine this region when you study the reproductive system.

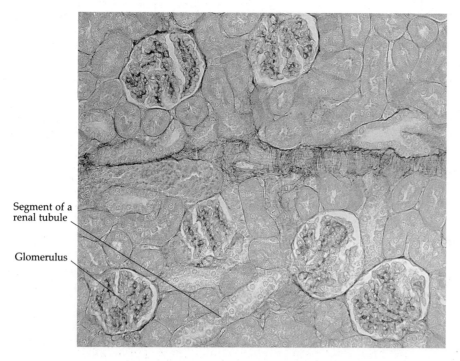

Segment of a renal tubule

Glomerulus

**FIGURE 25–5**

Light micrograph of a section through the kidney cortex showing glomeruli and short segments of renal (kidney) tubules of several nephrons. In this section of the kidney, it is not possible to distinguish between proximal and distal convoluted tubules; therefore, they are labelled generally as renal tubules. Notice the cuboidal epithelial cells, discussed in Topic 21, that compose the tubules.

In the male pig (see Fig. 27–2), only a short segment of the urethra is visible at the base of the bladder before it joins with two sperm ducts *(vasa deferentia)* of the reproductive system. The urethra continues posteriorly and becomes part of the penis where the urethra opens to the outside of the body. In male mammals, the urethra carries both urine and sperm, whereas the female urethra carries only urine. Do not follow the urethra through the pelvic region of the male pig at this time; you will examine this area when you study the reproductive system.

When you have finished with your fetal pig, wrap it in wet paper towels and return it to its storage bag for future use.

## POSTLAB QUESTIONS

1. Identify the structures of the mammalian kidney.

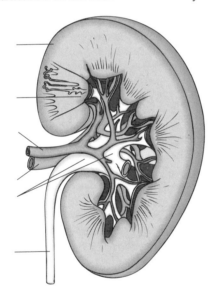

2. Why is the right kidney usually located slightly lower than the left kidney? (*Hint:* Review locations of organs of the digestive system.)

3. Infections of the bladder are much more common in females than males. What structural difference between males and females helps explain this phenomenon?

## FOR FURTHER READING

Eckert, R., and D. Randall. 1988. *Animal Physiology.* New York: Freeman.

Ganong, William F. 1991. *Review of Medical Physiology.* Norwalk, Conn.: Appleton & Lange.

Gordon, M. et al. 1982. *Animal Function: Principles and Adaptations.* New York, Macmillan.

Guyton, A. C. 1991. *Textbook of Medical Physiology.* Philadelphia: Saunders College Publishing.

Guyton, A. C. 1984. *Physiology of the Human Body.* Philadelphia: Saunders College Publishing.

Hole, J. W. 1981. *Human Anatomy and Physiology.* Dubuque, Iowa: William C. Brown.

Johnson, Leonard R., ed. 1992. *Essential Medical Physiology.* New York: Raven Press.

Krause, M. B., and L. K. Mahan. 1984. *Food, Nutrition, and Diet Therapy.* Philadelphia: W. B. Saunders Company.

Romer, A. S., and T. S. Parsons. 1986. *The Vertebrate Body.* Philadelphia: Saunders College Publishing.

Smith, H. M. 1960. *Evolution of Chordate Structure.* New York: Winston.

Spence, A. P., and E. B. Mason. 1987. *Human Anatomy and Physiology.* Menlo Park, Calif.: Benjamin-Cummings Publishing Co.

Thibodeau, Gary A., and Kevin T. Patton. 1992. *Structure and Function of the Body.* St. Louis: Mosby Year Book.

Tortora, Gerard J., and Nicholas P. Anagnostakos. 1990. *Principles of Anatomy and Physiology.* New York: Harper & Row.

Vander, A. et al. 1989. *Human Physiology: The Mechanisms of Body Function.* New York: McGraw-Hill.

❑

# Mammalian Heart and Circulatory System

## O B J E C T I V E S

1. Describe the principal components of the blood and give their functions.

2. Compare the structure and function of the different types of blood vessels, including arteries, capillaries, and veins.

3. Describe the structure and function of the parts of the heart and label them on a diagram.

4. Locate and identify the major blood vessels of the fetal pig.

5. Trace a drop of blood through the pulmonary and systemic circulations naming in proper sequence each structure through which it passes.

## INTRODUCTION

Most members of the animal kingdom have some type of **vascular system** that transports materials throughout the body. Some animals, such as insects, have an **open circulatory system** with open-ended blood vessels that release blood directly into the body cavity where an exchange of materials with the tissues occurs. The blood is propelled via heart and body muscle movement. Vertebrates have a closed circulatory system in which the blood never leaves the vessels. Exchange of substances between the blood and tissues occurs by simple diffusion through the thin vessel walls.

Fish have a heart that contains one atrium and one ventricle, a system that passes blood only once through the heart for each complete circuit through the body. This is called **single circulation.** The hearts of birds and mammals have two atria and two ventricles, one of each on both the right and left sides, that function as two side-by-side pumps. The right side

sends blood to the lungs where the blood is oxygenated. The oxygenated blood returns to the heart from the lungs where the blood is pumped by the left side of the heart to all parts of the body. This is **double circulation** because blood passes through the heart twice for each complete circuit through the body. This double circulation not only keeps oxygenated and deoxygenated blood separate but also keeps blood pressure high in the arteries and results in a more efficient transport of materials.

Vertebrate blood vessels are of three types: arteries, capillaries, and veins. **Arteries** carry blood away from the heart. Their thick walls are composed of smooth muscle, connective tissue, and elastic fibers all of which enable the vessel to withstand high blood pressure from the heart. Arteries branch and rebranch into all parts of the body until they divide into microscopic **capillaries,** the sites where exchange of materials actually occurs between blood and tissues. The capillary wall is composed of a single thin layer of endothelial cells through which diffusion of materials is easy.

288

Blood is carried back to the heart by **veins** formed from the joining of capillaries and smaller vessels leading from the tissues. Compared to arteries, veins are thin-walled vessels that are composed of connective tissue and smooth muscle. In general, veins have larger internal diameters than arteries having the same external circumference.

Throughout the mammalian circulatory system, **valves** occur in various locations and keep blood flowing in the appropriate direction. Valves are flaps of tissue that open when blood flows in the appropriate direction, but they close if blood begins to flow backward. The valves in the veins also facilitate low pressure blood flow against the pull of gravity. Valves between the heart chambers and in associated major vessels produce unidirectional circulation. Vessels of the lymphatic system also have valves that function similarly in preventing backward flow.

Transport of materials is accomplished by a definite circulation path throughout the body. Blood returns to the heart from the body by the large **venae cavae** veins that empty into the right atrium, which contracts and fills the right ventricle. The right ventricle contracts sending blood through the **pulmonary artery** to the lungs where carbon dioxide is removed and oxygen is received by capillaries around the alveoli. Oxygenated blood returns to the heart by **pulmonary veins** that empty into the left atrium. Then the left ventricle contracts sending oxygenated blood into the large **aorta** artery that branches to different body areas. Arteries divide into capillaries where oxygen and nutrients are delivered and carbon dioxide and wastes are received from the cells. Capillaries in the digestive system receive nutrients that are then distributed to the rest of the body. Capillaries from the tissues carry blood into larger vessels leading to veins and eventually to the venae cavae where blood is returned to the heart for another circulation.

Blood consists of water (50%), dissolved salts, plasma proteins, and three types of blood cells (red blood cells, white blood cells, and platelets). **Red blood cells** are the most numerous. Their main function is oxygen transport, which involves **hemoglobin.** **White blood cells** are important in fighting infection and producing immunity against diseases. **Platelets,** which are cell fragments, function in blood clotting.

We hope that you find the following exercises involving the mammalian heart and circulatory system interesting and useful in demonstrating the major aspects of this chapter.

## PRELAB QUESTIONS

1. What is a closed circulatory system?

2. What type of vessels carry blood away from the heart?

3. What is the function of valves?

4. What large vein returns blood to the heart?

5. What is the function of white blood cells?

## LAB 26.A
## ANATOMY OF THE MAMMALIAN HEART

### MATERIALS REQUIRED (per pair of students)

preserved sheep heart, partially sectioned

dissecting tray

blunt probes

### PROCEDURE

The heart is one of the most vital organs because of its function in maintaining blood flow throughout the body. If the heart ceases to function, tissues no longer receive a supply of oxygen and nutrients, and metabolic wastes are not removed from the cells. The structure of the heart reflects its function as a living pump that continually drives the blood stream.

1. Obtain a preserved sheep heart and observe its external structure. The sheep heart is similar to the human heart in size and shape, which makes it a good subject for study. As with all mammalian hearts, the sheep heart consists of four chambers — two thin-walled atria (atrium, singular) located above two large, muscular ventricles that form the bulk of the heart. Note the coronary arteries and veins over the surface of the heart that circulate blood to the heart muscle tissue. Note the pale fat deposits on the heart. Position the heart so that the large vessel arising from the top of the heart curves toward your right as in Figure 26–1a. (Figs. 26–1a and b are diagrams of the human heart, but they serve as a guide to the similar sheep heart. Refer to

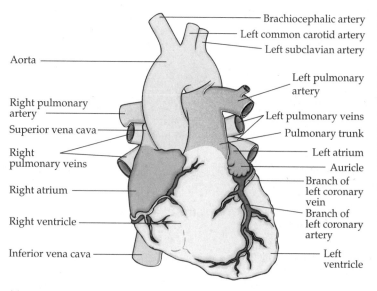

Aorta

Right pulmonary artery

Superior vena cava

Right pulmonary veins

Right atrium

Right ventricle

Inferior vena cava

Brachiocephalic artery
Left common carotid artery
Left subclavian artery

Left pulmonary artery

Left pulmonary veins
Pulmonary trunk
Left atrium
Auricle
Branch of left coronary vein
Branch of left coronary artery
Left ventricle

(a)

**FIGURE 26-1**

(a) Anterior view of the human heart as seen if facing another individual. This same view in mammals that walk on four legs is known as the ventral view, the side toward the ribs. (b) Posterior view of the human heart as seen if standing behind another individual. This same view in mammals that walk on four legs is known as the dorsal view, the side toward the spinal column.

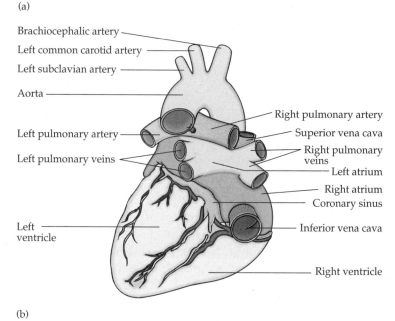

Brachiocephalic artery
Left common carotid artery
Left subclavian artery
Aorta

Left pulmonary artery

Left pulmonary veins

Left ventricle

Right pulmonary artery
Superior vena cava
Right pulmonary veins
Left atrium
Right atrium
Coronary sinus
Inferior vena cava

Right ventricle

(b)

them as you continue.) You are now looking at the ventral surface of the heart that would be facing the ribs in the sheep's body. Consequently, the *left* side of the heart is on your *right*; the heart is easier to study from this perspective.

2. Locate the right atrium, which receives blood from all of the body by two large veins—the anterior vena cava, which brings blood from the anterior body, and the posterior vena cava, which brings blood from the posterior body. (In humans, these vessels are called the superior and inferior vena

cavae as in Figs. 26–1a and b). Insert a blunt probe into these vessels to reach the right atrium.

3. Open the cut of the heart to expose its internal anatomy. Consult Figure 26–2 to locate the various heart structures as you follow the path of blood flow. Gently insert your probe into the described blood vessels to see where they lead. Blood from the right atrium enters the right ventricle through the tricuspid valve that separates these two chambers and keeps blood from flowing back into the atrium when the right ventricle contracts.

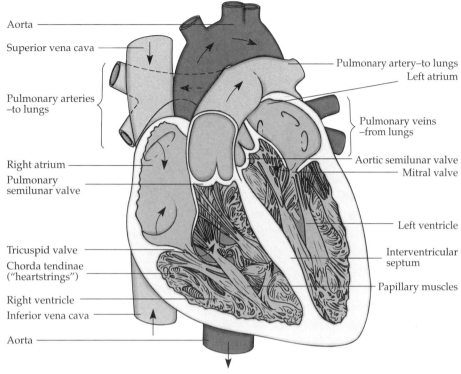

**FIGURE 26–2**

Internal view of the human heart.

Contraction of the right ventricle sends blood through the semilunar valve into the pulmonary artery that branches to each lung where the blood becomes oxygenated.

4. Locate the thin-walled pulmonary veins that return oxygenated blood from the lungs to the left atrium of the heart. Blood leaves the left atrium through the mitral (bicuspid) valve that separates the left atrium and left ventricle. The muscular left ventricle contracts, sending blood through another semilunar valve into the aorta which distributes blood to all regions of the body.

When finished, return the sheep heart to its container as directed by your instructor.

**POSTLAB QUESTIONS**

1. How do you suppose the semilunar valves received their names?

2. Label this mammalian heart.

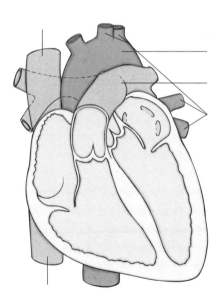

3. If the blood enters the heart by way of the inferior vena cava and leaves by way of the aorta, list, in order, the structures the blood passes through. Use

the following terms: pulmonary artery, pulmonary vein, left atrium, right atrium, left ventricle, right ventricle, mitral valve, semilunar valves, tricuspid valve.

4. Why are the sheep vena cavae anterior and posterior while the human vena cavae are superior and inferior?

5. Explain anatomically why the location of the heart lends itself to rhythmic compression of the heart in cases of cardiac arrest.

6. What organelle would you expect to be very abundant in a cardiac muscle cell?

## LAB 26.B
## MAJOR BLOOD VESSELS OF THE FETAL PIG

### MATERIALS REQUIRED (per pair of students)
dissected fetal pig from previous exercises

blunt probes

dissecting pins

### PROCEDURE
In the following exercise, arteries and veins will be considered together since corresponding vessels of an organ usually lie side by side. Arteries and veins are easily distinguishable in your pig specimen because the circulatory system has been injected with colored latex. Arteries are red, veins are blue. These colors distinguish the color differences of the blood carried in these vessels during life. Oxygenated blood, generally carried by the arteries, is bright red in color. Deoxygenated blood, typically carried in the veins, is darker red in color. Refer to Figures 26–3 and 26–4 as you proceed through the exercise.

1. Observe the heart in your fetal pig. The thin sac of membrane covering it is called the pericardium. Carefully remove the lungs from the thoracic cavity leaving the heart intact. Lift and push to one side the digestive organs to expose the major blood vessels that lie against the dorsal body wall.

2. Follow the large aorta, which arches posteriorly. Locate the first large artery that branches from the aorta toward the head. This is the short brachiocephalic artery that soon branches into the right subclavian artery, which feeds the right foreleg and chest. The corresponding vessel that drains the foreleg is the right brachial vein. It joins the right internal jugular vein leading from the head, which empties into the anterior vena cava.

3. Return to the aorta near the heart and locate the second branch, the left subclavian artery. This artery feeds the left foreleg and chest. The corresponding vein is the left brachial vein that joins the left internal jugular vein, which empties into the anterior vena cava.

4. Note how the aorta runs along the entire dorsal body wall parallel to the spine. It is known as the dorsal aorta in this location. Locate the point where the dorsal aorta passes through the diaphragm. The first artery to branch posterior to the diaphragm is the coeliac artery, which feeds the diaphragm, spleen, liver, and stomach. The large vein that corresponds to the dorsal aorta in this region is the posterior vena cava that drains the posterior body and hind legs. The posterior vena cava empties into the right atrium of the heart. Veins from the small intestine, large intestine, stomach, and spleen merge forming the hepatic portal vein that leads into a capillary network within the liver. Nutrients in the blood from the digestive organs are processed by the liver. Blood from the liver drains into the posterior vena cava by hepatic veins that are hidden within the liver tissue.

5. Locate the second branch of the dorsal aorta below the diaphragm, which is the large mesenteric ar-

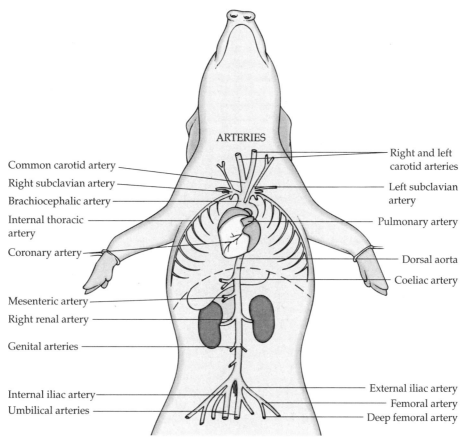

ARTERIES

Common carotid artery
Right subclavian artery
Brachiocephalic artery
Internal thoracic artery
Coronary artery

Mesenteric artery
Right renal artery

Genital arteries

Internal iliac artery
Umbilical arteries

Right and left carotid arteries
Left subclavian artery
Pulmonary artery

Dorsal aorta
Coeliac artery

External iliac artery
Femoral artery
Deep femoral artery

**FIGURE 26-3**

Major arteries of the fetal pig.

tery that feeds the intestine. Its branches can be seen throughout the transparent intestinal mesenteries.

6. Find the next branch of the aorta, which forms the right and left renal arteries that feed the kidneys. The right and left renal veins return blood to the posterior vena cava.

7. Continue posteriorly where the dorsal aorta divides into several major branches. The outermost branches are the right and left external iliac arteries that feed the hind legs. The two median branches are the umbilical arteries that lead into the umbilical cord. The right and left external iliac veins drain blood from the hind legs toward the posterior vena cava. The umbilical vein leading out of the umbilical cord enters the liver and joins the posterior vena cava.

There should be some mention here that the

fetal pig is an immature model of mammalian circulation. Adult structures are all present but there are some important differences in circulation since the oxygenated blood was supplied from the placenta. First, the ductus arteriosus is a small vessel that connects the pulmonary artery to the aorta. This shunts blood from nonfunctional lungs into systemic circulation. Second, the foramen ovale is an opening between the right and left atria; therefore, the blood bypasses the lungs. Third, the umbilical vessels connect the placenta, which functions in the fetus as lungs.

After you have located the arteries and veins mentioned above, cover your fetal pig with wet paper towels and return it to the storage bag for later use. Clean your dissecting tray and tools, and return them to their proper place.

**FIGURE 26-4**

Major veins of the fetal pig.

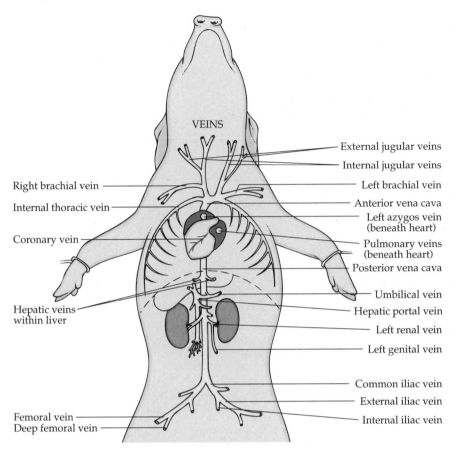

VEINS

External jugular veins

Internal jugular veins

Right brachial vein

Left brachial vein

Internal thoracic vein

Anterior vena cava

Left azygos vein
(beneath heart)

Coronary vein

Pulmonary veins
(beneath heart)

Posterior vena cava

Umbilical vein

Hepatic veins
within liver

Hepatic portal vein

Left renal vein

Left genital vein

Common iliac vein

External iliac vein

Femoral vein

Internal iliac vein

Deep femoral vein

## POSTLAB QUESTIONS

1. Identify the following arteries on the fetal pig.

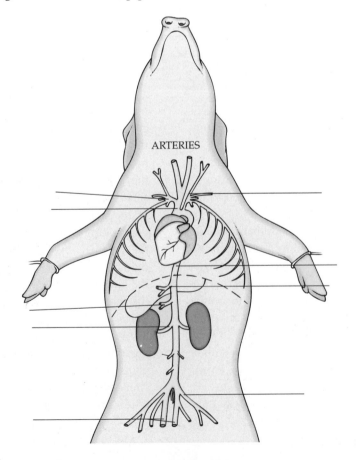

ARTERIES

2. Identify the following veins in the fetal pig.

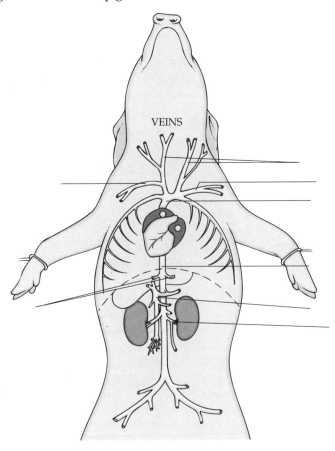

VEINS

3. Why does fetal blood circulation before birth differ from blood circulation after birth?

4. Why is almost all fetal blood a mixture of oxygenated and deoxygenated blood?

## LAB 26.C
## COMPARISON OF ARTERIES AND VEINS

### MATERIALS REQUIRED

prepared slide of artery and vein, cross section

compound microscope

### PROCEDURE

There are structural differences in arteries and veins due mainly to the differences in blood pressure that occur in these two types of blood vessels.

1. Observe the prepared slide of an artery and vein cross section under low and high power. Your slide was probably prepared from a section of an artery and its corresponding vein that supply and drain a particular region of a small mammal's body. Examine the artery, the vessel with a thicker wall and a proportionally smaller lumen, or cavity. The thick artery wall consists of elastic fibers and smooth muscle tissue. The elastic tissue allows the artery to expand with the surge of high blood pressure when the heart contracts. Steady contraction of the smooth muscle tissue strengthens the artery wall against overexpansion.

2. Examine the vein and notice the thinner wall and the proportionally larger lumen compared to that of the artery. Veins carry blood, which is typically low in pressure, toward the heart. Veins are composed of the same tissues as arteries but they lack thick walls due to the absence of high blood pressure in these blood vessels. Consequently, veins are flabby when empty, whereas arteries retain their circular shape.

## POSTLAB QUESTIONS

1. What are arterioles?

2. If the blood is typically low in pressure when it is in the veins, what keeps the blood flowing in the veins?

## LAB 26.D
## CAPILLARIES

### MATERIALS REQUIRED

compound microscope

small goldfish surrounded in wet cotton (demonstration prepared by instructor)

microscope slide

### PROCEDURE

In a sense, capillaries are the most important blood vessels because they are the site of exchange between cells and the blood stream. No exchange of materials occurs between the tissues and arteries or veins because their walls are much too thick for diffusion of materials to occur. The walls of capillaries consist of only one layer of thin epithelial cells, which easily permits diffusion of oxygen and nutrients into the tissues. You did not see capillaries in the fetal pig because capillaries are microscopic. However, you can see them with the aid of a compound microscope in thin living tissue such as the membranous tail of a goldfish.

1. Observe the demonstration of capillaries prepared by your instructor using a goldfish held in wet cotton. The wet cotton stabilizes the fish and maintains the animal while it is out of the aquarium.

2. Note the diameter of the capillaries relative to the red blood cells. Also note at what speed they pass through the capillaries.

## POSTLAB QUESTION

1. How do the structures of arteries, veins, and capillaries compare to their function?

## LAB 26.E
## BLOOD

### MATERIALS REQUIRED

prepared slide of human blood smear

compound microscope

### PROCEDURE

Of course, blood is a major component of the circulatory system. You know from Topic 21 that blood is a connective tissue. Blood consists of various types of cells and platelets suspended in plasma, the actual liquid portion of blood.

1. Examine the prepared blood smear slide under low and high power. The numerous cells throughout the slide are red blood cells, or erythrocytes, that carry the respiratory gases oxygen and some carbon dioxide. Erythrocytes contain hemoglobin, a red pigment that chemically binds with oxygen and carbon dioxide making transport of these gases possible. Focus carefully on an individual erythrocyte. The nucleus is absent, having been lost during development in the red bone marrow where erythrocytes are produced. Notice that erythrocytes are slightly concave on each side of these somewhat flattened cells (see Fig. 21–8, page 241). This shape increases the surface area of erythrocytes making them more efficient in transporting respiratory gases.

2. Return to low or scanning power and search the slide for larger, less common cells with large, darkly stained nuclei. These are white blood cells, generally known as leucocytes. When compared to erythrocytes, leucocytes are present at a ratio of about 1:700. There are various kinds of leucocytes, but they all function in some aspect of the immune system such as producing antibodies or engulfing foreign bacteria and viruses in amoeboid fashion.

3. Platelets, another component of blood, are not visible on your slide. Platelets are cell fragments of cytoplasm that pinch off from large cells in the bone marrow. Platelets are essential for the production of a blood clot that prevents excessive blood loss from a wound.

4. Plasma, not present on your slide, of course, is the liquid of blood in which platelets and red and white blood cells are suspended. Plasma is mostly water and contains a complex mixture of dissolved salts, gases, wastes, nutrients, hormones, and various proteins.

## POSTLAB QUESTIONS

1. Identify the components of this blood smear.

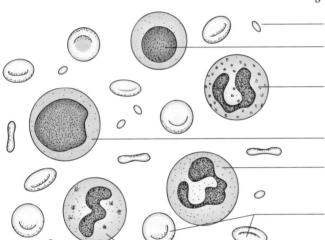

2. Anemia is a deficiency in hemoglobin. What type of blood cell could an anemic person be deficient in?

## FOR FURTHER READING

Bourne, G. H., ed. 1980. *Hearts and Heart-Like Organs.* New York: Academic Press.

Brooks, S. M. 1973. *Basic Facts of Body Water and Ions.* New York: Springer-Verlag, Inc.

Eckert, R., and D. Randall. 1988. *Animal Physiology.* New York: Freeman.

Guyton, A. C. 1984. *Physiology of the Human Body.* Philadelphia: Saunders College Publishing.

Hole, J. W. 1981. *Human Anatomy and Physiology.* Dubuque, Iowa: William C. Brown.

Kapff, C. T., and J. H. Jandl. 1991. *Blood: Atlas and Sourcebook of Hematology.* Boston: Little, Brown and Co.

Romer, A. S., and T. S. Parsons. 1986. *The Vertebrate Body.* Philadelphia: Saunders College Publishing.

Satchell, Geoffrey H. 1991. *Physiology and Form of Fish Circulation.* New York: Cambridge University Press.

Smith, H. M. 1960. *Evolution of Chordate Structure.* New York: Winston.

Spence, A. P., and E. B. Mason. 1987. *Human Anatomy and Physiology.* Menlo Park, Calif.: Benjamin-Cummings Publishing Co.

Vander, A. et al. 1989. *Human Physiology: The Mechanisms of Body Function.* New York: McGraw-Hill.

Warr, Gregory W., and Nicholas Cohen, eds. 1991. *Phylogenesis of Immune Functions.* Boca Raton, Fla.: CRC Press.

# Mammalian Reproductive System

## OBJECTIVES

1. Trace the development and fate of the ovum, labelling on a diagram each structure of the female reproductive system and the functions of those structures.

2. Trace the development of sperm cells and their passage through the male reproductive system, labelling on a diagram each male structure and giving its function.

3. Give the actions of testosterone in the male.

4. Describe the hormonal control of the menstrual cycle

and identify the timing of important events, such as ovulation and menstruation.

5. Describe the process of fertilization.

6. Describe the mode of action and give the advantages and disadvantages of each of the following methods of contraception: the Pill, IUD, spermicides, the sponge, condom, diaphragm, rhythm, douche, withdrawal, sterilization, implants and cervical cap.

---

## INTRODUCTION

The mammalian reproductive system can be illustrated by a study of the human reproductive system because the basic reproductive structures found in humans are similar to those found in other mammals.

The external sex organs of the woman are collectively known as the **vulva.** The **labia majora** and **labia minora** cover and protect the urinary opening, the entrance to the vagina, and the **clitoris,** which is the sensitive female counterpart of the male penis. The **vagina** receives the penis during **copulation** (sexual intercourse) and also functions as a birth canal during delivery of the fetus. Internally, the vagina leads to the **uterus,** which is where the embryo develops during pregnancy. The internal lining of the uterus is called the **endometrium,** which changes during the menstrual cycle. The opening of the uterus is the **cervix,** which protrudes into the upper end of the vagina. Its function is to support the fetus and uterine fluid during pregnancy. The **oviducts** (Fallopian tubes) branch from the upper right and left of the

uterus. They are tubes that carry the egg from the ovary toward the uterus. The **ovaries** (female **gonads**) are located at the free end of the oviducts. The ovaries produce **gametes** (eggs) and female hormones that maintain pregnancy and produce the **secondary sexual characteristics** of women.

The male gonads **(testes)** are carried externally in a sac, called the **scrotum.** The testes produce **testosterone,** the most important **androgen** (male sex hormone) responsible for male secondary sexual characteristics. Each testis is composed of many **seminiferous tubules** that produce male gametes **(sperm).** Sperm are stored in the **epididymis,** which is closely associated with each testis. Sperm are released from the epididymis during **orgasm,** a series of muscular contractions that occur in the reproductive tract of both sexes as a result of sufficient sexual stimulation. The contractions propel sperm through the **vasa deferentia** (sperm ducts) to the base of the urethra, where the sperm mix with secretions from the **seminal vesicles,** the **prostate** gland, and the **bulbourethral (Cowper's) glands** to form **semen.** Semen continues through the urethra and is expelled out the **penis,** which thereby introduces sperm into the vagina during copulation. The penis contains chambers of spongy vascular tissue that engorge with blood during sexual stimulation causing the penis to enlarge and stiffen. These changes in the penis facilitate placement of sperm into the upper vagina during intercourse. Sperm swim through the cervix, up the uterus, and into the oviducts where fertilization usually occurs.

**Fertilization,** (the union of haploid gametes) can occur in a woman only during a certain time of the **menstrual cycle.** The cycle is controlled by four interacting hormones: follicle stimulating hormone (FSH), estrogen, luteinizing hormone (LH), and progesterone. The cycle begins when the pituitary gland re-

leases **follicle stimulating hormone** (FSH), which causes maturation of an ovarian follicle and secretion of the female hormone **estrogen** from the ovarian follicle. Estrogen stimulates the pituitary to release **luteinizing hormone** (LH), which causes **ovulation,** the release of a mature egg from the ovary. This occurs on about the fourteenth day of the "typical" 28-day menstrual cycle. If sperm are present, fertilization usually occurs in the oviduct before the egg reaches the uterus. The egg disintegrates about 72 hours after ovulation if fertilization does not occur.

The ruptured follicle, under the influence of LH, becomes a **corpus luteum** ("yellow body") that secretes more estrogen and the female hormone **progesterone.** Both hormones cause thickening and a blood supply increase in the endometrium, which will

FEMALE

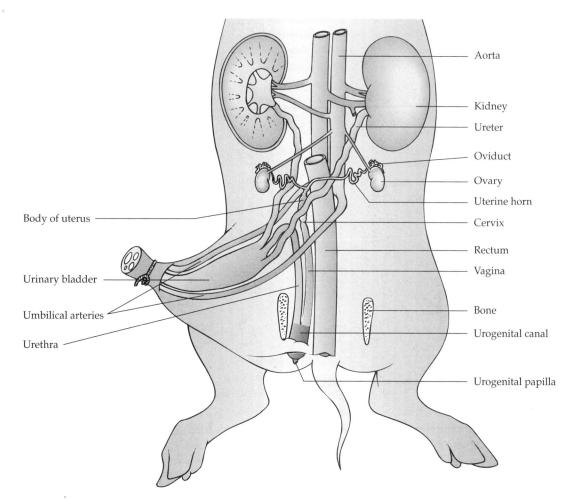

Aorta

Kidney

Ureter

Oviduct

Ovary

Uterine horn

Cervix

Rectum

Vagina

Bone

Urogenital canal

Urogenital papilla

Body of uterus

Urinary bladder

Umbilical arteries

Urethra

**FIGURE 27–1**

Major organs of the reproductive system in the female fetal pig.

facilitate implantation of the early embryo should fertilization occur. The high levels of estrogen and progesterone inhibit secretion of FSH and LH from the pituitary. Without these hormones, the corpus luteum produces much less estrogen and progesterone, which causes the endometrium to degenerate. This produces menstrual flow on about the twenty-eighth day. The pituitary again produces FSH without the corpus luteum secretions as inhibitors and the cycle repeats.

In order to avoid unwanted pregnancy, several **contraception** methods have been developed to prevent fertilization. These methods differ in a variety of ways, including which sex uses them, how effective the method is, and how the methods prevent fertilization of the egg by the sperm.

As with the other organ systems you have studied, the reproductive system of the fetal pig is representative in structure and function of most other mammals. Keep in mind as you observe the reproductive organs of the fetal pig that this is an immature specimen; therefore, most structures are proportionally much smaller than in an adult pig. As the pig matures, the reproductive system is stimulated by various hormones that cause growth and development of the organs as they become functional.

The female and male fetal pig reproductive systems are illustrated in Figures 27–1 and 27–2. Refer to these figures as you proceed through the following exercises. *Important:* After you have located all organs of your specimen, be sure to trade fetal pigs with an-

MALE

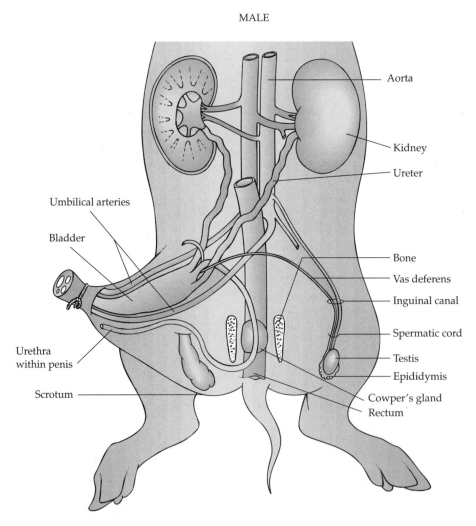

**FIGURE 27–2**

Major organs of the reproductive system in the male fetal pig.

other pair of students to examine the reproductive system of the opposite sex.

The following laboratory studies have been selected to demonstrate various important aspects of the mammalian reproductive system.

## PRELAB QUESTIONS

1. What is the function of ovaries?

2. What are the tubules called in which mammalian sperm are formed?

3. What is the function of luteinizing hormone in females?

4. What is fertilization?

5. What structure is the male counterpart of the female clitoris?

## LAB 27.A
## REPRODUCTIVE ORGANS OF THE FEMALE FETAL PIG

### MATERIALS REQUIRED
fetal pig specimen from previous exercises

dissecting tray

blunt probes

razor blade

forceps

scissors

dissecting pins

### PROCEDURE

1. Pin back the flaps of the abdominal wall. Expose the reproductive system of your fetal pig by pushing the digestive organs aside to reveal the kidneys and all posterior structures.

2. Locate the small ovaries; they are oval-shaped structures positioned just posterior to each kidney. The ovaries are the female gonads. They produce egg cells, or ova (ovum, singular), which is the female gamete. The ovaries also produce the female hormones estrogen and progesterone, which stimulate the female reproductive tract and cause the development of female traits in the adult animal.

3. Note the fine, convoluted tubule located on the anterior surface of each ovary. This is the oviduct, also known as the Fallopian tube, which transports egg cells released from the ovary. The egg cells are carried by the beating of cilia that line the oviduct. As in humans, the egg cells are fertilized in the oviduct if sperm are present. Early embryonic development occurs as the fertilized egg continues through the oviduct toward the uterus.

4. Observe the two horns of the uterus (uterine horns) located slightly posterior to each ovary. The tiny oviducts broaden and connect with the anterior end of the uterine horns. The forked uterus of the pig greatly differs from the single-chambered human uterus. The uterus holds and nourishes the developing pig embryos. In pigs and other mammals that give birth to offspring in litters, multiple embryos develop in each uterine horn. The uterine horns intersect at the body of the uterus.

5. Expose the lower reproductive tract by carefully using a razor to cut the tissue and pelvic bone at the midline of the hip region. Widely separate the hind legs of the pig after you cut through the pelvis.

6. Look for a constriction posteriorly on the body of the uterus. This is the location of the cervix, an internal ring of smooth muscle tissue that marks the end of the uterus body and the beginning of the vagina. The vagina is a tube of smooth muscle tissue that receives the penis during copulation and functions as a birth canal when muscular contractions of the uterus expel the offspring.

7. Follow the vagina posteriorly, carefully cutting tissue as necessary, and locate its intersection with the urethra from the urinary bladder near the terminal end of the vagina. The short tube between this intersection and the body opening is the urogenital canal. The urogenital canal exits to the outside of the body by the urogenital orifice, the opening of both the excretory and reproductive systems. The urogenital opening is surrounded by the urogenital papilla, the external genital organ of the female pig.

With the completion of this exercise, you may no longer need your fetal pig specimen. Clean and dry your dissecting tray and instruments, then follow your instructor's directions for disposal or storage of the fetal pig.

## POSTLAB QUESTIONS

1. Egg cells are to ovaries as sperm cells are to _____.

2. Identify the following reproductive structures of the female pig.

FEMALE

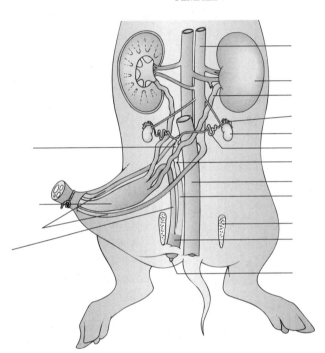

3. In a human female, how many openings are there from the pelvis to the exterior? How does this compare to the female fetal pig?

2. Locate the urinary bladder that leads from the attachment site of the umbilical cord inside the abdominal wall. Next, find the urethra leading posteriorly from the bladder. The vasa deferentia (vas deferens, singular) are two small tubules that intersect the urethra near the bladder. Do not confuse them with the ureters that lead to the bladder from the kidneys. The vasa deferentia carry sperm cells into the urethra. The urethra in male mammals carries both urine and sperm cells to the outside of the body, whereas the urethra in female mammals has no function regarding the reproductive system.

3. Notice that the urethra leads down posteriorly into the pig's pelvic area (hip region). Follow the urethra by carefully using a razor blade to make a midline cut through the muscle tissue and pelvic bone. The cut pelvis allows you to widely separate the hind legs for exposing structures within the pelvic area.

4. Use your finger or probe to touch the scrotum, an external sac ventral to the anus, to see if the testes have descended from the abdomen. Unless you have an older fetal pig, the testes will probably not

## LAB 27.B
## REPRODUCTIVE ORGANS OF THE MALE FETAL PIG

### MATERIALS REQUIRED

same as for Lab 27.A

### PROCEDURE

1. Pin back the flaps of the abdominal wall. Expose the reproductive system of your fetal pig by pushing the digestive organs aside to reveal the kidneys and all posterior structures.

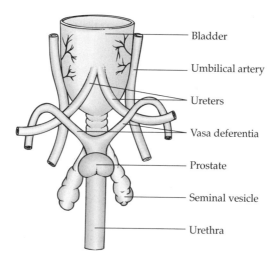

**FIGURE 27-3**

Dorsal view of the bladder and urethra region of the male fetal pig. Many ducts of the excretory and reproductive systems intersect at this point. The prostate and seminal vesicles secrete fluids into the urethra in this area.

yet be in the scrotum. If the scrotum is empty, locate a testis (singular) by following a vas deferens posteriorly from the bladder area. The vas deferens passes through the abdominal wall at the inguinal canal, an opening that also holds blood vessels leading to the testis. The vas deferens and combined vessels are called the spermatic cord, which continues down to the testis. Expose the testis by opening the membrane that covers it. The testis is the male gonad that produces sperm cells (gametes) and the male hormone testosterone, which matures the organs of the reproductive system and gives male characteristics to the adult animal.

5. Locate the epididymis, a mass of highly coiled small tubules on one side of the testis. The epididymis is continuous with the vas deferens and functions as a site of sperm storage after they are produced within the testis.

6. Return to the urethra near the bladder, the region illustrated in Figure 27–3. Locate the seminal vesicles, which are the light-colored glands near the junction of the vasa deferentia and urethra. At this same point, you may be able to distinguish the small, round prostate gland. Secretions from these glands mix with sperm cells in the urethra just before they are expelled from the body. This mixture is called semen. These secretions sustain the swimming sperm cells after they leave the male reproductive tract.

7. Follow the urethra through the pelvic area and find the bulbourethral (Cowper's) glands located on each side of the urethra. Fluid from bulbourethral glands also contributes to semen.

8. The urethra enters the penis just posterior to the bulbourethral glands. Use a blunt probe to separate the penis from the surrounding tissue. The penis opens to the outside of the body at the urogenital orifice located slightly posterior to the umbilical cord on the abdomen.

With the completion of this exercise, you may no longer need your fetal pig specimen. Clean and dry your dissecting tray and instruments, then follow your instructor's directions for disposal or storage of the fetal pig.

**POSTLAB QUESTIONS**

1. Vasa deferentia are to testes as _____ are to ovaries.

2. Identify the reproductive structures of a male pig on page 304.

MALE

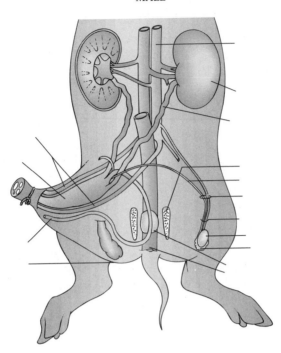

## LAB 27.C
### INTERNAL STRUCTURE OF THE FEMALE AND MALE GONADS

#### MATERIALS REQUIRED

prepared slides of the following:
a. mammalian testis, x.s.
b. mammalian ovary, x.s.

compound microscope

#### PROCEDURE

In Topic 10, you observed prepared slides of ovary and testis sections in the context of meiosis. Now is an appropriate time to review slides of these organs because they are the principal components of the female and male reproductive systems. In addition to producing gametes, the gonads also produce hormones that are essential to sexual development and to producing male and female characteristics that distinguish the sexes.

1. Observe under low power a prepared slide of a cross section of a mammalian testis. The numerous seminiferous tubules where sperm are produced appear as small circles throughout the slide.

2. Switch to high power and closely observe one of the seminiferous tubules. The larger cells nearer the outer edge of the tubule are diploid spermatogonia cells that develop into haploid sperm cells by spermatogenesis (meiosis and differentiation). The cells within the tubule wall closer to the lumen (opening) of the tubule are in various stages of spermatogenesis.

3. Examine closely under high power the lumen of the seminiferous tubule and look for extremely fine threads that are the tails of maturing sperm cells. The heads of the sperm cells are embedded in the tubule wall.

4. Look carefully for small cells located in the spaces between the round sections of seminiferous tubules. These are interstitial cells that secrete the male hormone testosterone, which produces male traits in the adult mammal. Compare what you see on your slide with the electron micrograph in Figure 10–14 (see page 105).

5. Obtain a prepared slide of an ovary cross section. Observe its general structure using the scanning objective or low power of your microscope. Notice the spherical pockets, called follicles, of various sizes scattered throughout the ovary. Each follicle originally contains one diploid oogonium cell that

gives rise to a single haploid egg cell by oogenesis (meiosis and differentiation). The larger follicles are those of more advanced development.

6. Search for a large, obvious follicle similar to the light micrograph in Figure 10–15 (see page 106). This is a mature follicle that should contain one large cell within its cavity. The cell within the large follicle is nearly a mature egg cell, or ovum. The large space within the follicle was filled with fluid before the slide was made. The ovum is released (ovulation) when the mature follicle ruptures and breaks through the ovary wall. Switch to high power and observe the developing ovum.

The cells of a developing follicle secrete the female hormone estrogen before ovulation. After ovulation, the ruptured follicle changes, becoming a corpus luteum that secretes progesterone, another female hormone. Both estrogen and progesterone are involved in maintaining the uterus during pregnancy, the human menstrual cycle, and producing female traits in the adult.

2. Identify the structures indicated on the above diagram.

3. What process is occurring in this structure?

4. If a woman's somatic cells have 46 chromosomes, how many chromosomes are contained in her mature ovum?

## POSTLAB QUESTIONS

1. Identify this structure.

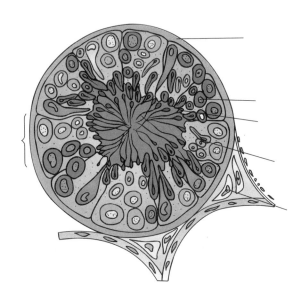

## LAB 27.D
## BIRTH CONTROL METHODS

### MATERIALS REQUIRED

Demonstration items of the following:
a. birth control pills
b. IUD
c. spermicidal foam and jelly
d. sponge
e. condom
f. diaphragm
g. cervical cap

### PROCEDURE

Several birth control or contraceptive methods are available, and most do not require a prescription or assistance from a physician for their use. The kind of birth control that is right for an individual is a matter of personal choice with consideration given to health concerns. Since most contraceptives function within the female, a woman should consult with her doctor when deciding upon a birth control method.

Contraceptive methods prevent pregnancy by interfering with fertilization, which is the fusion of a sperm and egg cell. A brief summary of common contraceptive methods is presented in Table 27–1. Observe the demonstration items of these methods on display as you proceed.

**Table 27–1**  Typical Reported Failure Rates During the First Year of Use of a Birth Control Method*

| Method of Birth Control | % of Women Experiencing an Accidental Pregnancy in the First Year of Use |
|---|---|
| Chance | 85 |
| Spermicides | 21 |
| Periodic abstinence | 20 |
| Withdrawal | 18 |
| Cervical cap | 18 |
| Sponge | |
|   Women who have borne children | 28 |
|   Women who have not borne children | 18 |
| Diaphragm | 18 |
| Condom | 12 |
| IUD | 3 |
| Pill | 3 |
| Female sterilization | 0.4 |
| Male sterilization | 0.15 |
| Implants | 0.04 |

*Data adapted from Hatcher et al., *Contraceptive Technology*, 1990–1992, 15th ed. (New York: Irvington Publishers, Inc., 1990).

1. *Birth control pill.* Birth control pills are available only by prescription. They contain synthetic female hormones (progesterone and estrogen) that interfere with the release of an ovum from the ovaries (ovulation) during the menstrual cycle. Taking one pill daily elevates the level of female hormones in the blood similar to what occurs during pregnancy. The female hormones inhibit the release of follicle stimulating hormone and luteinizing hormone from the pituitary gland that are essential for ovulation. In effect, the pill "tricks" the pituitary gland into sensing that the body is pregnant; therefore, no ovulation occurs and pregnancy is avoided. The pill is very effective if taken daily. It is also convenient because it does not require special procedures during inter-

course. Older women who use the pill have a somewhat higher risk of hypertension, blood clots, and strokes.

2. *IUD.* The intrauterine device (IUD) is a small device made of metal or plastic that is inserted by a physician into the uterus where it stays until removed. Unlike most birth control methods, an IUD does not interfere with fertilization; rather, it prevents implantation of the early embryo into the wall of the uterus and pregnancy is avoided. The IUD is effective and convenient because it is not bothersome during intercourse. However, the IUD has been linked to an increased risk of infertility and infection, which has decreased its usage.

3. *Spermicides.* As the name implies, spermicides prevent fertilization by killing sperm cells. Spermicides are inexpensive and available without a prescription. Spermicides are produced as a cream, as a foam, or as a jelly that is placed in the upper vagina with an applicator at the time of intercourse, which some may regard as inconvenient. Spermicides are most effective when used in combination with another contraceptive, such as a diaphragm or condom, described below.

4. *Sponge.* The sponge is just that—a disc-shaped foam sponge containing spermicide. It is available without a prescription. The sponge is placed over the cervix and can be inserted well before intercourse since the spermicide remains active for about 24 hours. A small strap is attached to the sponge for easy retrieval. The sponge is an effective contraceptive, but the additional use of a condom ensures safety from pregnancy.

5. *Condom.* The condom is the only contraceptive device designed for use by men. The condom is a thin, durable sheath of latex or animal membrane that fits closely over the penis. Semen is retained in the condom, thus preventing pregnancy. Condoms are the only contraceptive method that has the additional benefit of preventing the spread of sexually transmitted diseases. Condoms are inexpensive and widely available without a prescription. Some couples may consider condoms as an inconvenience because of their use at the time of intercourse. Use of a spermicide with the condom increases protection if the condom should break or leak. Petroleum-based or vegetable oil-based lubricants should be avoided since these cause breakdown of latex condoms.

6. *Diaphragm.* The diaphragm is a dome-shaped rubber shield that fits over the cervix. This prevents fertilization by blocking the site at which sperm enter the uterus. Diaphragms are designed for use with a spermicide and are reusable. Various sizes are available to accommodate most women. A diaphragm requires a prescription and must be fitted by a physician. The diaphragm requires some skill and practice for proper placement over the cervix. The diaphragm is inserted just before intercourse.

7. *Rhythm (periodic abstinence).* The rhythm method of contraception involves no chemicals or devices; it is simply the abstinence of sexual intercourse during the time of ovulation, which avoids pregnancy. Of course, this involves very careful observation of the menstrual cycle to identify ovulation, which usually (but not always) occurs at the midpoint of the cycle. Predicting ovulation strictly by the calender presents a high risk of pregnancy because the menstrual cycle can be variable. A more accurate prediction is possible by checking for a slight rise in body temperature that occurs just before ovulation. This method also is not totally reliable because health and emotional factors can affect body temperature. Some couples may not consider the rhythm method as a desirable choice of contraception because of its rather high pregnancy rate and the requirement of regular sexual abstinence.

8. *Douche.* Douching involves flushing the vagina with a solution to remove sperm from the reproductive tract. Douching has essentially no effect as a contraceptive because sperm rapidly swim through the cervix into the uterus where the douche cannot reach.

9. *Withdrawal.* This contraceptive practice involves removal of the penis from the vagina before the ejaculation of semen. Withdrawal is not a reliable means of contraception because extreme self-control is required on the part of the male, and lubricating fluids secreted from the penis prior to ejaculation may contain sperm cells.

10. *Sterilization.* Obviously, sterilization is the most effective means of contraception. Sterilization of the male is a relatively simple surgical procedure because the reproductive organs are external. The surgery, called a vasectomy, is usually performed in a doctor's office with local anesthesia. A vasectomy involves the cutting and tying off of each vas deferens that leads from the testes in the scrotum. As a result, the sperm cells never leave the testes, but are reabsorbed. A vasectomy does not affect the sexual function of a male in any way, and it gives a couple the peace of mind of permanent, convenient, and effective birth control. Attempts of vasectomy reversal have had limited success, so vasectomy should be considered only by those couples who plan to have no children in the future.

Because the female reproductive organs are internal, sterilization of the female requires hospitalization. The oviducts (Fallopian tubes) are cut and tied off, a procedure known as tubal ligation. This prevents egg cells from reaching the uterus. Like vasectomy, tubal ligation is very effective, permanent, and has no effect on the sexual function of the female. Limited success has been achieved in the reversal of tubal ligation; female sterilization should be considered only by those who no longer want children.

11. *Implants.* One of the newest forms of birth control is the surgical implantation of silicone capsules under the woman's skin (usually on the upper arm). These capsules are for long-term contraception and can inhibit ovulation for up to five years by a slow release of progesterone into the bloodstream. Contraception is reversible by surgical removal of the capsules.

12. *Cervical cap.* The cervical cap is a soft rubber cup-shaped device that fits directly over the cervix inside the vagina. The cap interior is coated with spermicide and inserted prior to intercourse and must be left in place six to eight hours afterwards. The similarities in structure and use with the diaphragm gives the cervical cap similar contraceptive effectiveness, but the smaller size may make the cap easier to handle.

## POSTLAB QUESTIONS

1. Identify the following contraceptive methods.

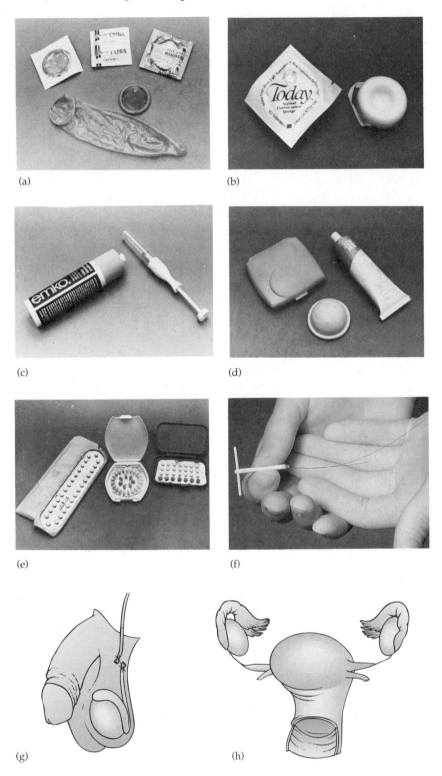

(a)                                      (b)

(c)                                      (d)

(e)                                      (f)

(g)                                      (h)

2. Complete Table 27–2.

**Table 27–2**  Contraceptive Methods

| Method | Function | Advantages | Disadvantages |
|---|---|---|---|
| Birth control pill | | | |
| Condom | | | |
| Diaphragm | | | |
| Douche | | | |
| IUD | | | |
| Rhythm | | | |
| Spermicides | | | |
| Sponge | | | |
| Sterilization | | | |
| Withdrawal | | | |
| Implants | | | |
| Cervical cap | | | |

## FOR FURTHER READING

Bell, G. 1982. *The Masterpiece of Nature: The Evolution and Genetics of Sexuality.* Berkeley: University of California Press.

Eckert, R., and D. Randall. 1988. *Animal Physiology.* New York: Freeman.

Gordon, M. et al. 1982. *Animal Function: Principles and Adaptations.* New York, Macmillan.

Guyton, A. C. 1984. *Physiology of the Human Body.* Philadelphia: Saunders College Publishing.

Hole, J. W. 1981. *Human Anatomy and Physiology.* Dubuque, Iowa: William C. Brown.

Masters, W. H. et al., eds. 1992. *Human Sexuality.* New York: Harper Collins Publishers.

Romer, A. S., and T. S. Parsons. 1986. *The Vertebrate Body.* Philadelphia: Saunders College Publishing.

Smith, H. M. 1960. *Evolution of Chordate Structure.* New York: Winston.

Solomon, E. P. et al. 1990. *Human Anatomy and Physiology.* Philadelphia: Saunders College Publishing.

Spence, A. P., and E. B. Mason. 1987. *Human Anatomy and Physiology.* Menlo Park, Calif.: Benjamin-Cummings Publishing Co.

Vander, A. et al. 1989. *Human Physiology: The Mechanisms of Body Function.* New York: McGraw-Hill.

Witters, W., and P. Witters. 1980. *Human Sexuality, A Biological Perspective.* New York: D. Van Nostrand.

❏

# Animal Development

1. Define growth, morphogenesis, and cellular differentiation, and describe the role of each process in the development of an organism.

2. Describe the principal events and characteristics of each

of the early stages of development: zygote, cleavage, blastula, and gastrula.

3. Define metamorphosis and describe how development continues after an animal is born or hatches.

## INTRODUCTION

Embryonic development is divided into four main stages: cleavage, gastrulation, neurulation, and organogenesis. The first stage, **cleavage,** is a period of rapid cell division of the zygote (fertilized egg) with no growth in cell size. Cleavage produces many new cells and segregates different cytoplasmic substances into different cells. Eventually, cleavage results in a **blastula,** or hollow ball of cells.

The second stage, **gastrulation,** occurs when the blastula undergoes an inpocketing of cells that form a **gastrula.** The cells eventually arrange into three **germ layers:** the ectoderm, the endoderm, and the mesoderm. Each germ layer gives rise to specific tissues and organs in the fully developed animal. The **ectoderm** (outermost layer) gives rise to the nervous system, the skin and its derivatives. The **endoderm** (innermost layer) produces digestive glands and the lining of the gut. The middle germ layer, the **mesoderm,** forms the skeleton, muscles, and various organs.

The third developmental stage, **neurulation,** occurs when the **neural tube** and head form by the joining of the parallel folds in the ectoderm. Neurulation proceeds after the three germ layers are established.

The fourth developmental stage, **organogenesis,** is a process of organ formation. This occurs as the germ layers give rise to their respective structures, producing embryonic organs and systems.

An organism grows from a zygote by an increase in cell number and by an increase in cell size. These cells are arranged into specific body forms by the process of **morphogenesis.** During embryonic development, seemingly similar cells give rise to all the different cell types in the fetus. This process is called **differentiation,** which results from the expression of different genes in different cells of the embryo. The course of development for a particular cell is somehow fixed, or determined. **Determination** occurs early in development and is not reversible. Determination is also heritable, as demonstrated by the fact that cells of one type give rise only to more cells of the same type.

Development of a young animal does not stop when it is born or hatched. Growth and cellular differentiation are continuous processes as the offspring matures into an adult of its species. This occurs as tissues and organs of the body are influenced by activated genes and various hormones that change the juvenile organism into a full-sized, sexually mature individual.

We consider the following investigation of various aspects of animal development, particularly early development, to be very useful in arriving at an understanding of the major concepts in this chapter.

## PRELAB QUESTIONS

1. What are the four stages of early embryonic development?

2. What embryonic stage is represented by a hollow ball of cells?

3. What process results from the expression of different genes in different cells of the embryo?

4. Is development completed at the time of birth or hatching?

## LAB 28.A

## CLEAVAGE AND EARLY EMBRYONIC DEVELOPMENT

### MATERIALS REQUIRED

prepared slide of sea star development

compound microscope

### PROCEDURE

The developing eggs of the sea star, a representative of phylum Echinodermata, are ideal for observing the events that occur in the early embryonic development.

1. Obtain a preserved slide of developing sea star eggs. Observe the slide under scanning and low power. Notice the slide contains many stages of development, from unfertilized eggs to sea star larvae.

2. Search the slide for a large cell with an obvious nucleus. This is an unfertilized sea star egg (Fig. 28–1a).

3. Continue by searching for a cell of similar size as the egg, but with no apparent nucleus. This cell is the fertilized egg, or zygote, formed when a sperm cell penetrates the egg and their nuclei fuse. The zygote is probably less common on your slide because it quickly begins development by cell division.

4. Scan the slide for small clusters of cells found in numbers of 2, 4, 8, 16 and so on (see Fig. 28–1b through 28–1d). These cell groups represent early developmental stages. They are produced when the zygote divides by mitosis. This early cell division is known as cleavage because the cells are continually divided, or cleaved, but there is no cell growth. Consequently, the cells double in number with each division, but they become progressively smaller as their cytoplasm is divided in half. At the 32-cell stage, the embryo is called a morula and is a solid ball of cells.

5. Search the slide for a hollow ball of cells known as the blastula (Fig. 28–1e), which develops from the morula as mitosis continues. The numerous cells of the blastula are produced by continual mitotic divisions. The fluid-filled cavity within the blastula is called the blastocoel.

6. Scan the slide for a gastrula, the next embryonic stage that develops from the blastula. The gastrula is easily identified by an inpocketing of cells that protrudes into the cavity of the developing gastrula. An early gastrula is identified by a slight inpocketing of cells as it begins to grow inward (Fig. 28–1f). A later gastrula has a conspicuous inpocketing. The internal cavity of the inpocketing is known as the archenteron, which gives rise to the digestive tract (Fig. 28–1g). The opening of the archenteron, called the blastopore (Fig. 28–1f and 28–1g), develops into the anus in echinoderms and chordate animals. Animals that exhibit this trait are known as deuterostomes. In other phyla, the blastopore gives rise to the mouth. These animals are known as protostomes.

7. Notice that formation of the archenteron results in two cell layers of the gastrula. The inner cell layer forming the wall of the archenteron is the endoderm, and the outer cell layer of the gastrula is the ectoderm. A middle layer of cells, the mesoderm, later develops between the endo- and ectoderm. Each of these cell layers will give rise to specific organs and tissues of the fully developed organism.

8. Search the slide for the odd, free-swimming sea star larva similar to Figure 28–1h. Much development has occurred between the gastrula and the sea star larva you see on the slide. The larva has a

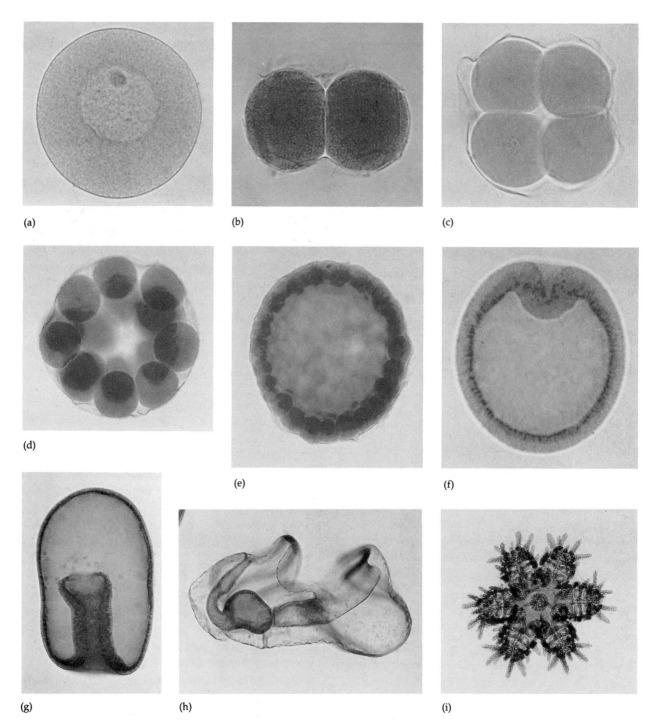

(a)

(b)

(c)

(d)

(e)

(f)

(g)

(h)

(i)

**FIGURE 28–1**

Sea star development. (a) Unfertilized egg. (b) Two-cell stage. (c) Four-cell stage. (d) Sixteen-cell stage. (e) A section through the blastula revealing the blastocoel. (f) A section through the early gastrula revealing the blastopore and the developing archenteron. (g) A section through the later gastrula revealing the archenteron. (h) A bilaterally symmetrical sea star larva. (i) A young radially symmetrical sea star.    *(Carolina Biological Supply Company)*

complete digestive tract with mouth and anus in addition to other tissues throughout its body that developed from the three germ layers. Further development occurs beyond the larval stage and the young sea star finally begins to resemble a miniature adult (Fig. 28–1i).

## POSTLAB QUESTIONS

1. List the proper development sequence from earliest to latest for an animal, using the following terms: blastula, gastrula, morula, and zygote.

2. You are looking at several slides that show different stages of animal development and you note that one structure has three germ layers. You are probably looking at a _____.

## LAB 28.B
## EARLY EMBRYONIC DEVELOPMENT OF THE CHICK

### MATERIALS REQUIRED on demonstration

prepared slides of chick embryos at 24, 33, 48, and 72 hours of development

### PROCEDURE

Chicken embryos have long been sources of developmental studies because they are easily obtained and observed. They also beautifully illustrate the remarkable and rapid changes that occur within an embryo as the germ layers continually divide and shift positions, resulting in a complex organism. In the following exercise, you will observe and compare chick embryos at various stages of development as indicated in hours after fertilization.

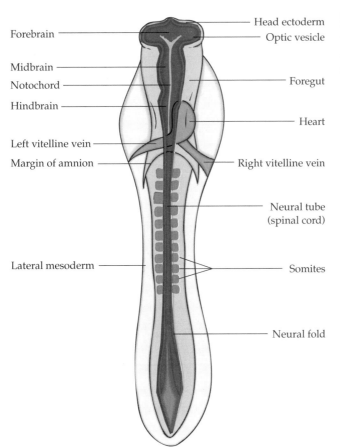

Forebrain
Midbrain
Notochord
Hindbrain
Left vitelline vein
Margin of amnion
Lateral mesoderm

Head ectoderm
Optic vesicle
Foregut
Heart
Right vitelline vein
Neural tube (spinal cord)
Somites
Neural fold

**FIGURE 28–2**

Whole mount of 33-hour chick embryo and some of its major structures.

1. Observe the prepared slide of a 24-hour chick embryo on the demonstration stereomicroscopes. Note the relatively simple structure of the 24-hour chick. The supportive notochord, common to all members of phylum Chordata at some stage of their life cycle, occurs lengthwise in the center line of the embryo. The notochord will be replaced by the vertebral column (backbone) as development continues. The head end of the embryo is somewhat visible, which consists partly of the developing brain. A few somites are faintly visible. Somites are blocks of mesodermal cells that give rise to the skeleton and skeletal muscle.

2. Notice the dramatic difference in the next prepared slide of a 33-hour chick embryo on demonstration. Much change has occurred in just 9 hours. The somites are more numerous and well defined. The head is better developed, with optic vesicles that will become eyes (Fig. 28–2). The growing brain is distinctly visible as three regions, forebrain, midbrain, and hindbrain. The neural tube, which will give rise to the spinal cord, has formed from the merging of neural folds found lengthwise along the midline of the embryo. The brain, neural tube, and all other nervous tissue arises from ectoderm.

The foregut, part of the digestive tract, is visible in the region of the midbrain. Notice the bulbous heart that protrudes on the right side of the embryo. The heart is connected to the vitelline (yolk) veins, which will soon contain blood that carries food to the embryo from the large yolk in the egg. The amnion is a membrane that surrounds the embryo of reptiles, birds, and mammals. The amnion forms a fluid-filled sac around the embryo that protects it from shock. The margin of the developing amnion may be seen in the region of the vitelline veins as it grows over the head and proceeds toward the tail.

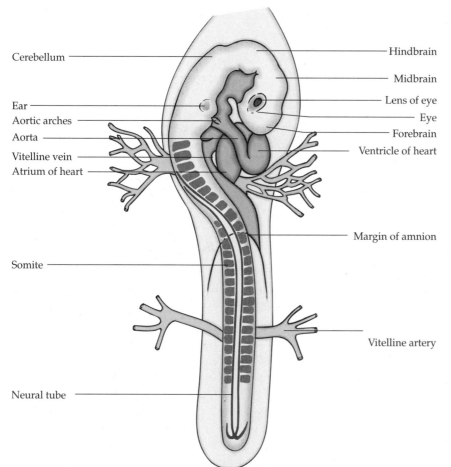

Cerebellum

Ear

Aortic arches

Aorta

Vitelline vein

Atrium of heart

Somite

Neural tube

Hindbrain

Midbrain

Lens of eye

Eye

Forebrain

Ventricle of heart

Margin of amnion

Vitelline artery

**FIGURE 28–3**

Whole mount of 48-hour chick embryo and some of its major structures.

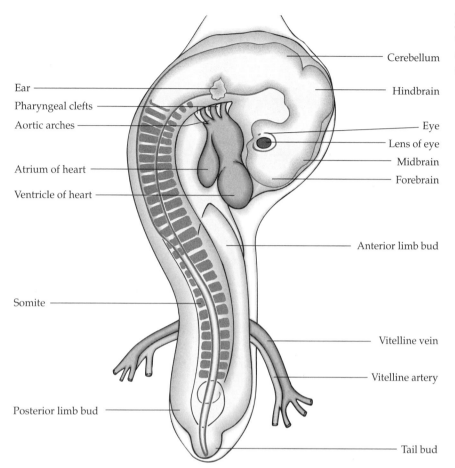

**FIGURE 28-4**

Whole mount of 72-hour chick embryo and some of its major structures.

Labels on figure: Ear, Pharyngeal clefts, Aortic arches, Atrium of heart, Ventricle of heart, Somite, Posterior limb bud, Cerebellum, Hindbrain, Eye, Lens of eye, Midbrain, Forebrain, Anterior limb bud, Vitelline vein, Vitelline artery, Tail bud

3. Examine the next demonstration slide of a 48-hour chick embryo. Many advances are evident in these 15 hours beyond the 33-hour specimen. The chick now lies on its left side and the body has curled slightly (Fig. 28–3). The developing eye contains a lens. The three regions of the brain have enlarged, giving much more form to the head. A developing ear may be visible posterior to the cerebellum, a portion of the hindbrain. The heart has enlarged and pumps a blood supply with two chambers: the atrium and ventricle. Notice the aorta, a major artery of all vertebrates, that branches into aortic arches. The vitelline blood vessels are well developed and extend over the yolk surface. The margin of the amnion has grown further over the embryo, and somites are more numerous.

4. Observe the preserved slide of a 72-hour chick embryo, which is now 24 hours older than the previous specimen. Notice the body has continued to curl (Fig. 28–4). The eye, regions of the brain, and ear have become more prominent, and somites continue to increase in number. The amnion entirely covers the embryo by this time of development and forms a protective fluid-filled sac around the chick. Look carefully in the region of the aortic arches for pharyngeal clefts (gill slits). These are an evolutionary remnant of the gills in fishes, the first vertebrates from which all others evolved. All vertebrates exhibit gill slits during their development. Limb buds begin to appear at this developmental stage. These will grow into wing and leg appendages in time.

Of the 21-day incubation period of chickens, only 6 days are required to establish all the organ systems within a chick embryo. This occurs by rapid division and differentiation of cells in the 3 germ layers. The embryonic organs grow and mature during the remainder of the incubation period, which enables the chick to hatch and function outside its egg.

## POSTLAB QUESTIONS

1. Identify the following structures of a 33-hour chick embryo.

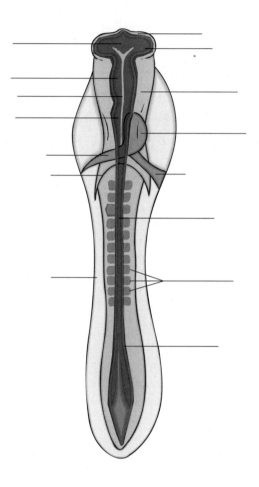

2. Identify the following structures of a 48-hour chick
   embryo.

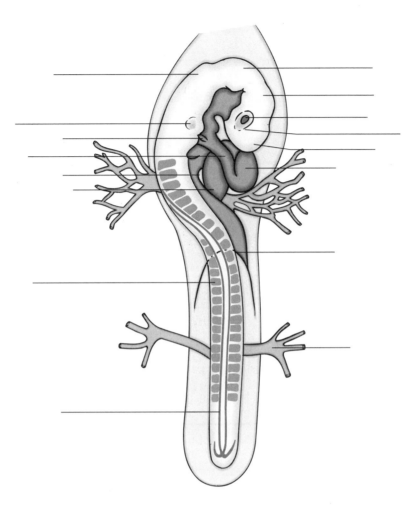

3. Identify the following structures of a 72-hour chick embryo.

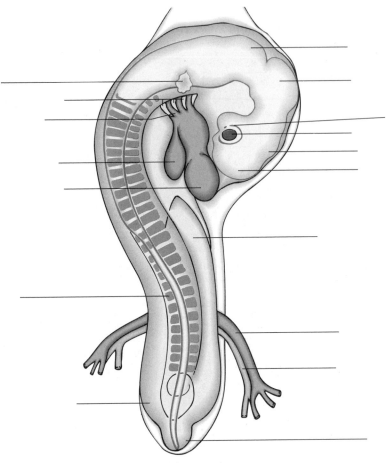

4. Identify the three distinguishing chordate features on the chick embryo.

5. Where are the blastula and gastrula stages in the chick embryo slides?

6. Can embryologic development be affected by environmental factors such as prenatal care? How?

7. Give a reason why some embryological stages of humans and chicks are similar. Would you expect the early or late stages to be most similar?

## LAB 28.C
## FROG METAMORPHOSIS

### MATERIALS REQUIRED
preserved tadpoles of five developmental stages, labelled (A) through (E)

hand lens

### PROCEDURE

The changes occurring in a frog tadpole as it matures are a dramatic example of development beyond the embryonic stages. This transition from a larva (tadpole) to an adult (frog) is known as metamorphosis, a process that commonly occurs in the life cycle of many kinds of animals in addition to amphibians.

1. Closely examine the preserved tadpole labelled (A). Note that it is somewhat fish like in appearance, with a prominent tail for swimming. Like

fish, tadpoles absorb oxygen from water by drawing it into the mouth, pumping it through their internal gills, and forcing it out a spiracle (opening) visible on the left side of the head region. Note the small mouth located toward the underside of the head (use a hand lens if necessary). Many tadpoles feed on algae they find on objects on the bottom of the pond.

2. Observe the tadpole labelled (B), representing the next stage of metamorphosis. Notice the reduced hind legs emerging from the body. The limbs first appear as small buds that gradually elongate into legs and feet with digits (toes).

3. Observe the next tadpole labelled (C). The shorter front limbs appear after the hind limbs. Look for any noticeable changes in the profile of the tadpole's body.

4. Next examine tadpole (D). Notice that the tail is becoming shorter as it is absorbed as a source of nutrients for the changing tadpole. The limbs are more pronounced and the body profile begins to appear more frog like. Look for changes in the size and shape of the mouth.

5. Examine the young frog in container (E). The tail is fully absorbed and the animal is distinctly a frog. The mouth is fully developed and is capable of swallowing insects instead of grazing on algae. Internal changes of the digestive organs would also reflect this change in diet. The frog now has lungs instead of gills and spends much of its time on land. When swimming, the frog surfaces occasionally to breathe, although considerable diffusion of oxygen from the water does occur through the highly vascular skin of amphibians. With the completion of metamorphosis, the juvenile frog will grow in size and eventually become a sexually mature adult.

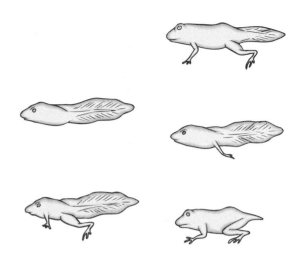

## POSTLAB QUESTIONS

1. Number the stages of frog metamorphosis in order of age.

## FOR FURTHER READING

Allen, James W. et al., eds. 1990. *Biology of Mammalian Germ Cell Mutagenesis.* Cold Spring Harbor, N.Y.: Cold Spring Harbor Laboratory Press.

Austin, C. R., and R. V. Short. 1972. *Embryonic and Fetal Development.* Cambridge: Cambridge University Press.

Bowder, L. 1984. *Developmental Biology.* Philadelphia: Saunders College Publishing.

Carlson, B. 1988. *Patten's Foundations of Embryology.* New York: McGraw-Hill.

De Pomerai, David. 1990. *From Gene to Animal: An Introduction to the Molecular Biology of Animal Development.* New York: Cambridge University Press.

Eckert, R., and D. Randall. 1988. *Animal Physiology.* New York: Freeman.

Gilbert, S. 1988. *Developmental Biology.* Sunderland, Mass.: Sinauer.

Gordon, M. et al. 1982. *Animal Function: Principles and Adaptations.* New York: Macmillan.

Guyton, A. C. 1984. *Physiology of the Human Body.* Philadelphia: Saunders College Publishing.

Hole, J. W. 1981. *Human Anatomy and Physiology.* Dubuque, Iowa: William C. Brown.

Romer, A. S., and T. S. Parsons. 1986. *The Vertebrate Body.* Philadelphia: Saunders College Publishing.

Rugh, R., and L. B. Shettles. 1971. *From Conception to Birth: The Drama of Life's Beginnings.* New York: Harper & Row.

Saunders, J. 1982. *Developmental Biology.* New York: Macmillan Publishing Co.

Smith, H. M. 1960. *Evolution of Chordate Structure.* New York: Winston.

Solomon, E. P. et al. 1990. *Human Anatomy and Physiology.* Philadelphia: Saunders College Publishing.

Spence, A. P., and E. B. Mason. 1987. *Human Anatomy and Physiology.* Menlo Park, Calif. Benjamin-Cummings Publishing Co.

Vander, A. et al. 1989. *Human Physiology: The Mechanisms of Body Function.* New York: McGraw-Hill.

# PART VIII

◻

# Evolution, Behavior, and Ecology

### TOPIC 29
Population Genetics and Natural Selection

### TOPIC 30
Behavior

### TOPIC 31
Population and Ecosystem Dynamics

### TOPIC 32
Human Ecology

TOPIC 29

❏

# Population Genetics and Natural Selection

OBJECTIVES

1. Explain natural selection as envisioned by Darwin.
2. State the Hardy-Weinberg law and discuss its significance in population genetics.
3. Explain how each of the following alters the gene fre-
quencies in large, random mating populations: mutation, gene flow, and natural selection.
4. Define the terms "selection coefficient" and "fitness" of a genotype.

## INTRODUCTION

Today's accepted theory of evolution was founded by Charles Darwin and Alfred Wallace who, working independently of one another in the mid-1800s, both proposed natural selection as the mechanism of evolution. **Natural selection** is the differential reproduction of genotypes from one generation to the next. Both Darwin and Wallace arrived at their conclusions after closely observing nature and after reading similar writings by other authors. When published, their theory of evolution was more convincing than those of earlier authors because they proposed a mechanism of evolution—natural selection.

Darwin illustrated the changing force of selection with selective breeding of domesticated plants and animals **(artificial selection),** which produces an organism with an increase of traits desired by agricultural breeders. Additional support for the theory of evolution comes from the fossil record, comparative anatomy, embryology, biogeography, and diverse other scientific disciplines.

The essence of evolution is descent with change

and, if change is significant enough, the establishment of new species. In order to see how new species occur (speciation), it is necessary to speak of populations. A **population** is all the members of a species that occupy a particular area at the same time. Populations, not individuals, are the units that may evolve into a new species. All the genes in a population are collectively called the population's **gene pool.** It is the gene pool that is affected by evolution. **Evolution** is change in the frequency of genes in a population's gene pool from one generation to the next. New species can arise by evolution from a population if the proper conditions occur.

One method to study the principles of evolution is to construct a simple theoretical population that does not change genetically from one generation to the next and, then, by subjecting the simple population to certain factors, see how those factors affect the population. The genetic model for the simple population is two alleles, $A$ and $a$, at an autosomal locus. The **Hardy-Weinberg law** follows from this model. We concern ourselves with changes in the frequencies of those two alleles in successive generations of the population.

The Hardy-Weinberg law holds that there will be no change in the frequencies of the two alleles given the following conditions:

1. There is no mutation. Mutation changes genes to other alleles, which automatically changes their frequency.
2. There must be no selection pressure (natural selection) so that no genotype has a reproductive advantage.
3. There must be no mating preferences among individuals; that is, mating is random.
4. The population must be isolated to avoid gene exchange (**gene flow**) with other populations.
5. The population must be of large size because the law is based on statistical probabilities, which are not accurate for small populations.

If any one of these factors is not satisfied, then the Hardy-Weinberg law is not applicable, the allelic frequencies in the population will not remain constant, and evolution will occur.

The effect of various evolutionary forces on the population can be examined by observing the nature of allelic frequency change caused by those forces within the population. Subsequently, the model can be expanded in virtually a limitless number of ways (more than two alleles at the given locus, nonautosomal modes of inheritance, multiple loci, combinations of factors mentioned, and so on) as the investigations become more complex and more similar to real-life situations. Such an expansion is beyond the scope of this manual, but we can investigate the simple model in some detail.

Let's begin by observing what happens when the *aa* genotype is lethal. In such a situation, there is complete selection against the *aa* genotype. As a result, the frequency of the *a* allele will decline, and the frequency of the *A* allele will increase. Contrary to what one might suspect, it will take some time to eliminate the *a* allele completely because of the fact that the allele will tend to be retained in the population due to the heterozygotes, the carriers. As a result, the lethal recessive allele will seldom be completely eliminated from a large population, but will be carried throughout many generations.

In the case of the lethal recessive allele, where all of the individuals of the genotype *aa* die or fail to reproduce, we say that the selection coefficient is 1. The selection coefficient of a particular genotype is the proportion (a number between 0 and 1 inclusive) of

individuals having that genotype that, for whatever reason, is eliminated from the breeding population that produces the next generation. The cause of the elimination of such individuals from the breeding population may be due to any of a variety of reasons —failure to survive infancy, failure to survive to sexual maturity, failure to be fertile, and so on, all of which, including other reasons as well, select against the given genotype. If, on average, half of the *aa* individuals die or fail to reproduce, then the selection coefficient, *s*, is 0.5. **Fitness,** *f,* is related to the selection by the formula $f = 1 - s$, from which it is seen that fitness is the proportion of individuals of a given genotype who survive and reproduce. If the fitness of a given genotype is 1, then the selection coefficient against individuals of that genotype is 0, and there is no selection against individuals of that genotype. Although the selection coefficient and the fitness of a given genotype are related by the mathematical relationship, the two concepts involved are quite different.

Individuals heterozygous for a particular gene may be more common in a population than predicted by the Hardy-Weinberg law. This is because there can be a selective advantage to the heterozygous condition, a situation termed **heterozygote advantage.** In such a case, there may even be selection against one or both of the homozygous genotypes.

Some interesting departures from the expectations of the Hardy-Weinberg law occur for the case we have been considering (two alleles at an autosomal locus) when one examines the consequences of varying levels of selection against one, two, or three of the genotypes involved, that is, populations where the fitnesses of the three genotypes are not all 1. Such effects are most easily demonstrated by using a simple computer simulation program such as the following one that we have written in BASIC for the microcomputer.

## PRELAB QUESTIONS

1. What is the differential reproduction of genotypes from one generation to the next?

2. What are the five criteria for the Hardy-Weinberg law to hold true?

3. What is the mathematical relationship between selection coefficient and fitness of a genotype?

# LAB 29.A

## HOW RAPIDLY WILL EQUILIBRIUM BE REACHED?

### MATERIALS REQUIRED

microcomputer with BASICA, GWBASIC, or equivalent

computer program (See Appendix B: "pgs" Computer Program)

printer

### PROCEDURE

1. Load the "pgs" program that has been provided, into the computer according to your lab instructor's directions.
2. Run a simulation of 15 generations with 0.3 as the initial frequency of the $a$ allele and a selection coefficient of 0 for each of the 3 genotypes $AA$, $Aa$, and $aa$. Simply answer the questions as prompted by the computer.
3. In the space provided below, use the data produced by the computer simulation and plot the frequencies of the $AA$, $Aa$, and $aa$ genotypes and of $p$ (the frequency of the $A$-allele), and of $q$ (the frequency of the $a$ allele).

### POSTLAB QUESTIONS

1. Did the genotypic and allelic frequencies change when all of the assumptions of the Hardy-Weinberg law were satisfied?

2. Perhaps the results depend upon the initial frequency of the $a$ allele. Design an experiment to investigate that question and repeat the simulation. Plot the results on the graph on page 325.

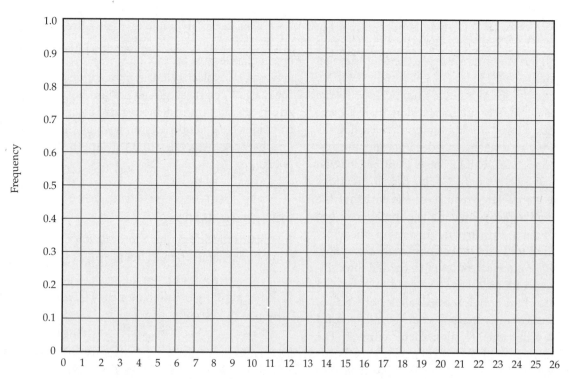

Generation

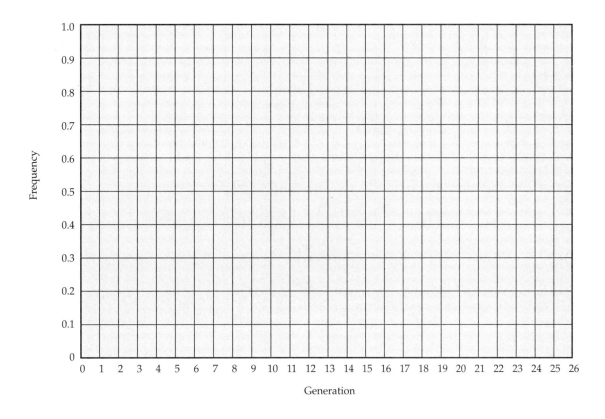

## LAB 29.B
## THE HARDY-WEINBERG EQUATION

### MATERIALS REQUIRED

calculator

### PROCEDURE

Let $p$ be the frequency of the $A$ allele, and let $q$ be the frequency of the $a$ allele. Given that a population satisfies the assumptions of the Hardy-Weinberg law, and given only the frequency of either the $A$ or $a$ allele, we can calculate the frequency of the other allele and the frequencies of the $AA$, $Aa$, and $aa$ genotypes.

To obtain the frequency of the other allele, given the frequency of the first allele, we need only subtract the frequency of the given first allele from 1, since the two allelic frequencies must add up to 1 ($p + q = 1$). For example, if we know that 0.6 is the frequency of the $A$ allele ($p$), then we can calculate the frequency of the $a$ allele ($q$) as $1.0 - 0.6 = 0.4$ ($1 - p = q$).

If we know the frequencies of the $A$ and $a$ allele, then we can calculate the frequencies of the $AA$, $Aa$, and $aa$ genotypes. The frequency of the $AA$ genotype will be the square of the frequency of the $A$ allele ($p^2$). The frequency of the $aa$ genotype will be the square of the $a$ allele ($q^2$). The frequency of the $Aa$ genotype will be twice the product of the $A$ allele and the $a$ allele ($2pq$). For example, suppose the frequency of the $A$ allele ($p$) is 0.4 and the frequency of the $a$ allele ($q$) is 0.6. The frequency of the $AA$ genotype will be $0.4 \times 0.4 = 0.16$ and the frequency of the $aa$ genotype will be $0.6 \times 0.6 = 0.36$.

There are two ways to calculate the frequency of the $Aa$ genotype. The first way is to calculate twice the product of the $A$ allele frequency and the $a$ allele frequency; that is, $2 \times 0.4 \times 0.6 = 0.48$.

The second way to calculate the frequency of the $Aa$ genotype is to use the relationship that the sum of the frequencies of the $AA$, $Aa$, and $aa$ genotypes is 1.0 or, $p^2 + 2pq + q^2 = 1$, which is the Hardy-Weinberg equation. Using that fact, for our example, the frequency of the $Aa$ genotype is $1.0 - (0.16 + 0.36) = 0.48$.

### POSTLAB QUESTIONS

Unless otherwise stated, all of these problems assume two alleles ($A$ and $a$) at an autosomal locus, a large, randomly mating population, no selection, no mutation, and no migration. For these problems, let $p$ be the frequency of the $A$ allele, let $q$ be the frequency of the $a$ allele, let $p^2$ be the frequency of the $AA$ genotype, let $2pq$ be the frequency of the $Aa$ genotype, and let $q^2$ be the frequency of the $aa$ genotype.

1. What would the frequency of the *A* allele be if 0.35 were the frequency of the *a* allele?

2. What would the frequency of the *Aa* genotype be if the frequencies of the *AA* and *aa* genotypes were 0.49 and 0.09, respectively?

3. What are the allelic and genotypic frequencies for the following population?

| Genotype | Number of Individuals |
|----------|----------------------|
| *AA* | 7200 |
| *Aa* | 9600 |
| *aa* | 3200 |

4. What are the genotypic frequencies if $p = 0.3$ and $q = 0.7$?

5. What are the allelic frequencies if 0.04, 0.32, and 0.64 are the respective frequencies of *AA*, *Aa*, and *aa*?

6. What are the allelic and genotypic frequencies if the frequency of the dominant phenotype (*AA* and *Aa*) is 0.64?

7. What are the allelic and genotypic frequencies if the frequency of the homozygous recessive genotype is 0.25?

8. What is the frequency of the carriers (*Aa*) in the population if 0.36 is the frequency of the *a* allele?

9. Is the population of Question 5 in Hardy-Weinberg equilibrium?

10. Is the following population in Hardy-Weinberg equilibrium? If not, what would be the allelic and genotypic frequencies when equilibrium is reached and how many generations would be required for the population to reach equilibrium?

| Genotype | Frequency |
|----------|-----------|
| *AA* | 0.75 |
| *Aa* | 0.10 |
| *aa* | 0.15 |

## LAB 29.C
## THE RELATIVE VALUE OF EUGENICS

### MATERIALS REQUIRED

microcomputer with BASICA, GWBASIC, or equivalent

computer program (See Appendix B: "pgs" Computer Program)

printer

### PROCEDURE

1. Run a simulation of 25 generations with 0.5 as the initial frequency of the *a* allele and a selection coefficient of 0.9 for the *aa* genotype and a fitness of 1.0 for the *AA* and *Aa* genotypes.
2. Plot the genotypic and allelic frequencies in the graph on page 327.

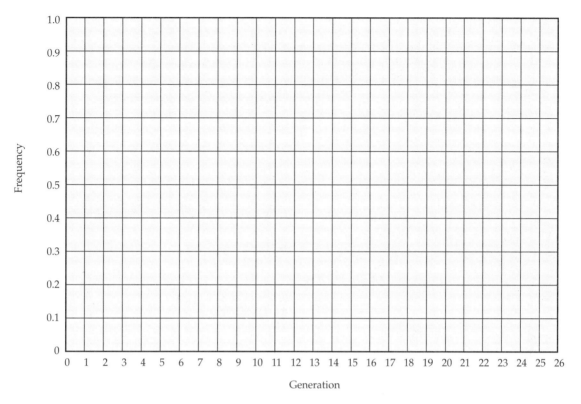

3. Run a simulation of 20 generations with 0.2 as the initial frequency of the *a* allele and a selection coefficient of 1.0 for the *aa* genotype and a fitness of 1.0 for the *AA* and *Aa* genotypes.

4. Plot the data of the simulation on the following graph.

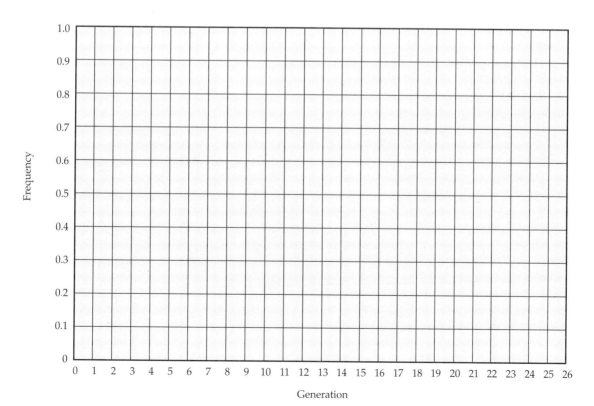

## POSTLAB QUESTIONS

1. What happened to the genotypic and allelic frequencies when there was almost complete selection against the *aa* genotype (selection coefficient equal to 0.9)?

2. What is eugenics?

3. Did the high rate of selection against the *aa* genotype in this laboratory exercise bring about the complete elimination of the *a* allele in this laboratory exercise?

4. What factor(s) acting in addition to complete selection against the *aa* genotype might offset the effect of such selection?

## LAB 29.D

### SELECTION AGAINST ALL THREE GENOTYPES, BUT MOST INTENSE SELECTION AGAINST THE HOMOZYGOUS RECESSIVE

### MATERIALS REQUIRED

microcomputer with BASICA, GWBASIC, or equivalent

computer program (See Appendix B: "pgs" Computer Program)

printer

### PROCEDURE

1. Run a simulation of 20 generations with 0.5 as the initial frequency of the *a* allele and a fitness of 0.6 for the *AA* and *Aa* genotypes and a fitness of 0.2 for the *aa* genotypes.
2. Plot the results of the simulation on the following graph.

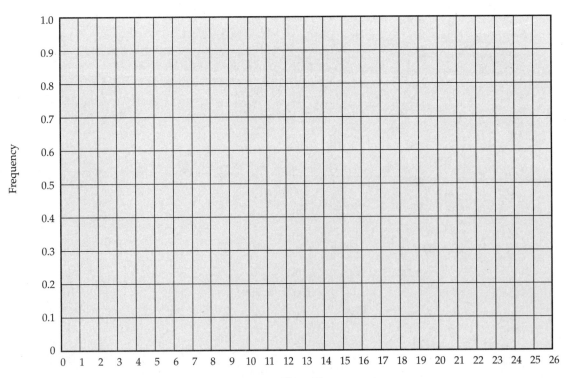

Frequency

Generation

## POSTLAB QUESTIONS

1. Overall, what is happening to the *A* allele?

2. What is happening to the *a* allele?

3. What will be the ultimate frequencies of those two alleles?

## LAB 29.E

## NO SELECTION AGAINST THE HETEROZYGOTE AND EQUAL SELECTION AGAINST HOMOZYGOUS GENOTYPES

### MATERIALS REQUIRED

microcomputer with BASICA, GWBASIC, or equivalent

computer program (See Appendix B: "pgs" Computer Program)

printer

### PROCEDURE

1. Run a simulation of 30 generations with 0.5 as the initial frequency of the *a* allele and with a fitness of the *Aa* genotype of 1.0. Both the *AA* genotype and *aa* genotype have a fitness of 0.3.

2. Plot the allelic and genotypic frequencies on the following graph.

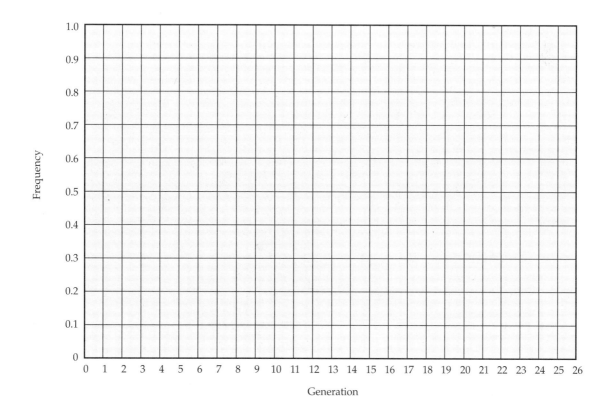

3. Repeat the simulation with the same coefficients but change the initial gene frequency to 0.2.

4. Plot the results on the following graph.

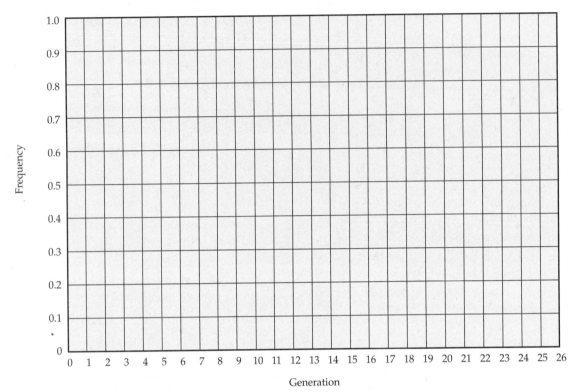

Generation

5. Repeat the last simulation again, with the same selection coefficients, but change the initial frequency of the *a* allele to 0.8.

6. Plot the results on the following graph.

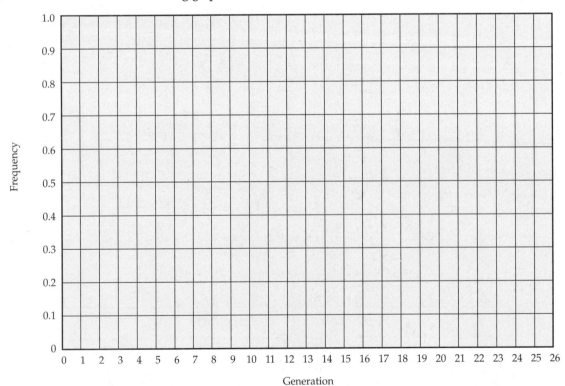

Generation

## POSTLAB QUESTIONS

1. What happened to the allelic and genotypic frequencies when 0.5 was the initial frequency of the *a* allele?

2. Compare the last three simulations and determine whether or not, ignoring generation times, the attainment of the equilibria frequencies reached by the allelic and genotypic frequencies are dependent upon initial frequency of the *a* allele.

## LAB 29.F

# SELECTION AGAINST ALL GENOTYPES BUT CIRCUMSTANCES FOR THE ASCENDANCY OF ONE ALLELE

## MATERIALS REQUIRED

microcomputer with BASICA, GWBASIC, or equivalent

computer program (See Appendix B: ''pgs'' Computer Program)

printer

## PROCEDURE

1. Perform 25 generations of the simulation with the initial frequency of the *a* allele being 0.5, and the selection coefficients of the *AA*, *Aa*, and *aa* genotypes being 0.7, 0.4, and 0.1, respectively. Can you predict the outcome?
2. Plot the results on the following graph.

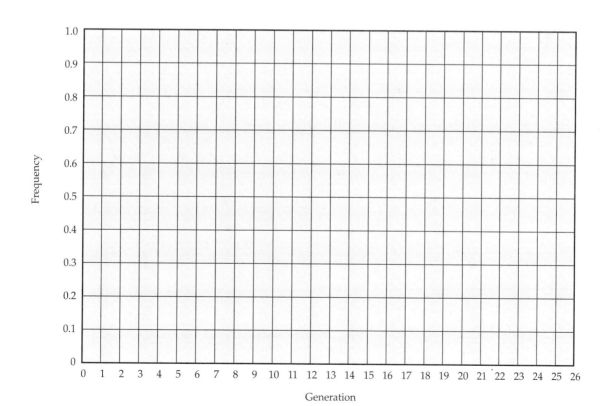

Generation

3. Repeat the last simulation again with 0.5 as the initial frequency of the *a* allele, but with the selection coefficients of the *AA*, *Aa*, and *aa* genotypes being 0.6, 0.7, and 0.9, respectively.

4. Plot the results of the simulation on the following graph.

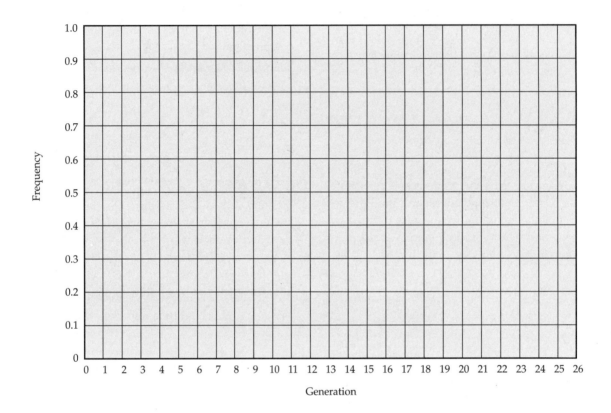

**POSTLAB QUESTION**

1. When an allelic frequency reaches 1.0, then that allele is said to be fixed in the population. How do you account for the fact that the *A* allele was fixed in this last simulation while the *a* allele was fixed in the first simulation of this last exercise?

# LAB 29.G
## HUMAN CHARACTERISTICS

### MATERIALS REQUIRED

strips (approximately 0.5 inch by 1.5 inch) of PTC paper

small hand mirror

calculator

**PROCEDURE**

There are a number of variations among humans that are based upon the inheritance of two alleles. Although many of these traits are characterized by variable expressivity (a range of degree of trait expression), for our purposes here, we shall treat the expression of these traits as an all-or-none phenomena. Furthermore, it should be stated that opinion is divided as to whether the traits we mention here are inherited in the simple fashions we have indicated. For simplicity we have assumed those simple modes of inheritance and we have used simple symbols for the alleles involved.

In this exercise, we will make a number of assumptions so that we may easily calculate the frequencies of the alleles involved based on their expression among the students of this course. For example, we shall regard the human population as a large, randomly mating population and the students in this class as being a random sample of that population.

Your instructor will announce which of the human traits listed below are to be studied in this exercise. For those traits to be studied, determine and record your phenotype and genotype. Share your genotypes with the rest of the students in the class so that frequencies of the alleles concerned can be calculated. If you show a dominant trait, for example, if you can taste PTC, record your phenotype as "taster" and record your genotype as *T* since you could either be homozygous or heterozygous for the dominant allele involved. If, however, you show the recessive trait, for example, if you cannot taste PTC, then record your phenotype as "nontaster" and record your genotype as *tt* since you can only be homozygous for the recessive allele.

These are the traits with which we shall be concerned:

1. *Ability to taste phenylthiocarbamide (PTC).* Phenylthiocarbamide (PTC) is a harmless chemical that, to some individuals, tastes bitter. Other individuals, however, cannot taste the compound at all. The ability to taste the chemical is inherited through a dominant allele (*T*). Individuals who can taste the chemical are of the genotype *TT* or *Tt*, while individuals who cannot taste the chemical are *tt*. There is considerable variation in the intensity of the ability of *TT* or *Tt* individuals to taste PTC. Place a piece of the PTC-impregnated filter paper on your tongue. If you cannot immediately taste the chemical, chew the paper. If you still cannot taste anything other than the paper itself, your genotype is *tt*. If you do taste the compound, your genotype is either *TT* or *Tt*.

   Your phenotype: _____

   Your genotype: _____

2. *Attached or free ear lobes.* The ear lobes of some people hang free, but some individuals have lobes that are completely attached to the side of the head (Fig. 29–1). The attached ear lobe condition is determined by homozygosity of a recessive allele; the genotype of such individuals may be designated as *ff*. Individuals with unattached or free ear lobes are of the genotype *FF* or *Ff*. Determine your phenotype for this trait and, if possible, your complete genotype.

   Your phenotype: _____

   Your genotype: _____

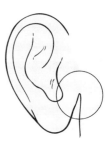

Attached ear lobes                    Free ear lobes

**FIGURE 29–1**

The ear lobe can hang free or be attached directly to the head.

3. *Presence or absence of hitchhiker's thumb.* Some individuals can bend the distal joint of the thumb back to nearly a 45° angle between the two joints of the thumb (Fig. 29–2). The ability to do this is produced by homozygosity for a recessive allele, *bb*. The condition is known as distal hyperextensibility of the thumb. Individuals of the genotype *BB* or *Bb* cannot bend their thumbs in this fashion. What is your phenotype and genotype for this trait?

   Your phenotype: _____

   Your genotype: _____

Hitchhiker's thumb                    Straight thumb

**FIGURE 29–2**

Hitchhiker's thumb is the ability to bend the distal joint of the thumb backwards at nearly a 45° angle.

4. *Presence or absence of cleft chin.* The actors Kirk Douglas and his son, Michael, are perhaps the best-known individuals with what is commonly called cleft chin. This trait is caused by a fissure in the bones that form the chin (Fig. 29–3). It is generally accepted that the condition is produced by a dominant allele, *C*. Individuals of the geno-

type *cc* do not show the trait, while *CC* and *Cc* individuals show cleft chins. Determine your phenotype and genotype for this trait. What are the phenotype and genotype of Michael Douglas, the son of Kirk Douglas?

Your phenotype: _____

Your genotype: _____

**FIGURE 29–3**

Cleft chin is caused by the fissure in the bones that form the chin.

5. *Presence or absence of bent little finger.* If you place your relaxed hands flat on the table and examine your little fingers, you will observe that they are either straight or bend slightly inwards at the last joint towards the fourth finger (Fig. 29–4). This bending of the little finger is apparently caused by a dominant allele. Individuals who are of the genotype *AA* or *Aa* show the trait, while individuals of the genotype *aa* do not show the trait. Determine your phenotype and genotype for this trait.

Your phenotype: _____

Your genotype: _____

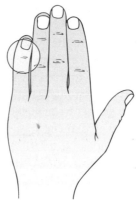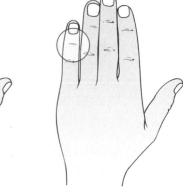

Bent little finger          Straight little finger

**FIGURE 29–4**

The third phalange of your little finger can be straight or bent toward your fourth finger.

6. *Presence or absence of widow's peak.* Certain individuals have a hairline that forms a definite downwards peak or point in the center of the forehead (Fig. 29–5). The condition is commonly known as widow's peak. Individuals who show this trait are of the genotype *WW* or *Ww*, the trait considered to be due to a dominant allele. Individuals who have a straight hair line are of the genotype *ww*. What is your phenotype and genotype for widow's peak?

Your phenotype: _____

Your genotype: _____

7. *Clasped hand configuration.* If you clasp your hands together and intertwine your fingers you will find that either your left thumb is on top of your right thumb or your right thumb is on top (Fig. 29–6). If you intentionally configure your clasped hands in the opposite arrangement, the configuration will feel slightly uncomfortable or unnatural. Individuals who naturally place the left thumb on top are thought to be of the genotype *LL* or *Ll*, the left thumb on top being inherited and due to a dominant allele. Individuals who interlock their hands with the right thumb on top are of the genotype *ll*. What is your phenotype and genotype for this finger interlocking trait?

Your phenotype: _____

Your genotype: _____

Widow's peak   Straight hairline

Right thumb over left  Left thumb over right

**FIGURE 29-6**

Either your left or right thumb will be on top when you clasp your hands and intertwine your fingers.

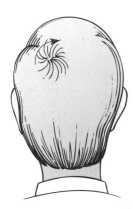

Clockwise   Counter-clockwise

**FIGURE 29-7**

Hair at the back of your head spirals in a clockwise or counterclockwise direction.

**FIGURE 29-5**

Widow's peak is a hairline that comes to a point in the center of the forehead.

8. *Configuration of hair whorl.* Some individuals have hair at the backs of their heads that spirals in a clockwise direction, while other individuals have hair spirals in a counterclockwise direction (Fig. 29-7). The clockwise configuration is dominant, and individuals who show that configuration are *DD* or *Dd*. Individuals who show the counterclockwise trait have the *dd* genotype. Have one of the other students in the class look at the back of your head to determine whether your hair shows the clockwise or counterclockwise whorl.

Your phenotype: _____

Your genotype: _____

9. *Presence or absence of ability to roll the tongue.* The ability to roll the tongue into a distinct U shape is apparently determined by a dominant gene (Fig. 29-8). Individuals who can do this are of either the *RR* or *Rr* genotype. Individuals who cannot roll the tongue are of the genotype *rr*. What is your genotype for this trait?

Your phenotype: _____

Your genotype: _____

10. *Presence of mid-digital hair.* Look carefully at the dorsal surface of your hands to see whether or not you have hair on the middle digits of your fingers

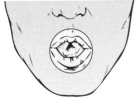

Tongue rolling                    Inability to roll tongue

**FIGURE 29-8**

Curling the tongue up at the sides is the ability to roll the tongue.

(Fig. 29–9). If you have mid-phalangeal hair, your genotype is either *HH* or *Hh* since the trait seems to be determined by a dominant gene. Individuals who do not show this trait are of the genotype *hh*. Determine your genotype for this trait.

Your phenotype: _____

Your genotype: _____

Calculate the frequencies of the dominant and recessive alleles of the traits studied, assuming that the assumptions of the Hardy-Weinberg law apply to the population concerned. For each trait, calculate the frequency of the recessive allele as the square root of the frequency of the recessive trait.

For example, suppose there are 35 individuals in the class. Of these, 15 (12 males and 3 females) do not show mid-digital hair. Of the 20 individuals who show mid-digital hair, 8 are males and 12 are females. These data have been recorded in the first line of Table 29–1. From the given data, the frequency of the re-

cessive trait is $15/35 = 0.43$. The frequency of the recessive allele is then the square root of 0.43 or 0.65, and the frequency of the dominant allele is $1.0 - 0.65$ or 0.35. One inference that might be drawn from the given data is that the frequency of the recessive trait is more common among males than females. That hypothesis can be tested; your instructor will show the class how that statistical test is applied to the data in question. The instructor will also determine, in class discussion, what additional hypotheses might be formulated and tested with the class data.

## POSTLAB QUESTIONS

1. What, if any, assumptions of the Hardy-Weinberg law were most likely not satisfied by the population sampled for the exercises of this laboratory?

2. Determine your genotype for each of the traits studied and, if possible, the corresponding genotypes of your siblings and parents.

3. Identify what, if any, might be the selective advantages or disadvantages of the traits studied.

4. Based upon the data obtained from this exercise, what would you expect to be the frequency of individuals showing both cleft chin and widow's peak in the general population?

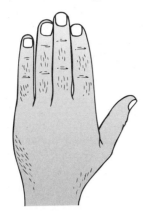

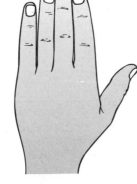

Mid-digital hair present          Mid-digital hair absent

**FIGURE 29-9**

The middle digit of each finger may or may not have hair.

**Table 29-1**

| Trait | Number showing Dominant Phenotype | | Number showing Recessive Phenotype | | Frequency of Recessive Allele | Frequency of Dominant Allele |
|---|---|---|---|---|---|---|
| | **Males** | **Females** | **Males** | **Females** | | |
| | **Total** | | **Total** | | | |
| Example | 8 | 12 | 12 | 3 | | |
| | 20 | | 15 | | 0.65 | 0.35 |
| 1. | | | | | | |
| 2. | | | | | | |
| 3. | | | | | | |
| 4. | | | | | | |
| 5. | | | | | | |
| 6. | | | | | | |
| 7. | | | | | | |
| 8. | | | | | | |
| 9. | | | | | | |
| 10. | | | | | | |

5. Is it to be expected that recessive alleles will be present in low frequencies in the overall population?

6. In addition to the sex of the individual, what other easily determined trait(s) might be considered to examine the possibility that the traits are associated?

7. How do you account for the fact that, in humans, the frequencies of sex-linked recessive traits are commonly much higher in males than in females?

## FOR FURTHER READING

Ayala, F. J., and J. W. Valentine. 1979. *Evolving.* Menlo Park, Calif.: Benjamin-Cummings Publishing Co.

Futuyma, D. 1987. *Evolutionary Biology.* Sunderland, Mass.: Sinauer.

Gould, S. J. 1977. *Ever Since Darwin.* New York: W. W. Norton.

Harvey, Paul H., and Mark D. Pagel. 1991. *The Comparative Method in Evolutionary Biology.* New York: Oxford University Press.

Langridge, John. 1991. *Molecular Genetics and Comparative Evolution.* New York: Wiley.

Lewin, R. 1982. *Thread of Life.* Washington, D. C.: Smithsonian Books.

Minkoff, E. C. 1983. *Evolutionary Biology.* Reading, Mass.: Addison-Wesley Publishing Co.

Nitecki, Matthew H., and Doris V. Nitecki, eds. 1990. *Evolutionary Innovations.* Chicago: University of Chicago Press.

Rose, Michael R. 1991. *Evolutionary Biology of Aging.* New York: Oxford University Press.

Strickberger, M. W. 1985. *Genetics.* New York: Macmillan.

Smartt, J. 1990. *Grain Legumes: Evolution and Genetic Resources.* New York: Cambridge University Press.

Takahata, Naoyuki, and James F. Crow, eds. 1990. *Population Biology of Genes and Molecules.* Tokyo, Japan: Baifukan.

Tuljapukar, Shripad. 1990. *Population Dynamics in Variable Environments.* New York: Springer.

White, M. J. D. 1978. *Modes of Speciation.* San Francisco: Freeman.

◻

# Behavior

## OBJECTIVES

1. Differentiate between innate and learned behavior and give two examples of each.

2. Define taxis and give examples of specific kinds of taxes.

3. Determine photo- and geotaxis in two invertebrate species.

4. Define reflex and identify two reflex responses in two invertebrate species.

5. Differentiate between trial-and-error and insight learning in human subjects.

---

## INTRODUCTION

Animal behavior is very diverse. Some behaviors are **innate** (inborn or hereditary), being genetically programmed into the nervous system. Examples are a perfect nest built by a bird at the very first try and unvarying courtship rituals among all males of a species. Other behavior is learned by the organism as it encounters various experiences in its environment. In actual fact, many ethologists (scientists who study animal behavior) hold that behaviors are affected by genetics and experience.

**Stereotyped behaviors** are acts performed in an identical pattern involving many muscles in a precisely timed sequence. Stereotyped behaviors are typically innate and involve few neurons in the central nervous system. Locomotion, sound production, and reflexes are examples of stereotyped behaviors. A **reflex** is an automatic, involuntary response to a given stimulus that is determined by the anatomical relationship of the neurons involved.

Learning occurs as a result of an individual organism's experiences. Learned behavior is classified into several categories based on different criteria. Two of these criteria—trial-and-error learning and insight learning—are presented in this section. **Trial-and-error learning** occurs when spontaneous activity produces a reward by chance. The organism then learns to repeat the same action in order to obtain the reward. **Insight learning** is quite the opposite; it requires reasoning that uses past experience to solve a new problem.

Animal response to stimuli in the environment is a major part of animal behavior. The orientation of an organism toward or away from a stimulus is called a **taxis** (plural, taxes [pronounced "tax-eez"]). A specific kind of taxis is named according to the type of stimulus that elicits the response. For example, in this section you will expose invertebrates to light as a stimulus. Therefore, you will determine if the organisms move toward light (positive phototaxis [photo = light]) or away from light (negative phototaxis).

## PRELAB QUESTIONS

1. What type of behavior is determined by your genetic makeup?

2. What type of stereotyped behavior is an automatic rapid response to a stimulus?

3. What type of response does an animal exhibit when the animal's body is oriented toward a light stimulus?

## LAB 30.A
## PHOTOTAXIS IN A PLANARIAN

### MATERIALS REQUIRED (per pair of students)

test tube

stopper

gooseneck lamp

wax pencil

plastic dropper, with enlarged opening

live planarian

spring or aquarium water

white sheet of paper

small metric ruler

paper towels

piece of aluminum foil

### PROCEDURE

Planarians are aquatic free-living flatworms (phylum Platyhelminthes) commonly found in slow streams and ponds. You will use a planarian as a subject to determine if this species exhibits positive or negative phototaxis in response to the stimulus of light.

A special note about handling planarians: planarians tend to adhere to the surface of the culture jar when you try to remove them from the container. To avoid injuring the worm during handling, first dislodge it with a burst of water from a dropper, then draw it into the dropper while it is free-floating in the water. Quickly transfer the planarian to a test tube before it has a chance to adhere to the interior of the dropper. If the worm does stick to the inside of the dropper, carefully insert a needle probe, nudge the worm loose (don't use the point!), then quickly transfer it to a test tube (See Fig. 30–1).

1. Use a ruler and wax pencil to draw a ring around the midpoint of a test tube.
2. Use a plastic dropper to transfer a planarian from the culture jar to the marked test tube. Completely fill the test tube with spring or aquarium water and insert a stopper, avoiding any bubbles. Wipe off any water overflow with a paper towel.
3. Mold the aluminum foil around the bottom half (opposite the stopper) of the tube to form a solid sheath that blocks light. Remove the foil sheath from the tube after it is formed.
4. Place the test tube horizontally on a white sheet of

paper. If the opening of the tube is lipped, it may be necessary to level the tube using stacked microscope slides, paper, or anything else that is handy. Place the lamp (light off) above the test tube, making sure it is far enough away to avoid warming the water when turned on.

5. Wait a few moments for the planarian to reach the center line while headed toward the stopper end of the tube.
6. Place the foil sheath over the test tube and turn on the lamp. Record the beginning time in Table 30–1. Watch the movement of the worm for 10 minutes. Record in Table 30–1 the total time the worm stays in the lighted and dark halves of the test tube. *Note:* After completing this exercise, retain the planarian in the test tube for Exercise 30.B, which follows.

**Table 30–1**  Phototaxis in a Planarian

Beginning time _____.    Total time in lighted tube half _____.
Total time in darkened tube half _____.

### POSTLAB QUESTIONS

1. Did the planarian exhibit a positive or negative phototaxis?

2. By observing the external features of the planarian, how do you think it perceives light?

## LAB 30.B
## GEOTAXIS IN A PLANARIAN

### MATERIALS REQUIRED (per pair of students)

live planarian in test tube from previous exercise

test tube rack

large brown paper bag

recovery container for "used" planarians

### PROCEDURE

In this exercise, you will use the same planarian in the test tube to test for geotaxis, the movement of the worm in response to the stimulus of gravity. Your observations will determine if the planarian exhibits

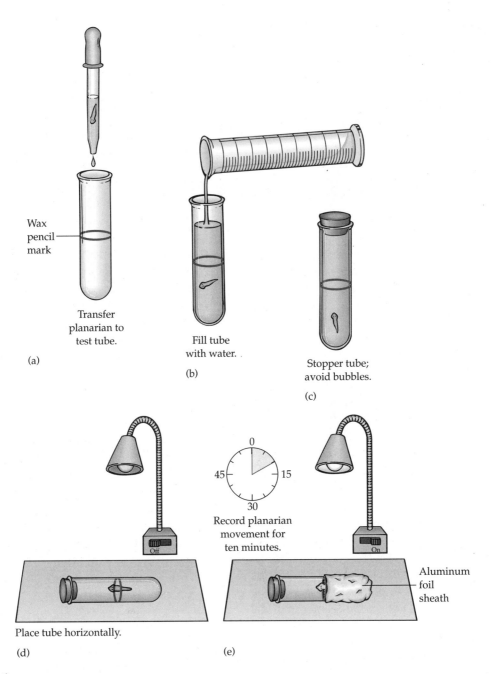

Wax
pencil
mark

Transfer
planarian to
test tube.

(a)

Fill tube
with water.

(b)

Stopper tube;
avoid bubbles.

(c)

0

45          15

30

Record planarian
movement for
ten minutes.

Off

On

Aluminum
foil
sheath

Place tube horizontally.

(d)                    (e)

**FIGURE 30–1**

Procedure to determine if planarians exhibit positive or negative phototaxis in response to the
stimulus of light.

negative geotaxis (movement away from gravity), or positive geotaxis (movement toward gravity) (See Fig. 30–2).

1. Open the paper bag and lay it on its widest side. Place the test tube rack deep inside the bag, where lighting is dim.
2. Hold the test tube horizontally until the planarian is positioned at the midpoint line. Immediately place the test tube in the rack with stopper end up. Be sure to place the tube in the rack where the planarian is easily visible by looking into the bag. Record the time in Table 30–2.
3. Observe the movements of the worm for 10 minutes. Record in Table 30–2 the total time the worm stays in the top half and bottom half of the test

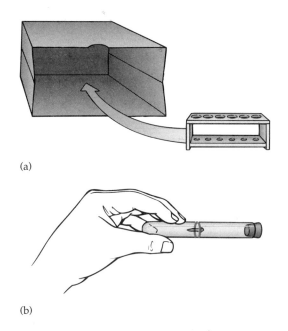

(a)

(b)

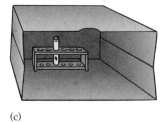

Observe planarian
movement for
ten minutes

(c)

**FIGURE 30-2**

Procedure to determine if planarians exhibit positive or negative geotaxis in response to the stimulus of gravity.

tube. *Note:* After completing the exercise, carefully transfer the planarian to a recovery container and return all materials to their original location.

**Table 30-2** Geotaxis in a Planarian

Beginning time ——. Total time in top half of tube ——.
Total time in bottom half of tube ——.

**POSTLAB QUESTIONS**

1. Did the planarian exhibit positive or negative geotaxis?

2. Design an experiment to test for an organism's taxis response to a stimulus.

## LAB 30.C
## PHOTOTAXIS IN FRUIT FLIES

**MATERIALS REQUIRED (per pair of students)**

T-shaped, stoppered tube containing 30 fruit flies

piece of aluminum foil

gooseneck lamp

**PROCEDURE**

In the following two exercises, you will test an organism that differs greatly in form and complexity from the planarian used previously for phototaxis and geotaxis. The common fruit fly (phylum Arthropoda, class Insecta) is a subject of many genetic studies. Here we use the fruit fly as another kind of invertebrate to observe its responses to light and gravity. *Note:* Be careful not to release any flies from the stoppered containers used in Exercises 30.C–D. Loose fruit flies in a laboratory are annoying (See Fig. 30–3).

1. Obtain a piece of aluminum foil and mold it over one of the arms of the T tube to form a solid sheath that completely blocks light. Make sure that the foil extends over the entire tube arm up to the stem of the T tube.

2. Move over to the lamp and turn it on so that it is ready. Keeping the T tube horizontal at all times, tap the tube with your finger to position all flies into the main stem of the tube. Quickly grasp the main stem to cover it with your hand. Hold the arms of the T tube horizontally in the bright light of the lamp.

3. After a few moments, count the number of flies (as close as possible) that move into the lighted and darkened arms of the T tube. Disregard any flies that remain in the main stem. Record your results for Trial 1 in Table 30–3.

4. Repeat the exercise and record results for Trial 2 in Table 30–3.

**Table 30-3** Phototaxis in Fruit Flies

Trial 1  Number of flies in light arm ——;
darkened arm ——.
Trial 2  Number of flies in light arm ——;
darkened arm ——.

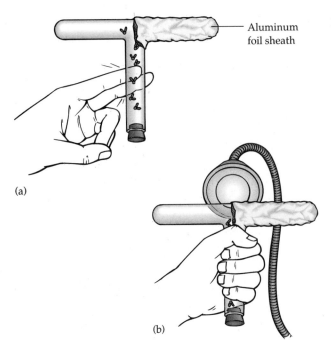

Aluminum foil sheath

(a)

(b)

**FIGURE 30-3**

Procedure to determine if fruit flies exhibit positive or negative phototaxis in response to the stimulus of light.

## POSTLAB QUESTIONS

1. Did the fruit fly exhibit a positive or negative phototaxis?

2. How does the fruit fly's response to light compare to a planarian's response? Explain the advantages of their responses.

3. Hold the vial upright and gently tap the vial with your finger to bring the flies to the lower end of the tube; quickly set it deep into the interior of the bag where lighting is dim.
4. After a few moments, count the number of flies (as close as possible) in the upper and lower half of the vial. Record the results for Trial 1 in Table 30-4.
5. Repeat the exercise a second time and record results for Trial 2 in Table 30-4.

## LAB 30.D
## GEOTAXIS IN FRUIT FLIES

### MATERIALS REQUIRED (per pair of students)

stoppered vial of 30 fruit flies

wax pencil

small metric ruler

large brown paper bag

### PROCEDURE

1. Use a ruler and wax pencil to draw a ring around the midpoint of the vial of fruit flies.
2. Open the paper bag and lay it on its widest side to have it ready nearby.

**Table 30-4**  Geotaxis in Fruit Flies

Trial 1  Number of flies in upper end of vial _____.
Trial 2  Number of flies in lower end of vial _____.

### POSTLAB QUESTIONS

1. Did the fruit fly exhibit negative or positive geotaxis?

2. How does the fruit fly's response to gravity compare to a planarian's response? Explain the advantages of their responses.

## LAB 30.E
## REFLEX BEHAVIOR: FEEDING RESPONSE IN A PLANARIAN

### MATERIALS REQUIRED (per pair of students)

live planarian (not used in previous exercises)

small watch glass

small sheet of white paper, if needed

aquarium or spring water

plastic dropper with enlarged opening

needle probe

stereomicroscope

fresh liver chopped in tiny bits

### PROCEDURE

Feeding behavior in planarian worms is apparently an automatic reflex response to the presence of food. When withdrawn, the tubular pharynx of the planarian is held within the pharyngeal chamber that leads to the mouth located on the midventral surface of the worm. If the stimulus of food is detected in the environment, the pharynx extends from the mouth and food is ingested (See Fig. 30–4).

1. Use a dropper to place a planarian in a watch glass filled with aquarium or spring water.
2. Place the watch glass on the stage of a stereomicroscope. If the stage of the microscope is not light in color, place a small sheet of white paper on the stage to help you observe the worm.
3. Allow the worm to become acclimated to its new container for five minutes. Observe the worm under low magnification during this time. Avoid moving or bumping the watch glass as much as possible.
4. Use a needle probe to place a few bits of fresh liver in the center of the watch glass. Watch the behavior of the worm for the next 15 minutes, noting any extension of the tubular pharynx in response to the liver.

### POSTLAB QUESTIONS

1. Describe the feeding behavior of planarians.

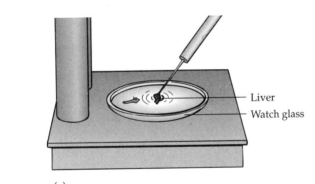

(a)

**FIGURE 30–4**

Procedure to demonstrate feeding behavior in planarians.

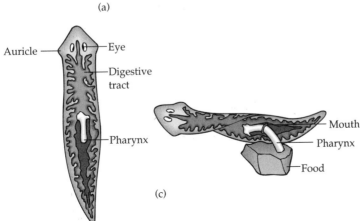

(b)

(c)

2. How successful is this method of feeding?

3. Design an experiment to observe the feeding behavior of fruit flies.

## LAB 30.F
### REFLEX BEHAVIOR: RESPONSE TO TOUCH IN A DECAPITATED COCKROACH

### MATERIALS REQUIRED (on demonstration)

decapitated cockroach mounted on cardboard support

small artist's brush

### PROCEDURE

The brain is typically not involved in many kinds of reflexes, which is well demonstrated in an organism with no head! The lab-reared cockroach on display was anesthetized, the head and wings were removed, then the resulting openings were covered with a sealant to prevent loss of body fluids. The hardy cockroach can live for an extended time in this state, perhaps two days or longer!

Use the small brush to lightly touch the abdomen, legs, and dorsal thorax of the roach. Note the kind of response when the roach is stimulated by touch. Wait for each response to cease before stimulating the roach again.

### POSTLAB QUESTIONS

1. Describe the cockroach's response to touch.

2. In reference to this response, why would your results vary somewhat from a fellow student?

## LAB 30.G
### LEARNED BEHAVIOR: TRIAL AND ERROR LEARNING

### MATERIALS REQUIRED (per pair of students)

8 jigsaw puzzle pieces, painted one color, in small bag

timepiece

### PROCEDURE

In Exercises 30.G–H, you will observe an organism very different from the invertebrates of previous exercises. You will observe a human! Learning is an important component of much behavior in humans and other vertebrates.

1. Determine which of you will be the subject and which will be the timekeeper. At the signal from the timekeeper, the subject spills the puzzle pieces from the bag and properly interlocks all puzzle pieces. Record the time required for assembly in the Trial 1 space of Table 30–5.
2. Disassemble the puzzle pieces and place them in the bag.
3. Repeat Steps 1 and 2 for a total of 3 trials, testing the same subject each trial. Record the time required for assembly in each trial in Table 30–5.

**Table 30–5**   Trial-and-Error Learning

| Time required for puzzle assembly: | Trial 1 _____ |
| | Trial 2 _____ |
| | Trial 3 _____ |

### POSTLAB QUESTIONS

1. What did you conclude from this experiment?

2. Design another experiment to demonstrate trial-and-error learning.

## LAB 30.H
## LEARNED BEHAVIOR: INSIGHT LEARNING

### MATERIALS REQUIRED (for class demonstration)

1 textbook of any kind

1 student subject (by volunteer or instructor's choice)

### PROCEDURE

Insight learning seems to be limited to humans and other primates. All of the technological achievements we enjoy as part of modern life are a result of this type of learned behavior.

Observe the subject faced with the following problem—how to place the textbook in a designated drawer on the opposite side of the room without moving from her/his seat. Record the subject's solution to the problem.

### POSTLAB QUESTIONS

1. Give examples of how communication changes the behavior of organisms. Do not use humans as examples.

2. How does trial-and-error learning compare to insight learning?

3. Why do planarians exhibit primarily innate behavior, whereas humans exhibit primarily learned behavior?

4. What type of behavior is exhibited when a nipple is placed in a newborn's mouth and the baby immediately begins to suck?

5. What type of behavior is exhibited when a toddler drinks from a cup?

6. How do innate and learned behaviors contribute to the survival of an organism?

### FOR FURTHER READING

Alcock, J. 1989. *Animal Behavior, an Evolutionary Approach.* Sunderland, Mass. Sinauer.

Bailey, Winston J. 1991. *Acoustic Behaviour of Insects: An Evolutionary Perspective.* London: Chapman and Hall.

Bonner, J. T. 1980. *The Evolution of Culture in Animals.* Princeton: Princeton University Press.

Brooks, Daniel R., and Deborah A. McLennan. 1991. *Phylogeny, Ecology, and Behavior.* Chicago: University of Chicago Press.

Evans, David L., and Justin O. Schmidt, eds. 1990. *Insect Defenses: Adaptive Mechanisms and Strategies of Prey and Predators.* Albany: State University of New York Press.

Gormley, Gerard. 1990. *Orcas of the Gulf.* San Francisco: Sierra Club Books.

Gould, J. L. 1982. *Ethology: The Mechanisms and Evolution of Behavior.* New York: Norton.

Grier, J. W. 1984. *Biology of Animal Behavior.* St. Louis, Mo.: Times Mirror/Mosby.

Kramer, Bernd. 1990. *Electrocommunication in Teleost Fishes.* New York: Springer-Verlag.

Krebs, J., and N. Davies. 1984. *Behavioral Ecology: An Evolutionary Approach.* Sunderland, Mass.: Sinauer.

Marler, P., and W. J. Hamilton. 1966. *Mechanisms of Animal Behavior.* New York: Wiley.

Oxnard, Charles E. et al. 1990. *Animal Lifestyles and Anatomies: The Case of the Prosimian Primates.* Seattle: University of Washington Press.

Ross, Kenneth G., and Robert W. Matthews, eds. 1991. *The Social Biology of Wasps.* Ithaca, N.Y.: Comstock Pub. Associates.

Trivers, R. 1985. *Social Evolution.* Menlo Park, Calif.: Benjamin-Cummings Publishing Co.

Vander Wall, Stephen B. 1990. *Food Hording in Animals.* Chicago: University of Chicago Press.

Von Frisch, K. 1974. *Animal Architecture.* New York: Harcourt Brace Jovanovich.

Williams, G. C. 1966. *Adaptation and Natural Selection.* Princeton, N.J.: Princeton University Press.

Wilson, E. O. 1975. *Sociobiology: The New Synthesis.* Cambridge, Mass.: Harvard University Press.

Wilson, M. F. 1984. *Vertebrate Natural History.* Philadelphia: Saunders College Publishing.

Wysocki, Charles J., and Morley R. Kare, eds. 1991. *Genetics of Perception and Communications.* New York: Dekker.

❏

# Population and Ecosystem Dynamics

1. Describe the effects of biotic potential and environmental resistance in keeping a population in balance.

2. Describe the effects of competition and predation on population size.

3. Define ecosystem, producer, consumer, food chain, food web, trophic level, and pyramid of energy.

4. Identify important nonbiotic factors and biotic factors in an ecosystem and explain how these factors affect the community.

5. Construct a pyramid of energy, food chain, and food web for an aquatic ecosystem, given the names and relative numbers of familiar organisms in the ecosystem. Identify levels of greatest and least biomass.

6. Summarize the energy relationships that are represented by a food web and a pyramid of energy.

---

## INTRODUCTION

For our purposes, a **population** is all the members of a given species occupying a given area at the same time. A population's **biotic potential** is its fastest possible growth rate under ideal conditions such as unlimited food supply. A population in nature seldom reaches its biotic potential because environmental conditions usually prevent such population growth.

The many factors that limit population size are collectively termed **environmental resistance.** Factors such as the availability of habitat, water, food, light, and oxygen are **nonbiotic** (nonliving) factors. Predators, competitors, and parasites are examples of **biotic** (living) factors of an environment. Ultimately, environmental resistance determines the maximum number of individuals of a species that can be supported indefinitely by the resources available. The importance of any one factor may change with time as conditions in the environment change.

**Competition** occurs when two or more organisms

use the same limited resource. Competition acts as a control of population size. When a species moves into an area of another species with a similar **niche** or function, one of them usually competes more strongly for resources and may cause extinction of the other species in that area. **Predation** is when one organism, the predator, kills and eats another organism, the prey. No matter what the environmental resistance is for the population, if the environmental resistance is greater than the biotic potential, then the population size will decrease. If the biotic potential is greater than the environmental resistance, then the population size will increase.

An **ecosystem** is the physical environment of a specified area and the **community** of all the different organisms living in that area. A **biome** is a large, distinct community of organisms determined by the interactions of climate, physical and biotic factors. There are a number of biomes that can be explored.

The sun is the ultimate energy source of nearly all ecosystems. **Producers** (green plants) are essential to an ecosystem because they have the ability to make their own food from inorganic substances by using light energy from the sun. By photosynthesis, food (and the energy available from it) becomes available for the **consumers,** which derive their food directly or indirectly from plants. **Decomposers,** also consumers, have the important function of breaking down dead plant and animal bodies, a process that releases nutrients back into the ecosystem. The released nutrients are then recycled in the ecosystem.

Energy in an ecosystem differs from the nutrients in an ecosystem in that energy flows one way, whereas nutrients are recycled. Except for the producers, the remaining organisms within an ecosystem obtain energy by consuming other organisms. There is thus formed a **food chain,** in which all organisms in the chain are each eaten by at least one other type of organism. Food chains can be interlinked in complex ways, the result being a **food web.**

Organisms of a food chain can be classified in terms of **trophic levels,** a classification that gives an indication of how far an organism is removed from green plants in obtaining its energy. Producers are the first trophic level, **primary consumers** which feed on producers are the second level, **secondary consumers** that feed on primary consumers are next, **tertiary consumers** feed on the secondary consumers, and the sequence continues up to higher trophic levels. Normally, organisms within even the most complex food chain can be classified into five or fewer trophic levels.

Trophic levels are limited to five steps because of energy loss that occurs when one organism consumes another. This explains why energy flow is one way in an ecosystem. This energy transfer can be represented graphically by a **pyramid of energy,** which shows the total amount of incoming energy at each trophic level. Energy is lost as heat from metabolism in the consumer's body. As a result, there is less food energy in a particular trophic level than in the one before it.

**Productivity** is the rate of new organic matter formation in organisms. Because energy is lost at each transfer, productivity decreases with each higher trophic level. Animals convert only about 10% of the energy they eat into new tissue by growth and reproduction. Therefore, lower trophic levels usually have a greater biomass than higher trophic levels. **Biomass** is simply the mass of all materials in organisms.

The laboratory exercises in this section have been selected to demonstrate some of the more important biological principles pertaining to biological populations and ecosystems.

## PRELAB QUESTIONS

1. What is a population's biotic potential?

2. What type of environmental factor is predation?

3. What occurs when one species moves into an area of another species with a similar niche?

4. What type of organism makes its own food from inorganic substances by using light energy from the sun?

5. What classification indicates how far an organism is removed from green plants in obtaining its energy?

6. Why is energy flow one way in an ecosystem?

7. Which trophic level will have the greatest biomass?

8. What is a pyramid of energy?

## LAB 31.A
## POPULATION DYNAMICS: PREDATOR–PREY

### MATERIALS REQUIRED ON DISPLAY

4 culture bowls individually labelled as A, B, C, and D

4 stereomicroscopes

### PROCEDURE

Two species of unicellular protozoa, *Paramecium* and *Didinium,* provide an example of predator and prey relationships in the microscopic world. *Paramecium* (Fig. 31–1) has an elongated oval shape and swims by numerous cilia that cover the surface of the cell. *Didinium* (Fig. 31–2) is more rounded in shape and swims by two bands of cilia that encircle the cell. One of these organisms is a predator of the other. In unfavorable conditions, such as the absence of prey, the predator becomes inactive and forms a dormant cyst.

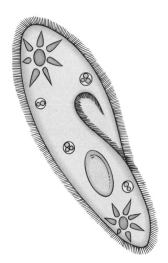

**FIGURE 31–1**

*Paramecium.*

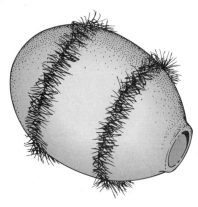

**FIGURE 31–2**

*Didinium.*

The predator becomes the active feeding form again when prey becomes available. In this exercise, you will examine mixed cultures of these two protozoa and determine which species is the predator and prey based on your observations.

Four cultures are set up on the lab bench for you to observe under the stereomicroscope. Culture A is a mixed population of approximately equal numbers of the protozoa *Paramecium* and *Didinium*. This mixed culture was prepared less than 48 hours ago. Culture B is a mixed population of *Paramecium* and *Didinium* that was prepared several days ago. Equal numbers of both kinds of protozoa were originally present in this culture. Culture C is a very old mixed culture. It was prepared over a week ago by adding equal numbers of

*Paramecium* and *Didinium*. Culture D is also a very old mixed culture of *Paramecium* and *Didinium*. In addition to equal numbers of the protozoa, fibers of cheesecloth were also added when the culture was prepared.

1. Carefully examine each of the four cultures under the highest power of the stereomicroscopes. Begin with culture A and continue through culture D. Although you will not see many details of the protozoa with the stereomicroscope, you can distinguish between *Paramecium* and *Didinium* by their general shape—*Didinium* is round, *Paramecium* is elongated. Estimate the relative numbers of *Paramecium* and *Didinium* in each culture. Look for four possible situations in the cultures: both species are present in approximately equal numbers, *Didinium* is more numerous, *Paramecium* is more numerous, or only cysts are present.

2. Record your observations for each culture, then present your findings on the appropriate line below.

Culture A _____

_____

Culture B _____

_____

Culture C _____

_____

Culture D _____

_____

**POSTLAB QUESTIONS**

1. Which organism is the predator?

2. Which organism is the prey?

3. How do you account for the relatively smaller number of *Paramecium* in Culture B?

4. What became of the *Paramecium* that were originally added to Culture C?

5. What was probably the single most important

factor responsible for the form of *Didinium* found in Culture C?

6. What would you predict would happen if a fresh supply of *Paramecium* were added to Culture C?

7. How do you explain the absence of the cyst form of *Didinium* in Culture D?

8. How do you account for the age of Culture D and the approximately equal numbers of the two protozoa in the culture?

9. Which culture(s) would you expect to remain essentially unchanged for the longest period of time?

10. What would you predict would happen to Culture D if the fibers were removed from the culture?

## LAB 31.B
## ECOSYSTEM DYNAMICS: ENERGETICS OF AN AQUATIC ECOSYSTEM

### MATERIALS REQUIRED

3–4 aquaria with signs identifying the contained freshwater organisms, plus the following, per group:

a. 1 thermometer

b. 1 meter stick

c. 1 light meter

d. 1 box of slides and coverslips

e. 1 dropper

f. 1 compound microscope

### PROCEDURE

For the purposes of this laboratory exercise, an aquarium has been set up for you that simulates a freshwater aquatic ecosystem. The plants and animals you see in the aquarium are the same or similar to those that you would see in a pond or small stream in your area. In observing this aquarium, you will study the energy relationships within this miniature ecosystem. Your instructor will divide the class into smaller groups that will examine one of the aquaria as a team.

### Nonbiotic Factors of the Ecosystem

1. An ecologist may begin a study of an ecosystem by examining the physical aspects of the environment. Begin your investigation by recording the water temperature in degrees Celsius. _____

2. Describe the physical objects lying on the bottom or floating in the aquarium. _____ _____ _____

3. Determine the dimensions of the ecosystem by measuring in centimeters and then figuring the volume of the aquarium in cubic centimeters.
   Width = ___cm   Height = ___cm
   Length = ___cm   Volume = ___cm$^3$
   $(h \times w \times 1 = \text{vol.})$

4. Describe the movement, aeration, and clarity of the water. _____ _____ _____ _____

5. Measure the light intensity falling upon the surface of the water. Use a light meter that measures in units called footcandles. A footcandle is the amount of light falling on one square foot from one standard candle at a distance of one foot. Be careful not to drop the light meter in the aquarium! Surface light in footcandles _____ .

### Biotic Factors of the Ecosystem

1. In addition to exploring qualities of the physical environment, an ecologist will also study the organisms present in the ecosystem. Begin by first observing the plants and animals visible with the unaided eye. Near the aquarium, there is a list of the common names and a brief description or illustration that identifies the larger aquatic organisms. Locate each species in the ecosystem and notice its relative abundance. List each species in Table 31–1, describe its appearance, and indicate its relative numbers with words such as abundant, many, some, few, or rare. If you can easily count the

number of a particular species, record that number instead.

**Table 31-1**

| Name | Description | Relative Number |
|------|-------------|-----------------|
|      |             |                 |

Of course, many of the organisms in the ecosystem cannot be seen with the unaided eye. Although they are small, the microscopic plants and animals of any ecosystem are of major importance. You are likely to observe four types of microorganisms in your aquarium: (a) phytoplankton (single- or multicelled microscopic plants usually with green pigment, generally do not move about); (b) zooplankton (single- or multicelled organisms lacking pigment, generally move about); (c) worms (multicellular, worm-like organisms); and (d) arthropods (insects or crustaceans that have noticeable eyes, legs, and scurry about). A fifth group of organisms that usually make up an ecosystem is decomposers. These organisms are mostly bacteria and fungi, and they feed on decaying organisms. Decomposers consume most of the energy remaining in decaying organisms and help return mineral components to the ecosystem. Are there any decomposers in this aquatic ecosystem? _____

2. Using a dropper, take a one-drop sample from near the surface of the ecosystem. Place the drop on a slide, add a coverslip, and examine the slide under the microscope. You do not have to identify the specific name of any organism. Instead, determine which of the above four groups it belongs to. List the organisms in Table 31–2. Assign a letter and subnumber to each different organism you observe. For example, assign $P_1$ to the first single-celled green organism you observe from the phy-

toplankton group. A new and different green microbe would be assigned $P_2$. The first organism from the zooplankton group would be $Z_1$, and so on. Repeat the sampling procedure at the middle of the aquarium and again on the bottom. Avoid drawing up excessive sand or debris with the dropper when sampling the bottom.

**Table 31-2**

| Level | Code | Description | Relative Number |
|-------|------|-------------|-----------------|
| S U R F A C E |  |  |  |
| M I D D E P T H |  |  |  |
| B O T T O M |  |  |  |

3. Figure 31–3 is a model of some common energy relationships in an ecosystem. This is frequently called a pyramid of energy and is divided into three trophic levels as indicated in the figure. Place the code name ($P_1$, $Z_1$, and so on) of each organism from Tables 31–1 and 31–2 into the box of Figure 31–3 which best represents the trophic level of that organism. Include with the name the word or number you used to describe its relative numbers in the ecosystem.

Now that you have collected information from the nonbiotic and biotic aspects of your ecosystem, you are ready to interpret your data by answering the following questions.

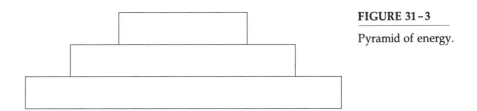

**FIGURE 31–3**
Pyramid of energy.

## POSTLAB QUESTIONS

1. What are the nonbiotic factors of the aquarium ecosystem on which all organisms in the ecosystem are very dependent?

2. Why would an ecologist collect data on the nonliving factors in a studied environment?

3. Which of the organisms from the groups of microorganisms would be the ones that photosynthesize?

4. What can you likely assume about the ecological role of the organisms that photosynthesize?

5. Why would biologists use a pyramid to illustrate biomass and energy relationships in an ecosystem?

6. A pyramid of energy illustrates energy utilization and loss. You know that when one organism consumes another, only part of the energy in the consumed organism is stored in the body of the consumer. What happens to the remainder of the energy in the consumed organism?

7. In Figure 31–3, the sizes of the boxes indicate the relative biomass of all organisms at each level. Examine the relative numbers you entered for each trophic level in Figure 31–3, and assume that your numbers are an accurate measure of biomass. Do the sizes of the boxes correspond to your perception of the total biomass at each level? If you answer yes, explain why this is apparent. If you answer no, explain the differences in proportions between the model and your data and why this might occur.

8. Study the food web shown in Figure 31–4, a food web for an entirely different ecosystem. Note that the web shows many different possible trophic relationships among organisms, including decomposers. You probably did not observe any of these decomposers in the aquarium. Why not?

9. Using all of the organisms from Figure 31–3, construct a food web for the aquarium ecosystem. Include decomposers in the web even though you did not observe them.

10. Are there any organisms in the ecosystem that are not dependent upon others?

11. Suppose that the entire population of producers in the aquarium ecosystem was killed by a poison. How would their absence affect the consumers?

12. How would the death of a population of dominant consumers affect the ecosystem?

13. What would be the results of the death or complete removal of all decomposers in the ecosystem?

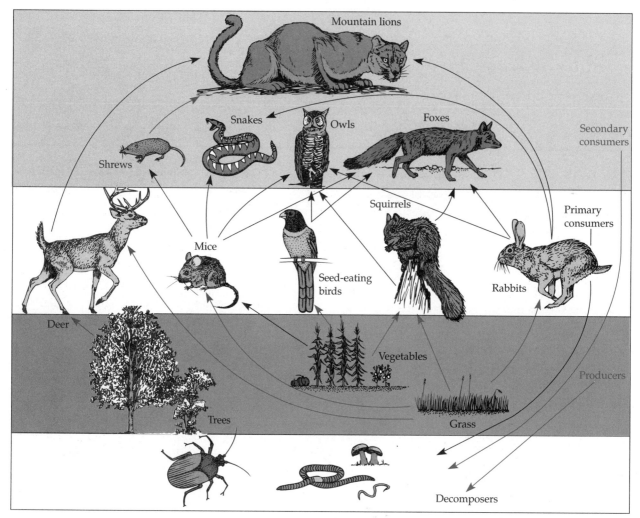

**FIGURE 31–4**

A simple terrestrial food web.

## FOR FURTHER READING

Andrews, John H. 1991. *Comparative Ecology of Microorganisms and Macroorganisms.* New York: Springer-Verlag.

Barnard, C. J., and J. M. Behnke, eds. 1991. *Parasitism and Host Behavior.* New York: Taylor and Francis.

Brewer, R. 1979. *Principles of Ecology.* San Francisco: W. H. Freeman.

Charles, Donald F., and Susan Christie, eds. 1991. *Acidic Deposition and Aquatic Ecosystems: Regional Case Studies.* New York: Springer-Verlag.

Clapham, W. B., Jr. 1983. *Natural Ecosystems.* New York: Macmillan.

Hunter, Christopher. 1991. *Better Trout Habitat: A Guide to Stream Restoration and Management.* Washington, D. C.: Island Press.

Keith, Lawrence H. 1991. *Environmental Sampling and Analysis: A Practical Guide.* Chelsea, Mich.: Lewis Publishers.

Krebs, C. J. 1985. *Ecology: The Experimental Analysis of Distribution and Abundance.* New York: Harper & Row.

Krebs, J. R., and N. B. Davies, eds. 1978. *Behavioral Ecology: An Evolutionary Approach.* New York: Sinauer.

Lott, Dale F. 1991. *Intraspecific Variation in the Social Systems of Wild Vertebrates.* New York: Cambridge University Press.

Nebel, B. J. 1987. *Environmental Science: The Way The World Works.*

Odum, E. P. 1983. *Basic Ecology.* Philadelphia: Saunders College Publishing.

Spellerberg, I. F. et al., eds. 1991. *The Scientific Management of Temperate Communities for Conservation: The 31st Symposium of the British Ecological Society, Southampton, 1989.* Boston: Blackwell Scientific Publications.

Turner, Monica G., and Robert H. Gardner, eds. 1991. *Quantitative Methods in Landscape Ecology: The Analysis and Interpretation of Landscape Heterogeneity.* New York: Springer-Verlag.

Whittaker, R. H. 1975. *Communities and Ecosystems.* New York: Macmillan.

◘

# Human Ecology

## OBJECTIVES

1. Review the development and impact of modern human life-styles upon the ecosystems of the earth.

2. Discuss the solutions to problems of inadequate food supply citing advantages, disadvantages, and examples of each.

3. Relate overpopulation to specific environmental problems and explain how humans can temporarily expand the carrying capacity of their habitat.

## INTRODUCTION

No other organism that has ever lived has affected the planet Earth as significantly and with such potential for harm as **Homo sapiens.** Through our industrial and agricultural technology, humans have been able to funnel the overall energy dynamics of the planet towards the attainment of one single objective: the production of human biomass. As our population increases, we have polluted the environment and, some would say, squandered our natural resources. It is a terrible realization that humans have the capacity to bring an end to our planet (Figs. 32–1 to 32–3).

Humans are not easy or kind towards our environment. Into our agricultural activities we channel energy, fertilizers, and pesticides, all of which would not normally enter a natural community. Our agricultural sites are generally characterized as being dedicated to primary production which is consumed at remote sites; precious little is recycled through the **primary production site.** Typically, our life-styles are highly consumptive and wasteful, which also contributes to **pollution.** Our rapid **population expansion** is usually at the cost of environmental destruction. Environmental destruction brings about wildlife endangerment as habitats are destroyed (Fig. 32–4).

**FIGURE 32–1**

Industrial waste is a major source of water pollution in the United States.   (© *Maurice E. Landre/NAS, Photo Researchers, Inc.*)

353

**FIGURE 32-2**

Combustion from industrial plants is a major source of air pollution. *(© Porterfield/Chickering, Photo Researchers, Inc.)*

**Overpopulation** is a serious threat to the human species. It is important that we limit human population through birth control (not by starvation and warfare) (Fig. 32-5). Limiting human population growth will also help stop the further deterioration of our environment. There are temporary steps that can be taken to expand the **carrying capacity** of the human habitat. These include doing more with recycling, exploiting genetic engineering to produce higher-yield crops with lowered dependence on fertilizers, encouraging developed nations to spend money more wisely and use their economic power to preserve threatened species and ecosystems, and continuing to be on the outlook for and to be sensitive to new threats to our ecosystems and environments.

We have designed the following experiment to demonstrate the factors that affect the growth of a population.

**FIGURE 32-3**

Oil spills, such as the one at Prince William Sound, have their effects on the environment for years or centuries to come. *(© Vanessa Vick, Photo Researchers, Inc.)*

**FIGURE 32-4**

The destruction of rainforests due to lumber harvesting eliminates many species of plants and animals. *(© Will and Deni McIntyre, Photo Researchers, Inc.)*

**FIGURE 32-5**

Modern technology has helped to increase the carrying capacity for the ever growing human population. *(Santi Visalli/ The Image Bank)*

**PRELAB QUESTIONS**

1. How has the human population impacted the environment?

2. What is the result of environmental destruction?

3. How can the carrying capacity of the human habitat be expanded?

# LAB 32.A
## ELEMENTARY ASPECTS OF POPULATION GROWTH

### MATERIALS REQUIRED

microcomputer with BASICA, GWBASIC, or equivalent

computer program included with this lab exercise, "sug"

printer attached to microcomputer

graph paper

### PROCEDURE

There are numerous factors that affect the growth of a population, factors such as the frequency of births, the longevity of members of the population, the abundance or scarcity of food supplies, and the interaction of members of the population with each other as well as with other organisms (predation, and so on). In actual fact, it may be shown that population growth is simply a direct function of population size. This is the model that we shall assume for these exercises. The model is a reasonable assumption for many populations, that is, for a population of bacterial cells growing in a culture media. In this exercise, we shall limit our consideration to two principal factors: the number of individuals comprising the population and what may be called a growth constant.

This BASIC computer program will allow us to investigate the effects, over time, of differing population sizes (N) and differing growth constants (K) on population size for populations undergoing exponential and unrestricted growth.

This is the computer program to be entered on a microcomputer for the associated laboratory exercises:

```
10 REM Program Name:  SUG
20 REM
30 REM Program to simulate unrestricted
growth of a population.
40 REM            *         *            *
50 REM            S         U            G
60 REM
70 CLS
80 INPUT ''What is the value for K, the
growth constant''; K
90 INPUT ''What is the initial population
size''; N
100 INPUT ''How many generations are to be
simulated''; TMAX
110 LPRINT ''Generation Population Size''
120 FOR T=0 TO TMAX
130 NT=N*EXP(K*T)
140 L PRINT TAB(5) T, TAB(20) INT(NT)
150 NEXT T
160 END
```

| Generation | Population Size |
| --- | --- |
| 0 | |
| 1 | 32 |
| 2 | 41 |
| 3 | 52 |
| 4 | 67 |
| 5 | 87 |
| 6 | 112 |
| 7 | 143 |
| 8 | 184 |
| 9 | 237 |
| 10 | 304 |

*Part 1.* In this portion of this exercise, you will investigate the affect that different growth constants have on population size over a short period of time. Make three separate runs of the program with $K = 0.1$, $K = 0.3$, and $K = 0.6$ and, for each run, 20 as the initial population size. Run each simulation for 10 generations. For each simulation, prepare a graph in which population size is plotted on the Y-axis and generation is plotted on the X-axis. In general, what is the affect on population size of increasing the growth constant?

*Part 2.* For this portion of the exercise, you will use the same program and hold $K$ constant while varying $N$. Run the program for 10 generations 3 times, each with $K = 0.3$, but with $N = 20, 30$, and $50$, for the first, second, and third run, respectively. Plot the data as before, with population size as the Y-axis and generation as the X-axis. Over the range of population sizes studied, what is the affect of initial population size on unrestricted population growth?

*Part 3.* Here are the data for a population experiencing unrestricted growth over 10 generations. The initial population size and growth constant for these data are the unknowns for you to identify. Plot these data as before and, by using the simulation program, identify the numerical values of the initial population size and the growth constant. If your values for $K$ and $N$ are correct, the simulation program will produce exactly the results given below.

## POSTLAB QUESTIONS

1. What would be the nature of the population growth being simulated with $K$ having a negative value? What would be the nature of the population growth if $K$ were 0?

2. Identify four factors that were not taken into account in the computer simulation used in the exercises, factors that have real effects on population growth. For each factor identified, write a brief statement concerning the nature of the effect of that factor on population growth.

3. In the simulation studied in these exercises, the unlimited growth of the population continues indefinitely. In reality, what factors might cause the population size to reach a plateau?

4. Make a sketch of the theoretical population growth curve demonstrated by the computer simulations of this exercise and, on the same figure, sketch what might be the more likely realized curve for an actual population. Identify the area between the two curves that represents "environmental resistance" as defined in Lab 31.

## FOR FURTHER READING

Anderson, S. 1985. *Managing Our Wildlife Resources.* Columbus, Ohio: Merrill.

Brewer, R. 1979. *Principles of Ecology.* Philadelphia: W. B. Saunders.

Colinvaux, P. 1986. *Ecology.* New York: John Wiley.

Ehrlich, P. R., and A. Ehrlich. 1981. *Extinction.* New York: Random House.

Ehrlich, P. R., and J. Roughgarden. 1987. *The Science of Ecology.* New York: Macmillan.

Gore, Albert. 1992. *Earth in the Balance.* Boston: Houghton Mifflin.

Hazen, W. E., ed. 1975. *Readings in Population and Community Ecology.* Philadelphia: W. B. Saunders.

Hussey, N. W., and N. Scopes, eds. 1985. *Biological Pest Control.* New York: Cornell University Press.

Huxley, A. 1985. *Green Inheritance.* Garden City, N.Y.: Anchor Press/Doubleday.

Jacobson, Harold K., and Martin F. Price. 1991. *A Framework for Research on the Human Dimensions of Global Environmental Change.* Paris: ISSC/UNESCO.

Mach, Mark, ed. 1991. *Healing the World and Me.* Indianapolis, Ind.: Knowledge Systems.

Mazur, Allan. 1991. *Global Social Problems.* Englewood Cliffs, N.J.: Prentice Hall.

Meadows, Donella H. 1991. *The Global Citizen.* Washington, D. C.: Island Press.

Metzger, Mary, and Cinthya P. Whittaker. 1991. *This Planet Is Mine: Teaching Environmental Awareness and Appreciation to Children.* New York: Simon & Schuster.

Odum, E. P. 1983. *Basic Ecology.* Philadelphia: Saunders College Publishing.

Orr, David M. 1992. *Ecological Literacy: Education and the Transition to a Postmodern World.* Albany, N.Y.: State University Press of New York.

Rothkrug, Paul, and Robert L. Olson, eds. 1991. *Mending the Earth: A World for Our Grandchildren.* Berkeley, Calif.: North Atlantic Books.

Simmons, Ian Gordon. 1991. *Earth, Air, and Water: Resources and Environment in the Late 20th Century.* New York: Edward Arnold.

Vajdi, Mehran. 1991. *The Human Jungle.* New York: Cornwall Books.

Western, D., and M. Pearl, eds. 1989. *Conservation for the Twenty-First Century.* New York: Oxford University Press.

Wilson, E. O. 1990. *Biodiversity.* Washington, D.C.: National Academy Press.

Wilson, E. O., and W. H. Bossert. 1971. *A Primer of Population Biology.* Sunderland, Mass.: Sinauer.

# APPENDIX A

❏

# Metric System

Biologists routinely use the metric system to measure length, mass, and volume. The units used to express these quantities are meter (m), gram (g), and liter (l), respectively. Within the metric system, the conversion from units of a particular measurement to other units is simply a matter of either multiplying by a multiple of 10 (for example, 1 centimeter (cm) = 10 millimeters (mm), 10 meters (m) = 10 × 100 (cm) = 1000 (cm) or by dividing by a multiple of 10 (for example, 150 cm = 150/100 = 1.5 m). (see Table 1–1)

**Table 1–1**  Relationships between Units in the Metric System

| Prefix | Multiplication Factor | Example |
|--------|----------------------|---------|
| *kilo* | 1000 | 1 kilogram = 1000 grams |
| *centi* | 0.01 | 1 centimeter = 0.01 meter |
| *milli* | 0.001 | 1 milliliter = 0.001 liter |

Table 1–2 gives the approximate American Standard equivalents of some of these metric units.

**Table 1–2**  Comparison of the American Standard and Metric Systems

| *Length* | 1.6 kilometers = 1 mile<br>1 meter = 1.1 yard<br>2.5 centimeters = 1 inch |
|----------|------------------------------------------------------------------------|
| *Mass* | 1 kilogram = 2.2 pounds<br>0.45 kilograms = 1 pound<br>28 grams = 1 ounce |
| *Volume* | 1 liter = 1.06 quarts<br>30 milliliters = 1 ounce |

It is important to become familiar with the metric system and develop a feel for the magnitude of these quantities.

## REVIEW QUESTIONS

1. How many millimeters (mm) are in a 10 cm ruler?
2. How many meters are in a 10 cm ruler?
3. Make the following conversions:
   - a.   26 cm =        mm
   - b.    1 km =        cm
   - c.  2800 cm =       m
   - d.   0.143 g =      mg
   - e.   246 mg =       kg
   - f.   892 g =        kg
   - g.   1.60 l =       ml
   - h.   532 ml =       l
4. Make the following conversions between the American Standard System and the metric system (refer to Table 1.2).
   - a.   14.2 in =      cm
   - b.  13.8 cm =       in
   - c.   2.1 mm =       in
   - d.   81.0 m =       yd
   - e.  1400 ml =       qt
   - f.   108 km =       mi
   - g.   1.00 lb =      g
   - h.   48.0 g =       oz
   - i.    72 mi =       km
   - j.     6 kg =       lb

◻

# "pgs" Computer Program

```
10 REM   Program Name:      PGS
20 REM   Population Genetics Simulations.
30 CLS
40 INPUT "Initial frequency of a-allele"; R
50 IF R>0 AND R>1 THEN GOTO 140
60 CLS
70 PRINT "Sorry, incorrect value for initial frequency of a-allele."
80 PRINT " "
90 PRINT "Correct initial frequency must be greater than"
100 PRINT "zero but less than one."
110 PRINT
120 PRINT "Strike any key to continue."
130 A$=INKEY$: IF A$="" THEN GOTO 130 ELSE GOTO 30
140 INPUT "Selection coefficient against the aa genotype"; SCR
150 IF SCR>=0 AND SCR<=1 THEN GOTO 240
160 PRINT " "
170 PRINT "Sorry, incorrect value for selection coefficient against"
180 PRINT "the aa genotype.  Correct value for selection coefficient"
190 PRINT "against aa genotype must be greater than or equal to zero"
200 PRINT "and less than or equal to one."
210 PRINT " "
220 PRINT "Strike any key to continue."
230 A$=INKEY$: IF A$="" THEN GOTO 230 ELSE GOTO 140
240 KR = 1 - SCR
250 INPUT "Selection coefficient against the AA genotype"; SCD
260 IF SCD>=0 AND SCD<=1 THEN GOTO 350
270 PRINT " "
280 PRINT "Sorry, incorrect value for selection coefficient against"
290 PRINT "the AA genotype.  Correct value for selection coefficient"
300 PRINT "against AA genotype must be greater than or equal to zero"
310 PRINT "and less than or equal to one."
320 PRINT " "
330 PRINT "Strike any key to continue."
340 A$=INKEY$: IF A$="" THEN GOTO 340 ELSE GOTO 250
350 KD = 1 - SCD
360 INPUT "Selection coefficient against the Aa genotype"; SCH
```

```
370 IF SCH>=0 AND SCH<=1 THEN GOTO 460
380 PRINT " "
390 PRINT "Sorry, incorrect value for selection coefficient against"
400 PRINT "the Aa genotype.  Correct value for selection coefficient"
410 PRINT "against Aa genotype must be greater than or equal to zero"
420 PRINT "and less than or equal to one."
430 PRINT " "
440 PRINT "Strike any key to continue."
450 A$=INKEY$: IF A$="" THEN GOTO 450 ELSE GOTO 360
460 KH = 1 - SCH
470 INPUT "How many generations to be simulated?"; GENS
480 GENS=ABS(INT(GENS))
490 PRINT " "
500 PRINT " "
510 PRINT "Get the printer ready, then strike any key to continue."
520 A$=INKEY$: IF A$="" THEN GOTO 520 ELSE GOTO 530
530 CLS
540 LPRINT "POPULATION GENETICS COMPUTER SIMULATION:"
550 LPRINT "Two alleles, A and a, at an autosomal locus."
560 LPRINT
570 LPRINT "Initial frequency of a-allele =" R
580 LPRINT " "
590 LPRINT "                              GENOTYPE"
600 LPRINT "                         AA        Aa        aa"
610 LPRINT "Selection Coefficient:";
620 LPRINT USING "          ##.##     ##.##     ##.##"; SCD,SCH,SCR
630 LPRINT "Fitness:";
640 LPRINT         "                         ";
650 LPRINT USING " ##.##     ##.##     ##.##"; KD, KH, KR
660 LPRINT " "
670 LPRINT "Number of generations to be simulated:", GENS
680 LPRINT
690 LPRINT       "GENERATION                 FREQUENCY "
700 LPRINT       "          ----------------------";
710 LPRINT       "--------------------"
720 LPRINT       "            AA      Aa      aa  ";
730 LPRINT       "        p         q     "
740 FOR I=0 TO GENS
750 S = R*R
760 T = 2*R*(1-R)
770 U = (1-R)*(1-R)
780 LPRINT USING "      #### ##.#### ##.#### ##.#### "; I,U,T,S;
790 LPRINT USING "     ##.####  ##.####"; 1-R, R
800 R= ((.5 * KH *T) + (KR * S)) / (  (KR*S) + (KH*T) + (KD*U) )
810 NEXT I
820 END
```

◘

# A Classification Summary of the Living World

The following is a capsule listing of the major taxonomic groups of organisms for quick reference in the use of this manual. A few major traits that characterize each group are also included along with some representatives. Groups are presented in phylogenetic order, i.e., older groups appear first, followed by those that have evolved more recently. Although most of these groups are currently recognized by most authorities, be aware that some texts may differ in their classification scheme. This is to be expected since there can be more than one approach to the classification of the overwhelming diversity of organisms that cover the planet.

## Kingdom Monera

Prokaryotic unicellular organisms; mostly heterotrophic saprobes, some autotrophic.

**Archaebacteria** - ancient anaerobic group adapted to extreme habitats

**Cyanobacteria** - photosynthetic group, ecologically important

**Eubacteria** - "true" bacteria, diverse group of great ecological and economic importance

## Kingdom Protista

Eukaryotic, mostly unicellular organisms; wide range of nutritional modes.

**Protozoa** - mostly animal-like, heterotrophic organisms

Phylum Sarcomastigophora - flagellate and amoeboid protozoa
*Amoeba, Trypanosoma*, foraminiferans
Phylum Ciliophora - ciliated protozoa
*Paramecium, Vorticella, Stentor*
Phylum Apicomplexa - parasitic, spore-forming protozoa
*Plasmodium*

**Algae** - autotrophic unicellular and multicellular organisms

Phylum Dinoflagellata - unicellular forms covered by cellulose plates, two flagella present
*Gonyaulax, Ceratium*
Phylum Bacillariophyta - mostly unicellular forms with cell walls of silica diatoms
Phylum Euglenophyta - unicellular flagellates
*Euglena*
Phylum Chlorophyta - the green algae, unicellular and multicellular forms
*Chlamydomonas, Spirogyra, Ulva*
Phylum Rhodophyta - the red algae, mostly multicellular forms
*Porphyra, Plumaria*
Phylum Phaeophyta - the brown algae, large multicellular forms
*Laminaria* (kelp), *Fucus, Sargassum*

**Fungus-like protists** - heterotrophic multicellular forms with superficial fungal-like appearance
Phylum Myxomycota - plasmodial slime molds
*Physarum*

Phylum Acrasiomycota - cellular slime molds
*Dictyostelium*
Phylum Oomycota - water molds
*Saprolegnia*

## Kingdom Fungi

Eukaryotic, unicellular and multicellular organisms; heterotrophic saprobes

Division Zygomycota - zygospore-forming fungi
*Rhizopus*
Division Ascomycota - sac fungi
*Morchella, Peziza,* yeasts
Division Basidiomycota - club fungi
*Agaricus* (common commercial mushroom), *Coprinus*
Division Deuteromycota - imperfect fungi
*Penicillium, Aspergillus*

## Kingdom Plantae

Eukaryotic, multicellular organisms; autotrophic

**Nonvascular plants** - xylem and phloem tissue absent, dominant gametophyte in life cycle
Division Bryophyta - mosses and similar plants
Class Bryopsida
mosses, *Mnium, Sphagnum*
Class Hepatopsida
liverworts, *Marchantia, Conocephalum*
Class Anthoceropsida
hornworts, *Anthoceros*

**Vascular seedless plants** - xylem and phloem present, dominant sporophyte in life cycle
Division Pterophyta - ferns
*Polypodium*
Division Psilophyta - whisk ferns
*Psilotum*
Division Sphenophyta - horsetails
*Equisetum*
Division Lycophyta - club mosses
*Lycopodium*

**Vascular seed plants, gymnosperms** - xylem and phloem present, dominant sporophyte in life cycle, "naked seeds" produced on bracts or other reproductive structures
Division Coniferophyta - conifers
pine, cedar, spruce
Division Cycadophyta - cycads
*Zamia*

Division Ginkgophyta - ginkgo
*Ginkgo*
Division Gnetophyta - gnetophytes
*Ephedra, Welwitschia*

**Vascular seed plants, angiosperms** - xylem and phloem present, dominant sporophyte in life cycle, seeds produced within a fruit resulting from a fertilized flower
Division Magnoliophyta - flowering plants
Class Magnoliopsida
dicot angiosperms, maple tree, grape vine, tomato plant
Class Liliopsida
monocot angiosperms, palm, iris, grass

## Kingdom Animalia

Eukaryotic, multicellular organisms; heterotrophic

**Acoelomate animals** - no true body cavity (coelom), body symmetry variable, invertebrates (animals without a vertebral column)

Phylum Porifera - sponges, mostly asymmetrical, tissue level of organization
natural bath sponge, *Grantia*
Phylum Cnidaria - cnidarians, radial symmetry, tentacles with cnidocytes (stinging cells)
Class Hydrozoa
hydra, Portuguese-man-of-war
Class Scyphozoa
jellyfish, *Aurelia*
Class Anthozoa
sea anemone, coral
Phylum Platyhelminthes - flatworms, organs derived from three germ layers, bilateral symmetry
Class Turbellaria
planarian
Class Trematoda
liver fluke
Class Cestoda
tapeworm
Phylum Nemertinea - proboscis worms, organ systems present, bilateral symmetry
*Tubularis*
Phylum Nematoda - roundworms, pseudocoelomate, bilateral symmetry
*Ascaris*
Phylum Rotifera - rotifers, pseudocoelomate, bilateral symmetry
*Stephanboceros*

**Coelomate protostomes** - animals with true body cavity lined with tissue of mesodermal origin, blastopore of gastrula develops into mouth of fully developed organism, bilaterally symmetrical invertebrates

Phylum Mollusca - mollusks typically with a shell of calcium carbonate, muscular foot
> Class Polyplacophora
>> chiton
>
> Class Gastropoda
>> snail
>
> Class Bivalvia
>> mussel
>
> Class Cephalopoda
>> squid

Phylum Annelida - segmented worms, first group to exhibit extensive segmentation
> Class Polychaeta
>> tubeworm, sandworm
>
> Class Oligochaeta
>> earthworm
>
> Class Hirudinea
>> leech

Phylum Onychophora - wormlike animals with both annelid and arthropod characteristics
> *Peripatus*

Phylum Arthropoda - segmented animals with jointed appendages, exoskeleton of chitin
> Subphylum Trilobitomorpha
>> Class Trilobita (extinct)
>>> trilobite
>
> Subphylum Chelicerata
>> Class Merostomata
>>> horseshoe crab
>>
>> Class Arachnida
>>> spider, mite
>
> Subphylum Crustacea
>> crab, barnacle
>
> Subphylum Uniramia
>> Class Chilopoda
>>> centipede
>>
>> Class Diplopoda
>>> millipede
>>
>> Class Insecta
>>> bee, grasshopper

**Coelomate deuterostomes** - animals with true body cavity lined with tissue of mesodermal origin, blastopore of gastrula develops into anus of fully developed organism, most bilaterally symmetrical (see Phylum Echinodermata); Subphylum Vertebrata contains the vertebrates (animals with a vertebral column)

Phylum Echinodermata - spiny-skinned animals, endoskeleton of calcium carbonate, pentaradial adults develop from bilaterally symmetrical larvae
> Class Crinoidea
>> sea lily
>
> Class Asteroidea
>> sea star
>
> Class Ophiuroidea
>> brittle star
>
> Class Echinoidea
>> sea urchin, sand dollar
>
> Class Holothuroidea
>> sea cucumber

Phylum Chordata - animals with a notochord, dorsal tubular nerve cord, pharyngeal gill slits
> Subphylum Urochordata
>> tunicate
>
> Subphylum Cephalochordata
>> sea lancelet
>
> Subphylum Vertebrata
>> Class Agnatha
>>> lamprey
>>
>> Class Chondrichthyes
>>> shark, ray
>>
>> Class Osteichthyes
>>> trout
>>
>> Class Amphibia
>>> salamander
>>
>> Class Reptilia
>>> lizard
>>
>> Class Mammalia
>>> mouse

# Glossary

**abscisic acid** A plant hormone involved in responses to stress and in dormancy.

**absorption spectrum** A measure of the amount of energy at specific wavelengths that has been absorbed as light passes through a substance. Each type of molecule has a characteristic absorption spectrum.

**acetyl CoA** A key intermediate compound in metabolism; consists of an acetyl group covalently bonded to coenzyme A.

**actin** A protein component of muscle fiber that, together with the protein myosin, is responsible for the ability of muscles to contract.

**action potential** The electrical activity developed in a muscle or nerve cell during activity; a neural impulse.

**active site** Area of an enzyme that accepts a substrate and catalyzes its reaction with another.

**active transport** Energy-requiring transport of a molecule across a membrane from a region of low concentration to a region of high concentration.

**adenosine triphosphate (ATP)** An organic compound containing adenine, ribose, and three phosphate groups; of prime importance for energy transfers in biological systems.

**adipose tissue** Tissue in which fat is stored.

**aerobes** Organisms that grow and metabolize in the presence of molecular oxygen.

**air spaces** Regions of air between loosely arranged mesophyll cells of leaves.

**alcoholic fermentation** An anaerobic pathway in yeast cells in which ethyl alcohol is produced.

**alleles** Genes governing variations of the same characteristic that occupy corresponding positions on homologous chromosomes; alternative forms of a single gene.

**allosteric enzyme** An enzyme that receives a substance on a region of the enzyme molecule other than the active site.

**allosteric interactions** An enzyme reaction involving a receptor site other than the active site of the enzyme.

**alternation of generations** A type of life cycle characteristic of plants. Plants spend part of their life in the haploid (gametophyte) stage and part in the diploid (sporophyte) stage.

**amino acid** An organic compound containing an amino group ($-NH_2$) and a carboxyl group ($-COOH$). Amino acids may be linked together to form the peptide chains of protein molecules.

**amino group** Group of atoms in a molecule arranged as $-NH_2$.

**amyloplasts** Plastids that store starch as a plant's reserve food supply.

**anaphase** The stage of mitosis between metaphase and telophase in which the chromatids separate and move toward opposite ends of the cell.

**androgen** Any substance that possesses masculinizing properties, such as sex hormones.

**antagonistic muscle** A muscle that produces the opposite movement of an agonist.

**anther** The part of the stamen in flowers that produces microspores and, ultimately, pollen.

**anticodon** A sequence of three nucleotides in transfer RNA that is complementary to, and combines with, the three nucleotide codon on messenger RNA, thus helping to specify the addition of a particular amino acid to the end of a growing peptide.

**aorta** The largest and main systemic artery of the body; arises from the left ventricle and branches to distribute blood to all parts of the body; main artery leaving the heart in vertebrates.

**apical meristem** An area of cell division located at the tips of plant stems and roots. Apical meristems produce primary tissues.

**arteries** Thick-walled blood vessels that carry blood away from the heart and toward the body organs.

**artificial selection** Selection by humans of traits that are desirable in plants or animals, and breeding only those individuals that possess the desired traits.

**atrioventricular node** Small mass of tissue located in the right atrium that spreads the cardiac impulse to the atrioventricular bundle.

**autonomic nervous system** The portion of the peripheral nervous system that controls the visceral functions of the body, e.g., regulates smooth muscle, cardiac muscle, and glands, thereby helping to maintain homeostasis.

**autosome**  A chromosome other than those involved in the determination of the sex of an individual.

**autotroph**  Organism that obtains organic molecules by synthesizing them from inorganic material.

**auxin**  A plant hormone involved in various aspects of plant growth and development.

**axillary bud**  A lateral bud located at the axils of leaves.

**axon**  The long, tubular extension of the neuron that transmits nerve impulses away from the cell body.

**bilateral symmetry**  A body form which can be divided into similar right and left halves.

**biomass**  The total weight of all the organisms in a particular habitat.

**biome**  A large, easily differentiated community unit arising as a result of complex interactions of climate, other physical factors, and biotic factors.

**biotic**  Living.

**biotic potential**  Inherent power of a population to increase in numbers when the age ratio is stable and all environmental conditions are optimal.

**blastula**  Usually a spherical structure produced by cleavage of a fertilized ovum; consists of a single layer of cells, the trophoblast, from which a small cluster, the inner cell mass, projects into a central cavity.

**blood**  Tissue made up of a liquid containing dissolved salts, plasma proteins, red blood cells, white blood cells, and platelets.

**bone**  Skeletal tissue of vertebrates.

**Bowman's capsule**  Double-walled, hollow sac of cells that surrounds the glomerulus at the end of each kidney tubule.

**bulbourethral glands**  Male gland located at the base of the penis that contributes secretions to the semen.

**capillaries**  Microscopic blood vessels occurring in the tissues which permit exchange of materials between tissues and blood. Lymph capillaries also drain away excess tissue fluid.

**capillarity**  The action by which water rises quickly in a thin tube.

**carbohydrate**  Compound containing carbon, hydrogen, and oxygen, in the ratio of 1C:2H:10, such as sugars, starches, and cellulose.

**carbon fixation**  The process of attaching a carbon dioxide molecule to an existing carbohydrate molecule.

**carboxyl group**  Group of atoms in a molecule arranged as $-COOH$.

**carcinogen**  A substance that causes cancer or accelerates its development.

**cardiac muscle**  Distinctive involuntary but striated type of muscle occurring only in the vertebrate heart.

**carpel**  The female structure in flowers that bears ovules.

**carrying capacity**  The ability of the habitat to support a population of organisms.

**cartilage**  Flexible skeletal tissue of vertebrates.

**catalyst**  A substance that regulates the speed at which a chemical reaction occurs without affecting the end point of the reaction and without being used up as a result of the reaction.

**cell cycle**  Cyclic series of events in the life of a dividing eukaryotic cell consisting of M (mitosis), cytokinesis, and the stages of interphase, which are the $G_1$ (first gap), S (DNA synthesis), and $G_2$ (second gap) phases.

**cell plate**  Forming cell wall that separates the two daughter cells produced by plant mitosis.

**cell theory**  A principle composed by Schleiden and Schwann stating that cells are the fundamental unit of life.

**cellular respiration**  The stepwise oxidation of high-energy food molecules to carbon dioxide and water, low-energy molecules.

**cell wall**  A thick, porous structure that lies just outside the plasma membrane of plant cells.

**central nervous system**  The brain and the spinal cord.

**centriole**  One of a pair of small dark-staining organelles lying near the nucleus in the cytoplasm of animal cells.

**centromere**  Specialized constricted region of a chromatid that serves as the site of spindle fiber attachment during cell division; sister chromatids are joined in the vicinity of their centromeres.

**cephalization**  The formation of a head

**cerebellum**  The deeply convoluted subdivision of the brain lying beneath the cerebrum which is concerned with the coordination of muscular movements; second largest part of the brain.

**cerebral hemispheres**  The two largest sections of the human brain which are involved in conscious awareness.

**cerebrum**  Largest subdivision of the brain; functions as the center for learning, voluntary movement, and interpretation of sensation.

**cervix**  The opening of the uterus leading into the vagina.

**chemiosmotic ATP synthesis**  A process of making ATP using a chemical ($H^+$) passing through a membrane.

**chlorophyll**  One of a group of light-trapping green pigments found in most photosynthetic organisms.

**chloroplast**  A chlorophyll-bearing intracellular organelle of plant cells; site of photosynthesis.

**chromosomes**  Filamentous or rod-shaped bodies in the cell nucleus that contain the hereditary units, the genes.

**cilium  pl. cilia**  One of many short, hairlike structures that project from the surface of some cells and are used for locomotion or movement of materials across the cell surface; structurally like flagella, including a cylinder of nine doublet microtubules and two central single microtubules, all covered by a plasma membrane.

**circulatory system**  A vascular system in which the transport fluid moves in a particular direction, usually because it is propelled by a muscular, pumping heart.

**citric acid cycle**  Aerobic series of chemical reactions in which food molecules are completely degraded to carbon dioxide and water with the release of metabolic energy; also known as the Krebs cycle.

**cleavage**  First of several cell divisions of an embryo which forms a multicellular blastula.

**cleavage furrow**  In animal cells, a groove resulting from constriction of microfilaments that divide the cyto-plasm into two cells.

**clitoris** A small, erectile body at the anterior part of the vulva, which is ho-mologous to the male penis.

**cnidocytes** Stinging cells of cnidar-ians.

**codon** A trio of mRNA bases that specifies an amino acid or a signal to terminate the polypeptide.

**coelom** Body cavity which forms within the mesoderm and is lined by mesoderm.

**coenzyme** A nonprotein substance that is required for a particular enzy-matic reaction to occur; participates in the reaction by donating or accepting some reactant; loosely bound to en-zyme. Most of the vitamins function as coenzymes.

**colchicine** A chemical that prevents formation of a mitotic spindle and so blocks mitosis and cell division.

**coleoptile** A protective sheath that encloses the stem in grass seeds.

**collagen** Protein in connective tissue fibers that is converted to gelatin by boiling.

**collecting duct** Part of the nephron where the actual concentration of urine takes place.

**colliculi** Midbrain centers for audi-tory and visual reflexes.

**colon** An organ that absorbs water, minerals and vitamin K and pushes the remaining fecal matter into the rectum.

**columnar** Shaped as columns.

**commensalism** A symbiotic rela-tionship in which an organism lives with another species without harming or benefiting its host.

**community** An assemblage of popu-lations that live in a defined area or habitat which can either be very large or quite small. The organisms consti-tuting the community interact in various ways with one another.

**competition** What occurs when two or more organisms attempt to exploit a limited resource, such as food or space.

**compound light microscope** An in-strument composed of lenses that functions in magnifying small objects or specimens by transmitting light through the specimen.

**concentration gradient** The gradual decrease in the concentration of a sub-stance.

**connective tissue** Vertebrate tissue consisting mostly of a matrix com-posed of cell products in which the cells are embedded, e.g., bone.

**consumers** Animals that eat plants or each other.

**contraception** Birth control methods that prevent fertilization.

**contract** To shorten.

**control group** Treatment group in an experiment that is subjected to the same conditions as the experimental treatment, except the factor(s) being investigated is not varied.

**copulation** Sexual union; act of physical joining of two animals during which sperm cells are transferred from one to the other.

**cork** Cells produced by the cork cambium. Cork is dead at maturity and functions for protection.

**cork cambium** Lateral meristem in plants that produces cork cells and cork parenchyma. Cork cambium and the tissues it produces make up the outer bark on a woody plant.

**corpus luteum** Pocket of endocrine tissue in the ovary that is derived from the follicle cells; secretes the hormone progesterone.

**cortex** The outer layer of an organ; in plants, the tissue beneath the epider-mis in many nonwoody plants.

**cotyledon** The seed leaf of the em-bryo of a plant, which may contain stored food for germination.

**Cowper's gland** Male gland that contributes secretions to the semen, located at the base of the penis.

**cranial nerves** Twelve pairs of nerves that connect the brain with various structures.

**crossing over** The breaking and re-joining of homologous (non-sister) chromatids during early meiotic pro-phase I, resulting in an exchange of genetic material.

**cuboidal** Shaped as cubes.

**cuticle** A waxy covering over the ep-idermis of the above-ground portion of land plants that reduces water loss from plant surfaces.

**cytokinesis** Stage of cell division in which the cytoplasm is divided to form two daughter cells.

**cytokinin** A plant hormone involved in various aspects of plant growth and development.

**cytoskeleton** Internal structure of microfilaments, intermediate fila-ments, and microtubules that gives shape and mechanical strength to ani-mal cells.

**day-neutral plant** A plant that does not flower in response to variations in day length which occur with changing seasons.

**decomposers** Microorganisms of decay.

**denature** To alter the physical prop-erties and three-dimensional structure of a protein, nucleic acid, or other macromolecule by mild treatment that does not break the primary structure, but which does destroy its activity.

**dendrite** Nerve fiber, typically branched, that conducts a nerve im-pulse toward the cell body.

**deoxyribonucleic acid** DNA; present in chromosomes; contains ge-netic information coded in specific se-quences of its constituent nucleotides.

**depolarization** A decrease in the po-tential difference across a membrane.

**determination** The progressive limi-tation of a cell line's potential fate dur-ing development.

**dialysis** Diffusion of a solute through a selectively permeable mem-brane.

**diaphragm** A respiratory muscle that extends across the bottom of the chest cavity, closing off the chest cav-ity from the abdominal cavity below.

**dicotyledon** A flowering plant with embryos having two seed leaves, or cotyledons; also known as a dicot.

**differentiation** Development toward a more mature state; a process changing a relatively unspecialized cell to a more specialized cell.

**diffusion** The net movement of mol-ecules from a region of high concen-tration to one of lower concentration, brought about by their kinetic energy.

**digestive system**  A system that ingests and digests food molecules and eliminates wastes.

**dihybrid cross**  A genetic cross which takes into account the behavior of two distinct pairs of genes.

**diploid**  A chromosome number twice that found in gametes; containing two sets of chromosomes.

**disaccharide**  A two-unit sugar, such as sucrose, that consists of a glucose and a fructose subunit.

**distal convoluted tubule**  Part of the nephron that functions in the secretion of potassium and hydrogen ions by active transport and in the secretion of ammonia by diffusion.

**DNA polymerase**  An enzyme that links free nucleotides into a new DNA polymer strand complementary to the template strand.

**dominant allele**  An allele that expresses itself.

**double circulation**  A type of circulation in which the blood passes through the heart twice in each complete circuit around the body.

**double helix**  The coiled arrangement of two complementary nucleotide chains in a DNA molecule.

**duodenum**  Portion of the small intestine directly adjacent to the stomach.

**ecosystem**  A natural unit of living and nonliving parts that interact to produce a stable system in which the exchange of materials between living and nonliving is recycled.

**ectoderm**  The outer of the three embryonic germ layers of the gastrula; gives rise to the skin and nervous system.

**effector**  A muscle or gland that contracts or secretes in direct response to nerve impulses.

**egestion**  A process of removing undigested and unabsorbed food from the body.

**electron microscope**  An instrument that uses a beam of electrons to magnify extremely small objects (e.g. organelles, cell membranes).

**electron transport system**  A series of about six chemical reactions during which hydrogens or their electrons are passed along from one acceptor molecule to another with the release of energy.

**embryo**  A young organism before it emerges from the egg, seed, or body of its mother; the developing human organism until the end of the second month, after which it is referred to as a fetus.

**embryo sac**  The female gametophyte generation in angiosperms.

**endocrine system**  A system of ductless glands that secrete hormones into the bloodstream.

**endocytosis**  The active transport of substances into a cell by the formation of invaginated regions of the plasma membrane which pinch off and become cytoplasmic vesicles.

**endoderm**  The inner germ layer of the gastrula lining the archenteron; becomes the digestive tract and its outgrowths—the liver, lungs, and pancreas.

**endodermis**  The innermost layer of the cortex in the plant root. Endodermis cells have a Casparian strip running around radial and transverse walls.

**endometrium**  Uterine lining.

**endoplasmic reticulum (ER)**  Structure composed of numerous internal membranes within eukaryotic cells.

**endosperm**  The nutritive tissue that is found at some point in all angiosperm seeds.

**energy of activation**  The amount of energy needed to raise molecules to an unstable, high-energy transition state.

**environmental resistance**  The sum of the physical and biological factors that prevent a species from reproducing at its maximum rate.

**enzyme**  A protein catalyst produced within a living organism that accelerates specific chemical reactions.

**epididymis**  A coiled tube that receives sperm from the testes and conveys it to the vas deferens.

**epithelial tissue**  The type of tissue that covers body surfaces, lines body cavities, and forms glands; also called epithelium.

**equilibrium**  The state in which the rate of reaction in one direction is the same as the rate of reaction in the opposite direction.

**esophagus**  The muscular tube extending from the pharynx to the stomach.

**estrogen**  Female sex hormone produced by the ovarian follicle; promotes the development and maintenance of female reproductive structures and of secondary sex characteristics.

**ethylene**  A plant hormone involved in various aspects of plant senescence.

**eukaryotic cells**  Cells of organisms that possess nuclei and cell organelles surrounded by membranes.

**evolution**  Genetic change in a population of organisms.

**excretion**  The discharge from the body of a waste product of metabolism (not to be confused with the elimination of undigested food materials).

**exoskeleton**  An external skeleton, such as the shell of mollusks or outer covering of arthropods; provides protection and sites of attachment for muscles.

**experiment**  A test designed to show that one or more of the hypotheses is more or less likely to be incorrect.

**experimental group**  A treatment group in an experiment that is subjected to the same conditions as the control group except the factor(s) being investigated is varied.

**extensor**  A muscle that serves to extend or straighten a limb.

**facilitated diffusion**  The passage of ions and molecules, bound to specific carrier proteins, across a cell membrane down their concentration gradients. Facilitated diffusion does not require a special source of energy.

**facultative anaerobes**  Organisms that perform aerobic or anaerobic respiration depending on the availability of oxygen.

**fatigue**  A condition caused by the accumulation of lactate in muscle cells.

**fermentation**  Anaerobic respiration that utilizes organic compounds both as electron donors and acceptors.

**fertilization**  Union of male and female gametes.

**fibrous root system** A root system in plants that has several main roots without a dominant root.

**filter feeders** Organisms that use cilia or muscles to set up water currents past (or through) their bodies and filter out or seize any food carried by the current.

**filtrate** Filtered fluid.

**first filial generation** The first generation of hybrid offspring resulting from a genetic cross.

**first foliage leaves** Initial photosynthetic leaf growth.

**fitness** A measure of the relative success of the phenotype; fitness is the complement of the selection coefficient (i.e., 1−S).

**flagellum** Long, whiplike movable structure of cells that is used in locomotion or in moving fluid past the tissue of which the cell is a part. Eukaryote flagella are composed of two single microtubules surrounded by nine double microtubules, but prokaryote flagellae are filaments rotated by special structures located in the plasma membrane and cell wall.

**flavin adenine dinucleotide (FAD)** A coenzyme that carries hydrogen atoms to an electron transport chain.

**flexor** A muscle that serves to bend a limb.

**flora** Plants.

**follicle stimulating hormone (FSH)** A gonadotropic pituitary hormone that helps to regulate sperm production in males and causes an ovarian follicle to mature and produce the steroid hormone estrogen in females.

**food chain** A sequence of organisms through which energy is transferred from its ultimate source in a plant; each organism eats the preceding and is eaten by the following member of the sequence.

**food web** Complex feeding relationships in a community of organisms.

**fruit** In angiosperms, a mature, ripened ovary. Fruits contain seeds and usually serve for seed protection and dispersal.

**fruiting bodies** Large, complex reproductive structures in fungi composed of many hyphae.

**gamete** A cell that functions in sexual reproduction; an egg or sperm whose union, in sexual reproduction, initiates the development of a new individual.

**gametophyte** The haploid or gamete-producing stage in the life cycle of a plant.

**gastrovascular cavity** A structure of cnidarians that serves as a digestive and circulatory system, distributing food around the body.

**gastrula** Early stage of embryonic development during which the three germ layers form.

**gastrulation** A process in embryonic development that results in the cells rearranging themselves into distinct layers.

**gene flow** The movement of alleles between local populations, or demes, due to the migration of individuals. Gene flow can have significant evolutionary consequences.

**gene pool** All the genes present in a species population.

**genetic engineering** Using chemical techniques to change genetic makeup, usually to produce a desired biological product.

**genetics** The study of patterns of inheritance as hereditary characteristics are passed on from parents to offspring.

**gene transplantation** Insertion of a gene into a DNA molecule.

**genotype** The complete genetic makeup of an organism.

**germinate** To begin growth in seeds or spores.

**germ layer** Primitive embryonic tissue layer; endoderm, mesoderm, or ectoderm.

**gibberellin** A plant hormone involved in various aspects of plant growth and development.

**gill** The respiratory organ of aquatic animals, usually a thin-walled projection from the body surface or from some part of the digestive tract.

**glial cell** Supporting cell of central nervous tissue, comparable to a connective tissue cell elsewhere in the body, but originating from ectoderm

and incapable of producing an extracellular matrix.

**glomerulus** A tuft of minute blood vessels; specifically, the knot of capillaries at the proximal end of a kidney tubule.

**glycerol** A three-carbon alcohol with three hydroxyl groups.

**glycolysis** The metabolic conversion of glucose into pyruvate or lactate with the production of ATP. Glycolysis is a metabolic pathway present in all living cells.

**Golgi complex** Cell organelle that modifies and sorts products of the endoplasmic reticulum.

**gonad** A gamete-producing gland; an ovary or testis.

**grafting** A means of artificial vegetative propagation.

**grana** Small bodies within chloroplasts that contain alternate layers of chlorophyll, protein, and lipid, and are the functional units of photosynthesis.

**guard cell** A cell in the epidermis of plant stems and leaves. Two guard cells form a pore for gas exchange, collectively called a stomate.

**guttation** The appearance of water droplets on leaves, forced out through leaf pores by root pressure.

**haploid** The chromosome number characteristic of gametes or spores; half the diploid number. In plants, the chromosome number of body cells of the gametophyte generation.

**Hardy-Weinberg law** The constancy of allelic frequencies at an autosomal locus in a large, random-mating population in the absence of selection, mutation, and migration.

**helix** Spiral.

**hemoglobin** The red iron-containing protein pigment of erythrocytes that transports oxygen and aids in regulation of pH.

**heterotrophs** Organisms that cannot synthesize their own food from inorganic materials and therefore must live either at the expense of autotrophs or upon decaying matter.

**heterozygote advantage** A phenomenon in which the heterozygous condition confers some special advantage on an individual that either homozygous condition does not, i.e., *Aa* has a higher degree of fitness than *AA* or *aa*.

**heterozygous**   Genetically mixed. Usually this term refers to a single pair of allelic genes that are unlike each other.

**homeothermic**   Using heat generated by metabolic activities to maintain a constant body temperature.

**homologous chromosomes**   Chromosomes that are similar in morphology and genetic constitution. In humans there are 23 pairs of homologous chromosomes, each containing one member from the mother and one member from the father.

*Homo sapiens*   Scientific name of humans.

**homozygous**   Possessing a pair of identical alleles.

**hormone**   A chemical messenger produced by an endocrine gland or by certain cells. In animals, hormones are usually transported in the blood and regulate some aspect of metabolism.

**hypertonic**   Having a greater concentration of solute molecules and a lower concentration of solvent (water) molecules, and hence an osmotic pressure greater than that of the solution with which it is compared.

**hypha**   One of the filaments composing the mycelium of a fungus.

**hypothalamus**   Part of the brain that functions in regulating the pituitary gland, the autonomic system, emotional responses, body temperature, water balance, and appetite; located below the thalamus.

**hypothesis**   A proposed answer to questions about what has been observed.

**hypotonic**   Having an osmotic pressure or solute content less than that of some standard of comparison.

**impermeable**   Preventing the passage of certain kinds of molecules.

**inducer molecules**   Molecules that function as signals which turn on the production of proteins.

**inhibitors**   Substances that decrease an enzyme's reaction rate.

**innate**   Behavior genetically programmed in the nervous system.

**insight learning**   The intuitive solution to a problem that has not previously been attempted.

**integumentary system**   The body's covering, including the skin and its nails, glands, hair, and other associated structures.

**intercalary meristems**   Areas of dividing cells in the stems, in stretches between the bases of leaves.

**interphase**   The period in the life cycle of a cell in which there is no visible mitotic division; period between mitotic divisions.

**invertebrates**   Animals without backbones.

**involuntary control**   Autonomic nervous system mechanism for adjusting function without conscious thought.

**isotonic**   Having identical concentrations of solute and solvent molecules, and hence the same osmotic pressure as the solution with which it is compared.

**kinetochore**   Portion of the chromosome centromere to which mitotic spindle fibers attach.

**labia majora**   In females, outer lips that cover and protect the urinary and genital openings.

**labia minora**   In females, inner lips that cover and protect the urinary and genital openings.

**lactate fermentation**   A pathway in which NADH transfers hydrogens to pyruvate forming lactate.

**large intestine**   An organ of the digestive system that absorbs sodium and water, harbors bacteria, and eliminates wastes.

**larva**   An immature free-living form in the life history of some animals in which it may be unlike the parent.

**lateral meristem**   An area of cell division located on the side of the plant. There are two lateral meristems, the vascular cambium and the cork cambium.

**law of independent assortment**   A principle that states that members of each gene pair are sorted into gametes independently of the members of the other gene pair.

**law of segregation**   A principle that states that homologous chromosomes are separated so that each member of the pair ends up in a separate gamete.

**ligament**   A connective tissue cable or strap that connects bones to each other or holds other organs in place.

**light-dependent reactions**   Phase of photosynthesis that absorbs light, splits water, and forms ATP and NADPH.

**light-independent reactions**   Phase of photosynthesis in which the energy of ATP and NADPH is used to manufacture carbohydrate.

**light microscope**   *See* compound light microscope.

**linkage group**   A group of genes located on the same chromosome that tend to be inherited together in successive generations.

**lipid**   Any of a group of organic compounds which are insoluble in water but soluble in fat solvents; lipids serve as a storage form of fuel and an important component of cell membranes.

**local potential**   Electrolyte imbalance in a localized area which produces an electrical charge.

**long-day plant**   A plant that flowers in response to long days (and short nights); compare with short-day plant.

**loop of Henle**   The U-shaped loop of a mammalian kidney tubule which extends down into the medulla.

**loose connective tissue**   Connective tissue consisting of diffusely arranged fibers in a semifluid matrix.

**lumen**   The cavity or channel within a tube or tubular organ, such as a blood vessel or the digestive tract.

**luteinizing hormone (LH)**   Anterior pituitary lobe hormone that stimulates ovulation and development of the corpus luteum; stimulates the production of progesterone; in males, stimulates testosterone secretion.

**lymphatic system**   Subsystem of the circulatory system that transports fats in the digestive system and immune system products.

**lysosome** Intracellular organelles present in many animal cells; contain a variety of hydrolytic enzymes which act when the lysosome ruptures or fuses with another vesicle. Lysosomes function in development and in phagocytosis.

**malignant** Uncontrolled or unregulated growth of cells such as in cancer.

**marrow** In bones, soft tissue with a number of different functions, including the production of blood cells.

**mass flow** In phloem transport, the movement of sucrose is created by the pressure of water entering the source to sink pathway.

**medulla** (1) The inner part of an organ, such as the medulla of the kidney; (2) the most posterior part of the brain, located next to the spinal cord.

**megaspore** The haploid spore in heterosporous plants that gives rise to a female gametophyte.

**meiosis** Division of the cell nucleus that produces haploid cells; produces gametes in animals and spores in plants.

**membrane carrier proteins** Proteins responsible for the transport of small molecules across a membrane.

**membrane potential** Electrical charge on a membrane.

**menstrual cycle** A monthly cycle of hormones and events that prepares a women's body to receive a fertilized ovum.

**meristem** A localized area of mitosis and growth in the plant body.

**mesoderm** The middle layer of the three basic tissue layers that develop in the early embryo; gives rise to connective tissue, muscle, bone, blood vessels, kidneys, and many other structures; lies between the ectoderm and the endoderm.

**messenger RNA** RNA that has been transcribed from DNA that specifies the amino acid sequence of a protein in eukaryotes and prokaryotes.

**metabolic pathways** A series of enzyme-mediated reactions in metabolism.

**metabolic rate** Rate at which energy is released from the breakdown of food molecules.

**metabolism** The sum of all the physical and chemical processes by which living organized substance is produced and maintained; the transformations by which energy and matter are made available for use by the organism.

**metaphase** The middle stage of mitosis and meiosis during which the chromosomes line up in the equatorial plate.

**metastasize** To transfer a disease such as cancer from one organ or part of the body to another not directly connected to it.

**microspore** The haploid spore in heterosporous plants that gives rise to a male gametophyte.

**microtubules** Hollow, cytoplasmic cylinders, composed mainly of tubulin protein, that comprise such organelles as flagella and centrioles, and serve as a skeletal component of the cell.

**microvilli** Minute projections of the cell membrane which increase the surface area of the cell; found mainly in cells concerned with absorption or secretion, such as those lining the intestine or kidney tubules.

**mitochondria** Spherical or elongate intracellular organelles that contain the electron transport system and certain other enzymes; site of oxidative phosphorylation. Sometimes referred to as the powerhouses of the cell.

**mitosis** Division of the cell nucleus resulting in the distribution of a complete set of chromosomes to each daughter cell; cytokinesis (actual division of the cell itself) usually occurs during the telophase stage of mitosis. Mitosis consists of four phases: prophase, metaphase, anaphase, and telophase.

**mitotic spindle** Structure consisting mainly of microtubules that provides the framework for chromosome movement during cell division.

**molt** To shed and replace an outer covering such as hair, feathers, or exoskeleton.

**monocotyledon** A flowering plant with embryos having one seed, leaf, or cotyledon; also known as a monocot.

**monohybrid cross** Genetic cross involving only one trait of the parents.

**monomer** A simple molecule of relatively low molecular weight that can be linked with others to form a polymer.

**monosaccharide** A simple hexose sugar; one that cannot be degraded by hydrolysis to a simpler sugar.

**morphogenesis** The development of the form and structures of the body and its parts by precise movements of its cells.

**motor unit** All the skeletal muscle fibers that are stimulated by a single motor neuron.

**multinucleated** A cell with two or more nuclei.

**muscle fiber** Fusion of several embryonic muscle cells to form an elongated multinucleated cell.

**muscle tissue** Cells that have the ability to contract.

**muscular system** A system of tissues whose contraction produces movement.

**mutualism** An association whereby two organisms of different species each gain from being together and are often unable to survive separately.

**mycelium** The vegetative body of fungi and certain protists (water molds); consists of a branched network of hyphae.

**mycorrhizae** Mutalistic associations of fungi and plant roots that aid in the absorption of materials.

**myelin sheath** The fatty material that forms around the axons of nerve cells in the central nervous system and in certain peripheral nerves.

**myofibrils** Tiny threadlike organelles found in the cytoplasm of striated and cardiac muscles that are responsible for contractions of the cell.

**myosin** A protein which, together with actin, is responsible for muscle contraction.

**natural selection** The differential reproduction of genotypes from one generation to the next.

**negative feedback** Regulation of the rate of a process by the concentration of its products.

**nephron** The functional, microscopic unit of the vertebrate kidney.

**nerve impulse**  The form in which information travels in the nervous system.

**nervous system**  A system using electrical impulses to convey information.

**nervous tissue**  Cells specializing in conducting electrochemical impulses.

**neural tube**  Structure in the developing embryo from which all the nervous system develops.

**neuron**  A nerve cell; a conducting cell of the nervous system which typically consists of a cell body, dendrites, and an axon.

**neurotransmitter**  Substance released by neurons to transmit impulses across a synapse.

**neurulation**  Process in embryonic development when the neural tube and head begin to develop.

**niche**  Functional role of a species or population in an ecosystem.

**nicotinamide adenine dinucleotide (NAD⁺)**  Coenzyme that carries hydrogen atoms to an electron transport chain.

**nitrifying bacteria**  Chemosynthetic bacteria that oxidize ammonia or ammonium to nitrites and nitrites to nitrates.

**nitrogen fixation**  Complex set of reactions that reduce nitrogen gas ($N_2$) from the air to ammonia ($NH_3$).

**nodes of Ranvier**  Naked areas of a myelinated axon.

**nonbiotic**  Non-living.

**notochord**  The flexible, longitudinal rod in the anteroposterior axis that serves as an internal skeleton in the embryos of all chordates and in the adults of some.

**nuclear area**  Region of nucleic acids not surrounded by a membrane.

**nuclear envelope**  The double membrane system that surrounds the cell nucleus of eukaryotes.

**nuclear transplantation**  The transfer of a nucleus from one cell to another devoid of a nucleus.

**nucleic acid**  DNA or RNA. A polymer composed of nucleotides that contain the purine bases adenine and guanine and/or the pyrimidine bases cytosine, thymine, and uracil.

**nucleolus**  Specialized structure in the nucleus formed from regions of several chromosomes; site of ribosome synthesis.

**nucleotide**  A molecule composed of a phosphate group, a 5-carbon sugar (ribose or deoxyribose), and a nitrogenous base (purine or pyrimidine); one of the subunits into which nucleic acids are split by the action of nucleases.

**nucleus**  A cellular organelle containing DNA and serving as the control center of the cell.

**obligate anaerobe**  An anaerobic organism that is killed by oxygen.

**oncogene**  Any of a number of genes that usually play an essential role in cell growth or division, and that cause the formation of a cancer cell when mutated.

**oogenesis**  Production of female gametes (eggs).

**open circulatory system**  A circulatory system in which there are open-ended blood vessels.

**operator**  A sequence of bases in DNA that controls the production of mRNA.

**operon**  In prokaryotes, a group of structural genes that are transcribed as a single message plus their associated regulatory elements. An operon is controlled by a single repressor.

**organ**  A differentiated part of the body made up of tissues and adapted to perform a specific function or group of functions, such as the heart or liver.

**organic compounds**  Molecules composed of carbon atoms.

**organogenesis**  The development of organs.

**organ systems**  Network of organs that produce a coordinated function or functions.

**orgasm**  Increase in respiratory rate, heart rate, and blood pressure followed by vascular engorgement of tissues, which finally result in an explosive burst of involuntary muscular contractions.

**osmosis**  Diffusion of water (the principal solvent in biological systems) through a selectively permeable membrane from a region of higher concentration of water to a region of lower concentration of water.

**osmotic potential**  Tendency, or potential of a solution to gain water by osmosis when separated from pure water by a selectively permeable membrane.

**ovary**  (1) In animals, one of the paired female gonads; responsible for producing eggs and sex hormones; (2) in flowering plants, the base of the carpel that contains ovules. Ovaries develop into fruits after fertilization.

**over-population**  Population growth above the carrying capacity of the environment.

**oviduct**  Tube that carries ova from the ovary to the uterus, cloaca, or body exterior.

**ovulation**  Rupturing of a mature follicle and the release of an egg into the coelom.

**ovule**  The part (i.e., megasporangium) which develops into the seed after fertilization.

**oxygen debt**  The accumulation of lactic acid in muscles during strenuous exercise followed by rapid breathing that provides the extra oxygen necessary to metabolize the lactic acid.

**palisade mesophyll**  The vertically stacked photosynthetic cells near the upper epidermis in dicot leaves.

**parasitism**  An intimate living relationship between organisms of two different species in which one benefits and the other is harmed.

**parenchyma**  Plant cells that are relatively unspecialized, are thin-walled, may contain chlorophyll, and are typically rather loosely packed; they function in photosynthesis and in the storage of nutrients.

**parental generation**  The original parents of a genetic cross.

**pathogen**  An organism capable of producing disease.

**penis**  External sex organ that introduces semen into the vagina during sexual intercourse.

**pentaradial symmetry**  A body form that is divided into five parts around a central area.

**pepsin**  The chief enzyme of gastric

juice; hydrolyzes proteins, particularly collagen.

**peptide bond** A distinctive covalent carbon-to-nitrogen bond that links amino acids in peptides and proteins.

**pericycle** A layer of meristematic cells in roots that gives rise to branch roots.

**peripheral nervous system** The nerves and ganglia that lie outside the central nervous system.

**peristalsis** Powerful, rhythmic waves of muscular contraction and relaxation in the walls of hollow tubular organs, such as the ureter or parts of the digestive tract, that serve to move the contents through the tube.

**petals** The colored cluster of modified leaves that constitute the next-to-outermost portion of a flower.

**phagocytosis** Literally, "cell eating"; a type of endocytosis by which certain cells engulf food particles, microorganisms, foreign matter, or other cells.

**pharyngeal gill slits** Embryologic gill remnants in chordates.

**pharynx** That part of the digestive tract from which the gill pouches or slits develop; in higher vertebrates it is bounded anteriorly by the mouth and nasal cavities, and posteriorly by the esophagus and larynx; the throat region in humans.

**phenotype** The physical or chemical expression of an organism's genes.

**phloem** Vascular tissue that conducts food in plants.

**photoperiodism** The physiological response of animals and plants to variations of light and darkness.

**phytochrome** A blue-green, proteinaceous pigment that is the photoreceptor for a wide variety of physiological responses, including initiation of flowering in certain plants.

**phytoplankton** Microscopic floating algae and plants which are distributed throughout oceans or lakes; autotrophic plankton.

**pith** Large, thin-walled parenchyma cells found as the innermost tissue in many plants.

**plankton** Free-floating, mainly microscopic aquatic organisms found in the upper layers of the water; includes phytoplankton, which are photosynthetic organisms, and zooplankton, which are heterotrophic organisms.

**plasma membrane** A living, functional part of the cell through which all nutrients entering the cell and all waste products or secretions leaving it must pass; the surface membrane of the cell which acts as a selective barrier to passage of molecules and ions into the cell.

**plasmids** Small circular DNA molecules that carry genes separate from the main bacterial chromosome.

**plastids** A family of membrane-bounded organelles occurring in photosynthetic eukaryotic cells; examples are chloroplasts and amyloplasts.

**platelets** Cell fragments in the blood that function in clotting.

**polar body** Small cell that consists almost entirely of a nucleus that is formed during oogenesis.

**polarized** Separation of electrical charges, usually across a membrane.

**pollen** The mass of microspores of seed plants which produce haploid nuclei capable of fertilization.

**pollen grain** Immature male gametophyte.

**pollination** In seed plants, the transfer of pollen from the male to the female part of the plant.

**pollution** An undesirable change in an ecosystem's physical, chemical or biological characteristics.

**polymer** A molecule built up from repeating units of the same general type, such as a protein, nucleic acid, or polysaccharide.

**polypeptide** A chain of many amino acids linked by peptide bonds.

**polysaccharide** A carbohydrate consisting of many monosaccharide units; examples are starch, glycogen, and cellulose.

**pons** The convex white bulge that is the part of the brainstem between the medulla and midbrain; connects various parts of the brain.

**population** All the members of a species occupying an area at the same time.

**population expansion** Increase in population growth, usually in response to environmental changes favorable to the organisms.

**positive geotropism** The growth response of an organism toward the direction of gravity.

**positive phototropism** The growth response of an organism toward the direction of light.

**postsynaptic membrane** Plasma membrane of the cell receiving an impulse; site of neurotransmitter function.

**predation** Relationship in which a species kills and devours other species.

**pressure flow** Mechanism of phloem transport involving a sucrose gradient.

**primary consumer** Herbivore, or plant-eating animal.

**primary growth** An increase in the length of a plant. This growth occurs at the tips of the stems and roots due to the activity of apical meristems.

**primary production site** The energy stored in organic matter by photosynthesis.

**producers** Organisms, such as plants, that produce food materials from simple inorganic substances.

**productivity** Rate at which energy is stored in organic matter.

**progesterone** Steroid hormone secreted by the ovaries that prepares the endometrium for implantation of the embryo and stimulates enlargement of breasts during pregnancy.

**prokaryotic cells** Cells of organisms that lack membrane-bounded nuclei, and other membrane-bounded organelles; the bacteria and archaebacteria.

**promoter site** Sites along DNA to which RNA polymerase attaches to begin transcription.

**prophase** The first stage in mitosis, during which the chromatin threads condense, distinct chromosomes become evident, and a spindle forms.

**prostate** The largest accessory sex gland of male mammals; surrounds the urethra at the point where the vasa deferentia enter the urethra, and secretes a large portion of the seminal fluid.

**protein** A large, complex organic

compound composed of chemically linked amino acid subunits; contains carbon, hydrogen, oxygen, nitrogen, and sulfur; proteins are the principal structural constituents of cells.

**protein channels**  Transmural membrane proteins which function in membrane transport.

**proximal convoluted tubule**  Part of a nephron that functions mainly in resorption of such substances as proteins, glucose, and ions.

**pseudopodia**  Temporary extensions of an ameboid cell, which the cell uses for feeding and locomotion.

**pulmonary artery**  In mammals, a vessel that carries blood from the heart to the lungs.

**pulmonary veins**  In mammals, vessels that carry blood from the lungs to the heart.

**Punnett square**  A tool used to determine the genotypes and genotypic ratios of the offspring when the genotypes of the parents are known.

**pyramid of energy**  A diagram of the total amount of incoming energy from successive trophic levels.

**pyruvate**  A three-carbon molecule that is the end product of glycolysis.

**radial symmetry**  A body form that can be divided into mirror image halves by several different planes passing through its long axis.

**reactants**  Materials acted upon in a chemical reaction.

**receptor**  (1) A specialized neural structure that is excited by a specific type of stimulus; (2) a site on the cell surface specialized to combine with a specific substance such as a hormone or neurotransmitter substance.

**recessive allele**  An allele that is masked and only expressed in the homozygous condition.

**reciprocal inhibition**  Any stimulus that causes a muscle to contract also inhibits the contraction of its antagonist, by means of a reflex arc.

**recombinant DNA**  Any DNA molecule made by combining genes from different organisms.

**rectum**  An organ that holds fecal matter until it is voided.

**red blood cells**  Cells of blood that carry the pigment hemoglobin.

**reflected light**  Light energy bounced off a surface.

**reflex**  An inborn, automatic, involuntary response to a given stimulus which is determined by the anatomic relations of the involved neurons; generally functions to restore homeostasis.

**reflex arc**  Pathway of sensory and motor neurons and usually one or more interneurons between them; under little control from higher centers.

**regulatory genes**  Special genes that provide codes for the synthesis of repressor or activator proteins.

**renal artery**  Vessel that carries blood from the aorta into the kidney.

**renal vein**  Vessel that carries purified blood from the kidney.

**replication**  Duplication of DNA.

**repressor protein**  The substance produced by a regulator gene that represses protein synthesis in a specific gene.

**reproductive system**  Organ system functioning to produce gametes and provides a means for production of offspring.

**resorbed**  Reuptake of materials through a membrane or other structure.

**respiratory medium**  An animal's immediate source of oxygen.

**respiratory surface**  Area where the respiratory medium gives up the oxygen to the animal's body.

**respiratory system**  In mammals, lungs function in supplying oxygen and excreting carbon dioxide.

**resting potential**  The membrane potential (difference in electric charge) of an inactive neuron (about 70 millivolts).

**restriction enzymes**  Bacterial proteins which cut DNA molecules at specific sites based on the nucleotide sequence.

**reverse transcriptase**  Enzyme produced by retroviruses which mediates the transcription of DNA from the viral RNA in the host cell.

**rhizoids**  Colorless, hairlike, absorptive filaments analogous to roots that extend from the base of the stem of mosses, liverworts, and fern prothallia.

**rhizome**  A horizontal underground stem that gives rise to above-ground leaves and stems.

**ribonucleic acid**  A family of single-stranded nucleic acids that function mainly in protein synthesis.

**ribosomes**  Organelles that are part of the protein synthesis machinery; consist of a larger and a smaller subunit each composed of ribosomal RNA (rRNA) and ribosomal proteins.

**ribulose bisphosphate**  Five-carbon sugar with two phosphate groups.

**RNA polymerase**  Family of enzymes that catalyze the synthesis of RNA molecules from DNA templates.

**root cap**  A covering of cells over the root tip that protects the delicate meristematic tissue directly behind it.

**root hair**  An extension of an epidermal cell in roots. Root hairs increase the absorptive capacity of roots.

**root pressure**  The positive pressure of the sap in the roots of plants, generated by the hypertonicity of the sap with respect to the water in the surrounding wall.

**saliva**  A substance released into the mouth in order to moisten the food, and it also contains a starch-digesting enzyme.

**saprobes**  Organisms that feed on dead organic matter.

**sarcolemma**  The muscle cell plasma membrane.

**sarcomere**  A segment of a striated muscle cell located between adjacent Z-lines that serves as a unit of contraction.

**sarcoplasmic reticulum**  System of vesicles in a skeletal or cardiac muscle cell which surrounds the myofibrils and releases calcium in muscle contraction; a modified endoplasmic reticulum.

**scientific law**  An accurate, comprehensive, and thoroughly accepted account of a natural phenomenon.

**scientific method**  A way of answering questions about cause and effect in the natural world.

**scientific principle** A theory that has stood the test of time and, therefore, is regarded as a nearly universally accepted explanation of a natural phenomenon.

**scion** Twig or bud of a desirable plant used in a graft.

**scrotum** Loose sac outside the male mammalian body that contains the testes.

**secondary consumers** Animals that feed on other animals.

**secondary growth** An increase in the width of a plant due to the activity of lateral meristems, vascular cambium, and cork cambium.

**secondary sexual characteristics** Features that are not directly related to reproduction but identify an individual as a sexually mature adult.

**secondary xylem** Cells produced inside the ring of vascular cambium.

**second filial generation** The offspring of the first filial $F_1$ generation.

**secrete** To actively release substances from cells.

**seed** A plant reproductive body that is composed of a young, multicellular plant and nutritive tissue (food).

**seed coat** "Skin" of a seed.

**segmentation** Division of an organism by partitions into repeated sections.

**selectively permeable membrane** A membrane that allows some substances to cross it more easily than others. Biological membranes are generally permeable to water but restrict the passage of many solutes.

**semen** Product of the male ejaculation, containing sperm and fluid and nutrients from the seminal vesicles, prostate gland and Cowper's gland.

**seminal vesicles** Male glands located behind and above the urinary bladder that contribute secretions to the semen.

**seminiferous tubules** Tubules within the testes where sperm are produced.

**sepals** The outermost parts of a flower, usually leaflike in appearance, that protect the flower as a bud.

**septa** Partitions.

**sessile** Permanently attached to one location.

**sex-linked genes** Genes located on the sex chromosome. In animals, having the XY (male) and XX (female) pattern of sex determination, nearly all sex-linked genes are carried on the X chromosome.

**shoot system** Plant part made up of one or more stems with leaves.

**short-day plant** A plant that flowers in response to short days and long nights; compare with long-day plants.

**simple epithelium** Epithelial tissue that is one cell layer thick.

**single circulation** A type of circulation in which the blood passes through the heart only once in a complete circuit around the body.

**sinoatrial node** Region of the heart in which the impulse triggering the heartbeat originates; the pacemaker of the heart.

**sister chromatids** A chromosome and its copy attached at the centromere.

**skeletal muscle** Voluntary or striated muscle of vertebrates, so-called because it usually is directly or indirectly attached to some part of the skeleton.

**skeletal system** The structural system of the body that functions in support.

**sliding filament** Protein structures in muscle cells that move past each other, producing shortening and lengthening of the muscle fiber.

**smooth muscle** Tissue that lines the walls of many of the internal organs, and the blood vessels.

**sodium-potassium pump** Cellular active transport mechanism that transports sodium out of, and potassium into, cells.

**soma** The part of the neuron containing the nucleus and most of the cell's organelles.

**somatic cell** A cell of the body not involved in sexual reproduction.

**somatic nervous system** That part of the nervous system that keeps the body in adjustment with the external environment; includes the sensory receptors on the body surface and within the muscles, and the nerves that link them with the central nervous system.

**sperm** Mature male gamete.

**spermatogenesis** The production of sperm by meiosis.

**sphincter** A group of circularly arranged muscle fibers, the contractions of which close an opening, such as the pyloric sphincter at the end of the stomach.

**spinal nerves** Thirty-one pairs of nerves extending from the spinal cord.

**spongy mesophyll** The irregularly arranged, photosynthetic tissue closest to the lower epidermis in leaves.

**sporangium** A spore case, found in plants and certain protists and fungi.

**spore** A reproductive cell that gives rise to individual offspring in plants, algae, fungi, and certain protozoa.

**sporophyte** The diploid portion of a plant life cycle which produces spores by meiosis.

**squamous** Flat and scalelike, as in squamous epithelium.

**stamen** The male part of flowers that produces pollen.

**stereomicroscope** Magnifying system using reflected light and binocular eyepieces, for examination of large specimens.

**stereotyped behaviors** Acts, involving the use of many muscles in a precisely timed sequence, that are always performed in an essentially identical pattern.

**stigma** (1) That portion of the carpel where the pollen lands prior to fertilization; (2) sometimes used to mean an eyespot.

**stock** Root system or stem of a plant, from which a twig or bud similar to the scion has just been removed.

**stomach** Anterior part of the intestine that is specialized to store food and secrete digestive enzymes.

**stomate** Small pore flanked by specialized cells (i.e., guard cells) which are located in the epidermis of land plants; stomata allow for gas exchange necessary for photosynthesis.

**stratified epithelium**  Epithelial tissue that is composed of two or more cell layers.

**striations**  Alternating dark and light areas of cardiac and skeletal muscle tissues.

**stroma**  The fluid region of the chloroplast which surrounds the grana and thylakoids.

**structural genes**  Portion of chromosome that contains codes for proteins.

**style**  The neck connecting the stigma to the ovary of a carpel.

**substrate**  A substance on which an enzyme acts; a reactant in an enzymatically catalyzed reaction.

**sucrose sinks**  Areas in a plant where sucrose is consumed.

**symbiosis**  An intimate relationship between two or more organisms of different species.

**synapse**  The junction between two neurons or between a neuron and an effector.

**synapsis**  The pairing and side-by-side union of homologous chromosomes during Prophase I of meiosis.

**synaptic cleft**  The tiny gap between neurons.

**tap root system**  A root system in plants that has one main root with smaller roots branching off it.

**taxis**  An orientation movement of a motile organism's response to a stimulus.

**telophase**  The last stage of mitosis and meiosis when, having reached the poles, the chromosomes become decondensed and a nuclear envelope forms around each group.

**tendon**  A connective tissue structure that joins a muscle to another muscle, or a muscle to a bone. Tendons transmit the force generated by a muscle.

**tension-cohesion**  Physical phenomenon based on the attraction between water molecules so that the evaporation of water in the leaf, results in water being pulled up from the root via the xylem.

**tertiary consumers**  Animals that eat or prey on secondary consumers of the food chain.

**testis**  Male gonad that produces sperm and secretes androgens.

**testosterone**  Steroid hormone secreted primarily by the testes that promotes sperm formation and stimulates primary and secondary sexual characteristics.

**tetanus**  Sustained, steady maximal contraction of a muscle, without distinct twitching, resulting from a rapid succession of nerve impulses.

**thalamus**  The part of the brain that serves as a main relay center transmitting information between the spinal cord and the cerebrum.

**theory**  A hypothesis supported by many different lines of evidence from repeated experiments.

**thick filaments**  In muscle tissue, protein filaments which are centered in the sarcomere.

**thigmotropism**  Plant growth in response to contact with a solid object, such as plant tendrils.

**thin filaments**  In muscle tissue, protein filaments attached to each side of a Z-line which extend less than halfway to the center of the sarcomere.

**thylakoids**  Stacks of flat membranous sacs inside the chloroplast where light energy is converted into ATP and NADPH used in carbohydrate synthesis.

**tissue**  A group of closely associated, similar cells that work together to carry out specific functions.

**totipotent**  Ability of a cell (or nucleus) to provide information for the development of an entire organism.

**transcription**  The synthesis of messenger RNA from a DNA template.

**transfer RNA molecules**  A form of RNA composed of about 70 nucleotides which serve as adaptor molecules in the synthesis of proteins. An amino acid is bound to a specific kind of tRNA and then arranged in order by the complementary pairing of the nucleotide triplet (codon) in template or mRNA and the triplet anticodon of tRNA.

**translation**  In protein synthesis, the processing of messenger RNA by ribosomes leading to the incorporation of specific amino acids into a polypeptide chain based upon nucleotide sequences of the messenger RNA.

**transmitted light**  Light energy passed through a medium.

**transpiration pull**  Negative water pressure produced within a plant by the evaporation of water into the air.

**trial-and-error learning**  A behavior pattern produced by success or failure by random attempts.

**trophic level**  The distance of an organism in a food chain from the primary producers of a community.

**tropism**  A growth response in plants that is elicited by an external stimulus.

**tropomyosin**  A muscle protein involved in the control of contraction.

**troponin**  In muscle tissue, a protein that binds to tropomyosin during contraction.

**trypsin**  Digestive enzyme produced in the pancreas, which breaks down polypeptides.

**tumor**  Mass of tissue that is growing in an uncontrolled manner; a neoplasm.

**turgor**  The building up of pressure within a cell.

**urinary system**  The excretory system of the body which regulates blood chemistry.

**uterus**  Highly elastic muscular organ which functions to hold a developing embryo and expel it during childbirth.

**vacuole**  A cavity enclosed by a membrane and found within the cytoplasm; may function in storage, digestion, or water elimination.

**vagina**  Receptacle for penis during copulation and the pathway to the exterior for the baby during childbirth.

**valves**  Flaps of tissue that help keep blood flowing in one direction.

**variable**  Changeable component of an experiment.

**Vasa deferentia**  Ducts between the testes and urethra which carry semen.

**vascular cambium**  A lateral meristem in plants that produces secondary xylem (wood) and secondary phloem (inner bark).

**vascular system** Fluid system that transports substances in most members of the animal kingdom.

**vascular tissue** Tissue that conducts food and water within the shoot system of some plants.

**vegetative reproduction** The production of new individuals with genetic makeup identical to the parent's.

**vein** A blood vessel that carries blood from the tissues toward the heart.

**venae cavae** Two large veins that return blood from the body to the heart.

**ventilation** A process that moves the respiratory medium and brings a constant fresh supply of oxygen past the respiratory surface.

**vertebrates** Chordates that possess a bony vertebral column; fish, amphibians, reptiles, birds, and mammals.

**villus** A minute, elongated projection from the surface of a membrane.

**voluntary control** Nervous system mechanism for adjusting function with conscious thought.

**vulva** External sex organs of a woman composed of folds of skin covering the vagina and urethra.

**white blood cells** Five different kinds of cells found in the blood that function in combating disease.

**xylem** Vascular tissue that conducts water and dissolved minerals in certain plants.

**Z lines** Protein-containing structures extending across the myofibril.

**zone of elongation** Cells in the root that have stopped dividing and are growing longer, pushing the root cap and apical meristem through the soil.

**zone of maturation** Cells in the root that have reached full size and are developing into specialized cells that will perform particular functions.

**zygote** The diploid ($2n$) cell that results from the union of two haploid gametes; a fertilized egg.

# Index